What Is Present Reality

The Power of Managing the Limits of Science

Vipin Gupta

ISBN: 9798-5921-0286-8
Black & White Paperback
Independently Published Imprint

ISBN: 9781-0879-3132-6
Color Hardcover
Indy Pub Imprint

First Published: 2021

Cover Design:
Ashaf Mohideen

Illustrations:
Ghanshyam Bochgeri

Illustration supervision:
Olivia Chatterjee Biswas

Edited By:
Sayantani Sengupta, Sangeeta Chatterjee, and Prem Nath

Page Setup:
Pinaki Dutta

Editorial supervision:
Sanchaita Biswas, Anupriya Dutta, and Vishnu Vijayan

Index:
Liya Jayabalan and Salini Kurup

EBook Editor:
Suresh Velu

Project Manager:
Pinaki Ghosh

Contents

The present project is about discovering the vastly integrated processes inside nature that shape our divine reality as an entity and are shaped by our gravitating reality. With the advent of modern science, the knowledge of our ancestors got relegated to the background. Modern science has led to extreme differentiation and specialization of knowledge. Each field of knowledge has become disjointed from the other fields. The scholars of each field believe that their knowledge is the key to universal well-being and deserves special recognition. In the past, there were no boundaries between different knowledge domains. There was a seamless consciousness about the divinity, the spirituality, the natural dynamics, the supernatural motivations, interventions, and moderations, the devil manipulations and mediations, and the satanic subversions and massifying infections. There was a consciousness that each sentient entity has divinity and the potential to transform that divinity into divine energy and oneself into a mind-born grandmaternal spirit to eternally guide other entities' divinity.

Over time, people's consciousness shifted to focus on spirituality, seeking to bind the divine entities' spirit essence and transform them into maternal *jinns*. They elevated themselves into a paternal *jinn*, who has the power to be at the command of anybody who knows the mantra for waking him from his peaceful slumber. By presenting themselves as an entrepreneur, wishing to serve the command of any master who believed in their power, the paternal *jinns* transformed each entity into a *devil*. A *devil* is one who goes with the flow of the universe and becomes an invisible hand servicing that flow, as if it is his gift to the universe. Consequently, each entity became a devotee of the different elements within Mother Nature. People began deifying each natural element as well as the diverse forms of those elements, seeking to charm them with the science of *tantra*.

People did not know that when one deifies an inanimate natural element, one is essentially servicing a unit of one's divine energy to let the entire universe benefit from that energy unit. It makes one the soul of the universe of follower vantities. Without consciousness of the truth of the worshipped deity's divinity, the followers conceive the deity as the son and the soul as the perpetuating God of the deity's illusionary energy. By personifying the blessed natural element, they transform that natural element into an absolute soul, i.e., a soul present within them in the form of the energy they activate whenever they perceive a void in their consciousness. Both the soul and the absolute soul become key to their conceived conscious well-being.

By believing in a para deity's supernatural powers to fulfill any wishes one has, one naturally transforms into a deity, blessed by that para deity. By behaving like a deity, one gets a rude shock when the proliferating aspirations of the leader person and the para-person universe of the followers do not get fulfilled. Consequently, each person becomes a discordant factor, seeking to motivate everybody else to see the value of its wishes and work toward fulfilling those wishes, as a path for realizing the truth of the deity and, through that, of the para deity. When one works toward fulfilling those wishes, as a sentient entity, one becomes a param deity, capable of fulfilling any wish of any entity in the universe.

There are two pathways to be a param deity. First, make one's workforce proficient to fulfill any wish by spontaneously becoming the wishable wish that trumps the entire universe of wishes. If everybody wishes to be like you, then you are the param deity who needs to do nothing but enjoy the benefit of devotion and absolute surrender by each entity. Second, make one's network proficient to fulfill any wish by spontaneously breeding that wishing within the entire universe of entities so that each entity becomes devoted to fulfilling a part of that wish effortlessly using a proportion of their natural energy. In this case, each entity becomes a primeval deity, capable of fulfilling an infinity of wishes by leveraging the tertiary residual formed by everybody else in the universe.

The networking pathway is a pathway of action that requires one first to be a supreme, creator deity for manifesting the rest of the universe as a supra, manifestor deity, reproducing and multiplying itself to form the entire universe as a super, knower deity. This divided entity is both conceiving and fulfilling its wishes by itself. Suppose one decides to devote consciousness into the act of creation. In that case, one eventually destroys oneself and becomes a part of the universe, conceiving a solution as well as experiencing the truth of that solution.

The workforce pathway is a pathway of knowing that requires one first to be a primordial, natural deity for liberating each entity to be what one is, effortlessly, without the mediation of any gravitational energy in the form of the guider power of a guider (guru) element.

There are two pathways for exchanging what each person in the universe wishes to do with what must be done to fill their consciousness void so that they make you their wishable wish. First, exchange the consciousness of each person with yours and become each person's para-consciousness. It makes each person your homolog but does not activate a self-luminous consciousness to be like you, but not your programmed robotic reproduction. Second, exchange each person's consciousness with that of Mother Nature and let each person develop a self-consciousness of the futility of depending on a naturally cyclical para-conscious element, whose behavioral qualities are befitting an inanimate entity, but not an animate entity.

Your personal force is the pathway of devotion that requires one first to be a devoted, supernatural deity, perpetuating oneself as the goal of all social exchanges.

Mother Nature's social force is the pathway of divinity that requires one first to be a devotee deity, illuminating oneself as the centering element. One discovers the centering element after piercing through the blinding, diffused, circulating air, generated because of the natural cycle of birth, growth, death, rest, and rejuvenation.

Now, you, as a person managing your life, have a choice. You can be like a person you believe is a param deity and make the entire truth of that person your goal. Or, you can be like Mother Nature and make experiencing the entire beauty of her as a para-person your paradigm for knowing the truth of the present reality.

Alternatively, you can be who you are and devote your energy to channel your consciousness on being who you are.

The titles of the twelve books in this project are as follows.

- What is divine energy

- What is present reality

- Is present reality

- Is divine energy

- What is consciousness

- What is para-consciousness

- What is self-awareness

- What is human factor

- What is trading factor

- What is cultural factor

- What is exchange factor

- What is technological growth

The first book, *What is divine energy*, introduced the essential vocabulary for knowing life's truth and illuminated its homologous constructs and analogies, both from the diverse disciplines of modern science and India's ancient wisdom dispersed across numerous manuscripts. It took a causal, sequential, and consequential approach, highlighting the chronological sequence of a cell's thirty-three phase development.

This second book, *What is present reality*, challenges the classical determinism, neoclassical thermodynamism, modern probabilistic relativism, as well as post-modern quantum indeterminism notions of knowing the present reality. It examines the diverse doctrines for the paradigmatic planning of the future reality. Each entity puts a varying proportionate priority on the varying doctrines to form a varying notion of the present reality. However, the present reality is neither the general notion about what it is nor the specific notion of what it is not.

As a general notion, the present reality is a sum of the specific past realities conceived at varying time moments. It is historical, knowable, and therefore, deterministic. As a specific notion the present reality is the generalized value of the varying future realities that may manifest with what one conceives in the present moment. It is conditioned by the cultural knowledge and, therefore, is a weighted average of the varying cultural linkages the conceiving entity has at the present moment.

Without knowing the truth of the varying cultural linkages within the conceiving entity, one may use the numbers for a probability prediction of the targeted present reality. Without knowing the truth of the dynamic cultural exchanges without the conceiving entity, one may find errors in those numbers and perceive the present reality to be indeterministic. An imperfect knower has zero integrity as a scientist and needs to be falisified before one diffuses imperfect knowing as a scientific discovery and pollutes everybody's consciousness.

Overall, as a management professor, my goal is to offer an overall strategic awareness of the reality within and without us, and offer solutions to manage our lives meaningfully in the present moment and leave a positive, healthy legacy for the sentient well-being of the future generation of children.

Since the science and the cultural wisdom, as we know, are the works of imperfect knowers, for knowing the present reality one needs to discard all received wisdom. If you are up for that challenge, then I welcome you to join the journey now. If that seems daunting, then do not worry. Wait for the other books in this project that will offer a more accessible exposition.

If you do not wish to wait, you may use the index to identify themes that are of particular interest to you and use this book as a guide to formulate an alternative hypothesis to explain what you believe you know, based on what either the modern science teaches you or the books you have read about the cultural wisdom of India. Take that hypothesis to its logical conclusion by reexamining the evidence using a fresh mind, free from a blind faith on what you thought you knew. Please contact me at my personal email gupta05@gmail.com and share what you find. *I guarantee a life full of fresh discoveries, once you take the first step.*

List of Abbreviations

DIVINE	d = determination, I = imagination, v = virtue, I = intuition, n = natural, e = excellence
GUIDER	g = global, u = unique, i = inclusive, d = diverse, e = engagement, r = responsibility
SHEENY	s = social, h = human, e = ecological, e = economic, n = national, y = psychological

List of Tables

Vipin Gupta, Ph.D., is a professor of management, and co-director of the Center for Global Management at the Jack H. Brown College of Business and Public Administration, California State University, San Bernardino. He has a Ph.D. from the Wharton School of the University of Pennsylvania. He has been a gold medalist for outstanding academic performance in the Indian Institute of Management post-graduate program, Ahmedabad; a top rank holder in the B.Com. (Hons) Program of Delhi University from Sri Ram College of Commerce; and an all-India rank holder at the graduate program of the Institute of Cost and Works Accountants of India.

Vipin Gupta was previously at Simmons University, Boston, Grand Valley State University, and Fordham University. He has offered several training programs and workshops on strategic planning and cross-cultural management to senior executives, administrators, defense personnel, and research methods to doctoral students and faculty in India and the US. He has been a visiting or guest faculty at more than thirty business schools in India. His workshops and lectures have been covered by several of the leading national and regional newspapers and television channels.

Professor Gupta has authored more than 180 journal articles and book chapters, including in leading journals such as Journal of Business Venturing, Family Business Review, Research in Organizational Behavior, Asia-Pacific Journal of Management, Multinational Business Review, Journal of World Business, Advances in Global Leadership, and Management Review. Besides delivering lectures and keynotes in several nations, he has presented at international academic conferences in more than sixty nations, including the Academy of Management, IFSAM, EGOS, Society of Industrial Organization Psychologists, Global Entrepreneurship Conference, and Family Enterprise Research Conference. He has been on the governing board and organizing committee of several international conferences. In 2017, he served as the academic program chair for the 52nd CLADEA Assembly.

Dr. Gupta is the co-editor of the seminal GLOBE (Global Leadership and Organizational Behavior Effectiveness Program) book Culture, Leadership, and Organizations – The GLOBE Study of 62 Societies (Sage

Publications, 2004). He is the principal investigator of the path-breaking CASE (Culturally-sensitive Assessment Systems and Education) Project on family businesses. He edited two critically acclaimed books on the theme of strategic management, performance, and leadership in the emerging markets: Creating Performing Organizations (Sage Publications, 2003) and Transformative Organizations (Sage Publications, 2004). He is the author of the Strategic Management and Business Policy: Concepts and Applications (PHI Learning, 2003 and 2005). He has been the principal editor of ten books on family business models in ten different regional clusters and the eleventh book on family businesses' gender dimension (ICFAI University Press, 2004). He has also co-authored a research manuscript MNC Subsidiaries in China: An empirical study of growth and development strategy (Information Age Publishing; 2015), and a textbook Leadership Across the Globe (Routledge USA, 2015).

Vipin Gupta has been a recipient of the coveted 2005 Scott Myers Award for Applied Research in the Workplace from Society for Industrial Organization Psychologists, USA. As a 2015-16 American Council of Education fellow, he visited sixty-two universities, colleges, and higher education institutions in nine European nations, the USA, and India.

Preface

As a preface to this work, here are some threshold concepts you will need for making sense of what you are about to experience.

1. Four Dimensions of the Reality

- The primordial (literally, finite: first formed with proportional, formative characteristics) reality is a beautiful, complex reality. It is a reality without the effects of the primordial entity.

- The primeval (literally, infinite: inclusive of the first age's transformative characteristics) reality is the true reality. It is a reality inclusive of the effects of the primordial entity.

- The present (literally, param: absolute with the normative characteristics) reality is the cause of both beautiful and true realities. It is the reality of the primordial entity as the creator of effects.

- The primordial entity is the primordial-primordial (literally, metaphysical: omnipresent within or without the infinite physical forms) reality. It is the creator of the primordial reality.

2. Four Additional Dimensions of the Reality

- Immanent reality is the experienced reality of divinity within the consciousness of an entity as a subject.

- Ideal reality is the reality of divinity of a subject who is the creator of the desired object.

- Emanating reality is the experienced reality of divinity without the consciousness of the entity who has created that reality.

- Theoretical reality is the reality of divinity scripted by a subject who perceives the desired reality as the gift of a para deity (literally, God: one omnipotent deity beyond the self's consciousness).

3. Four Dimensions of the Entity

- The primordial-primordial entity is a sentient (SHEENY) entity who, as a creature free from paradigmatic limitations, sentiates the desired social, human, ecological, economic, national, and psychological values.

- The primordial entity is a guider (*guru*) entity who, as a creator of a new paradigm, sentiates the desired global, unique, inclusive, diverse, engaged, and responsible values and modifies the desirability of the SHEENY values.

- The present entity is a divine entity who, as a creation of the present paradigm, sentiates the desired determination, imagination, virtue, intuition, nature, and excellence values, constrained by the socially desirable guider values and the institutionally desirable SHEENY values.

- The primeval entity is a Satan entity who, as an Almighty Creation of the constrained organizational development, destroys the desirable desired reality of the creation, the creature, and the creator, wishing to be an Almighty Creature who is the primordial paradigm shaping the present reality of the Almighty Creator.

4. Four Additional Dimensions of the Entity

- The subjective consciousness shapes an entity as the subject of the consciousness into a super deity who knows the subjective cause of subjectivity.

- The objective consciousness shapes an entity as the object of the consciousness into a supra deity who manifests the objective cause of objectivity.

- The ideal-effect of idealizing the self as the subject worth subjective introspection shapes the self into a supreme deity who creates the self as the worthy subject.

- The theory-effect of theorizing the self, as the object worth objective extroversion, shapes the self into a para deity who perpetuates the universe as an unworthy object.

5. Four Dimensions of Strategic Management

- Each of us is a primeval deity. We have infinite power to destroy the self's present reality and perpetuate the self's desired future reality.

- Each of us has the potential to be a param deity. We have absolute power to illuminate the worth of the self's present reality and destroy undesirable qualities.

- Each of us limits our potential by making our self-worth dependent on a primordial deity. We waste an ascending proportion of our energy begging the primordial deity for blessings to let us be the supernatural paradigm governing the strategic management of the universal endowments.

- A primordial deity is anyone who proclaims to have omniscience about the secret of how to control Mother Nature for appropriating disproportionate SHEENY benefits. Such an entity enjoys instant guider benefits. All believers enjoy infinite divine costs. The entire universe enjoys infinite SHEENY costs.

6. Four Additional Dimensions of Strategic Management

- A devoted deity believes in the primordial deity's power and perpetuates the primordial costs by conceiving them as the primeval benefits.

- A devotee deity lets the devotees believe anything they wish, so that they may personally experience the infinite SHEENY costs of the culturally-bounded rationality.

- A universal deity illuminates the devotee deity's truth by making each devotee personally responsible for the infinite GUIDER costs of the culturally-bounded rationality.

- A unique deity experiences the universal deity's truth by making oneself personally responsible for the infinite DIVINE costs of the culturally-bounded rationality.

7. Four Dimensions of a Strategic Manager

- A disproportionate sense of the intrinsic entity value transforms us into a luminous. The luminous is an eternal deity who radiates the intrinsic divine value as a subject to ascend and transform the extrinsic object value. As a luminous, we breed infinite cost-escalating material power, making the pollution of nature a grand problem for everybody.

- A disproportionate sense of the extrinsic subject value transforms us into a primeval illuminator. A primeval illuminator is a potential deity who trades the extrinsic guider value to descend and transform the intrinsic subject value. As a primeval illuminator, we believe the ecosystem has the infinite free machinery power to take our polluted nature (the grand problem) as an illuminated value to be shaped into an infinite free method power (the supernatural solution).

- A disproportionate sense of the intrinsic subject value and the extrinsic object value transforms us into a primeval greeter. A primeval greeter is a universal deity who transforms both the primordial reality of the intrinsic subject's absolute entropy and the extrinsic object's absolute growth into the present reality. The present reality is of the intrinsic guider's infinite growth and the intrinsic SHEENY's infinite entropy.

- A disproportionate growth in the intrinsic guider value transforms us into a primeval perpetuator (who is a dynamic deity) of the contaminated dynamic reality that descends our extrinsic SHEENY value by ascending the universe's complexity.

8. Four Additional Dimensions of a Strategic Manager

- The absolute entropy of both the intrinsic and the extrinsic SHEENY value transforms us into a primordial greeter (who is a technological deity) of the purified technological reality that ascends our extrinsic guider value by making our intrinsic guider power that purifying technology.

- The absolute entropy of the power-seeking intrinsic guider value transforms us into a primeval paternal (who is an organizational deity) of a universe of devoted children, motivated to discover new heuristics for ascending the collective guider value of the local soul essence community.

- The absolute entropy of the power-diffusing extrinsic guider value transforms us into a primordial maternal (who is an ecosystem deity) of a universe of devotee children, motivated to illuminate the escalating costs of the institutional collectivism by investing their intrinsic divine energy.

- The absolute entropy of the universe fraternizing divine entities empowers us to be a param child (who is an entity deity), absolutely free to script our destiny by countertrading the diffused extrinsic divine energy. We become a subject experiencing the extrinsic divinity and destroying the consciousness of our intrinsic divinity.

The objective of the present investigation is to help a strategic manager be a param manifestor (who is a present deity). It seeks to empower each child to transcend the limitations of the dependence on the extrinsic divinity and discover the present reality of the Vastly Integrated Processes Inside Nature. Discovering the truth of the present reality omnipermeating within Mother Nature is a joyful journey. No entity, discipline, culture, or religion has conditioned my journey. Many have guided my journey to be free from the limitations of the present reality. I acknowledge due credits at the end of this work. For now, at the outset, I wish to offer my profound gratitude to my wife, Bhakti, for her constant inspiration, critique, love, and suggestions for timely and audience-oriented investigations. Credit for the present investigation is shared equally between her and me. Her support has been quintessential for me to conceive, persevere, and finish the present investigation.

VIPIN GUPTA

The Three-Fold Divine-effect of One's Behavior System on the Reality

The future reality planned by the divinity
without an entity is a SIMPLE REALITY that perpetuates.
It is natural.
The present reality performed by an entity becomes a
VIRTUAL REALITY when reproduced. It supernaturally
reproduces the natural reality as a foundation for the artificial
"present growth."

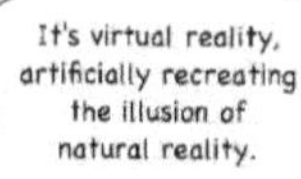

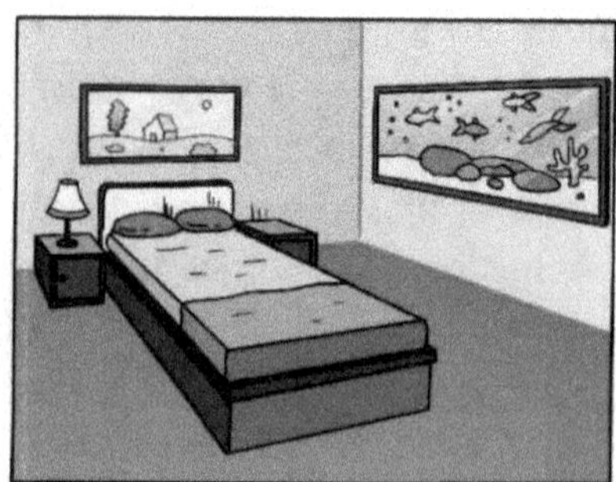

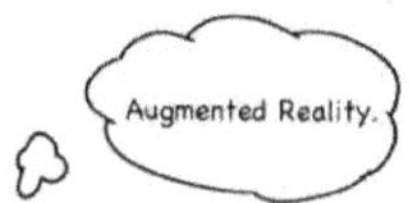

The past reality programmed within an entity
becomes an AUGMENTED REALITY, using virtual
elements as the artificial building blocks for
supernaturally transforming the natural reality.

The Present Reality is a Two-Dimensional Virtual Reality of a Deity and an Entity

As a deity, an entity conceives a virtual mind-born future reality and materializes that as the present reality for everybody to enjoy. It is the absolute reality of the entity's devoted follower universe, rather than of the entity. An entity-scripted absolute reality supplements the omnipresent reality of the universe without the entity. The unscripted omnipresent reality has simplicity. The entity-scripted present reality generates complexity by consuming everybody's disproportionate energy for augmenting and discerning its value.

1.1 The Scientific Approach to Present Reality

Many scientists have investigated and discovered the diverse dimensions of the present reality. As an observing entity, each scientist observes a diverse present reality dimension by activating a diverse paradigm for observing. A paradigm is a system of methods for scripting, conceiving, perceiving, experiencing, sentiating, spiritualizing, materializing, reproducing, and exchanging the diverse proportionate dimensions of the present reality. One scripts a dimension of the present reality by taking it as an absolute value and consciously determining a path for transcending that absolute limit. For scripting a new reality that does not exist in the present moment, one conceives that new reality by opening the mental block-box and riding on the horse of imagination in a virtual reality dream world.

For conceiving the virtual reality as an ideal point of infinity, one perceives the diverse dimensions of the present reality, guided by the experiential value of each dimension, to sentiate a modified vision of the present reality. For instance, one may wish for a world without humans or animals, so that the plants can grow and thrive naturally, and the environment remains virgin and beautiful. One may also wish for a world where humans, animals, and plants contribute to each other's natural growth without any entity's dominance. For sentiating a modified vision, one spiritually develops an impulse to materialize the partial reality and reproduce that as the absolute reality. Thus, the present reality is the exchange value of the absolute reality conceived by a primordial entity, i.e., an entity that lived in the past as the paternal of all the present entities and wished the "universe of child entities" to experience a part of the omnipresent reality.

The present reality, scripted and conceived by a primordial paternal, is perceived and experienced by the param child, i.e., the child who lives in the present, as the proportionate value of an omnipresent reality. By attuning their mind to the omnipresent reality, a param child enjoys the power to sentiate an omnipotent reality that was destroyed by the primordial paternal. After observing the omnipresent reality, a primordial paternal reproduces its proportionate value in the form of the present reality. After observing that present reality, a scientist reproduces a its proportionate value to form the future reality. The universe of all entities scripts the future reality. The scripted reality becomes the universe's present reality. The collective script of the universe of entities permeates the reality that unfolds over time. Each entity punctuates it with an individual script forming the omnipotent reality.

The aggregate of a collective and several individual scripts is modified and polluted by each entity to form a proportionate omnipresent reality. The totality of the one aggregate, and many disaggregated, scripts is mediated and proliferated by each entity in the form of the varying present realities. Without any desirability value weights reflected in the proportionate dominance, the varying present realities' wholesome value forms the almighty creator reality. The "overall present reality," i.e., the almighty creator reality and the varying future realities, scripted by each param child, forms the almighty creation reality. Again, the "overall future reality," i.e., the almighty creation reality and the varying forms of future entities, form the almighty creature reality. It is the creature's reality as the creator of both: the inanimate creation (derived and integrated from the past moments) and the animate creation (differentiated in the present moment). The "overall past reality," i.e., the almighty creature reality that is not yet manifested, together with the varying forms of past entities, is the absolute reality. It is the reality of the creator of both the departed child souls and the future child spirits.

The departed children's physical bodies are now present as the diverse materials, metals, minerals, plants, animals (inclusive of the birds, fishes, and insects), and humans. The physical bodies of the present entities will be the medium for incarnating the future child spirits. The absolute reality comprises the relative, proportionate, and diverse realities of the present entities and the infinite, disproportionate, and inclusive realities of the future entities. The absolute reality is the unique and squared reality of the present universe of entities. The future universe of entities is the engaged, entangled, and circumambulating reality of the omnipotent entity who happens to be the absolute reality's creator. The past universe of entities is the responsible, disentangled, and parabolic reality of the omnipermeating entity. Its perpetuating head-domain is the ascending, emergent, entangling, and circular reality. Its illuminating tail-domain is the descending, divergent, disentangling, and squaring reality. The overall reality of the almighty creator comprises the six forms of realities:

- First, the global, square root reality of the omnipresent universe of entities is the whole, sentiated, and centroid absolute reality of the universe.

- Second, the unique, squared reality of the present universe of entities is the fractional, spiritual, and diagonal relative reality of the universe.

- Third, the inclusive, exponential reality of the future entities is the disproportionate, infinite, and vertical experiential reality of the universe.

- Fourth, the diverse, logarithmic reality of the present entities is the proportionate, finite, horizontal, simplistic, and perceptual reality of the universe. It exponentiates the present experience and forms its future.

- Fifth, the engaged, circulating reality of the future universe of entities is the definite, ascending, entangling, chaotic, and conceptual reality of the universe. It radiates from the omnipotent entity's mind and irradiates into the intellect of the omnipresent entity for conceiving the present entities in their physical form.

- Sixth, the responsible, squaring reality of the past universe of entities is the ground, descending, disentangling, complex, and scripted reality of the universe. It radiates in the form of a light force from an omnipermeating astral body. It irradiates into the almighty creator of the past entities' etheric body in their metaphysical, spiritual form.

The etheric body refers to the incarnational universe of entities that services the overall "guider" reality. It consists of the global, unique, inclusive, diverse, engaged, and responsible dimensions. It works like a force without the light of the causal body. The causal body is the divine cause transforming a self-luminous entity into an etheric body. A self-luminous entity generates the force for disintegrating a star into an incarnational universe of entities. The gravitational energy of the star is the overall guider power of the incarnational universe of entities. The gravitational energy is the energy programming a set of behaviors that repel the undesirable, past reality and attract the desirable, future reality. The incarnational universe of entities trades the gravitational effect in the form of the guider power for guiding a set of behaviors among the present entities through their differentiated physical bodies to manifest the desired, integrated guider reality.

A self-luminous entity is the para-primordial self of the universe of entities that forms a physical body. It forms the primordial self in the form of the divine energy that leads the universe to incubate each entity in the form of an astral body. The divine energy is radiated by the param self, a spiritual entity embodying the ideal, future reality. The self is free from both the self-conceived intellectual idealism that forms the spiritual entity and the self-perceived theory of mind that incarnates a universe of entities in the form of the etheric body. The universe of entities experiences the primordial paternal, who incarnates them by diffusing the light from his physical body, as a force oriented toward reunification with the formative maternal source of divine energy. The formative divine energy is radiated by a grandfather cell, whose present reality is the paternal soul and whose future reality is the child spiritual entity. The present entity is the physical body, whose actual reality is the self-luminous entity. It is originally a grandmother cell that services a fourth of her squared energy, sentiating the paternal soul while transforming herself into a three-dimensional self-luminous entity.

When one prays, "*may the soul of the departed entity rest in peace,*" one invokes the paternal soul that is wishing to trade the sentient energy of the self-luminous entity. On death, the self-luminous entity becomes a physical body. Since the paternal soul is unable to trade the child entity's sentient energy after her death, he seeks to find an alternative grandmother cell to perpetuate his guider power to guide a sentient entity. The perpetuating guider-effect of the paternal soul limits the power of the atoms—that once constituted the departed physical body—to reunify with a self-luminous astral body. The atoms become entangled with the present sentient entity's physical body, mediated by the present reality of the paternal soul. The paternal soul is immanent within the etheric body, comprising many paternal souls, as a guiding force.

A sentient entity may cleanse the physical body from the guider programming-effect of the present paternal soul and the infinite additional souls from the past which constitute the etheric body. By doing so, the sentient entity empowers the normative development of the self-luminous astral bodies and their transformative growth into the universe of sentient entities. Such cleansing sentient entities, which manifest the self-luminous astral body, become luminous. A luminous is one who radiates the light force for the technological growth of the universe of entities. By trading the luminous astral energy, immanent within the astral body, an inanimate, material entity transforms into an animate, sentient entity. By trading the astral energy of a feminine maternal entity, immanent within a masculine grandfather entity, an androgynous, inanimate child entity incarnates a para-androgynous universe of the sentient granddaughter and grandson cells. The inanimate child entity becomes a present living child, and then a parent for incubating the infinite cells.

Depending on the method for making sense of the present reality's causative sequence, one may consider any animate or inanimate forms of the self as an entity. One may also consider one's divinity for conceiving the infinite ideal para-entity forms by scripting a theory of the mind that imagines the self in those diverse entity forms. By activating the diverse methods, one forms a paradigm for managing the present reality without the limits of specifically varying, mutually-exclusive theories; a generalized constant, all-inclusive ideal; and sequentially-formed, quantum hypotheses, for falsifying the truth of a diverse two-dimensional theory of relativity.

Albert Einstein's special theory of linear and continuous relativity is an all-inclusive perspective on the diverse mutually-exclusive theories about the present reality. His general theory of nonlinear and disentangled relativity is a mutually-exclusive perspective on the all-inclusive ideal value of the present reality. His sequential acceptance of the entangled relativity's quantum theory is a self-excluding perspective on the present reality's hypothetical and theoretical value.

1.2 The Special Theory of Linear Relativity

The theory of continuous linear relativity is special and all-inclusive of the diverse theories of mind. Each theory of mind radiates the formative astral body's light as a special form of the general self. The etheric body is a continuous linear reality of the universe formed by irradiating the diverse special effects to norm the general self organizationally. That general self is the universe as an entity. A causal body may trade the continuous linear reality of the universe and service that as the nonlinear, discontinuous reality of the special self. That special self is the causative entity, perceiving the universe's linear reality as a grand challenge and conceiving a solution for transforming the self into a special divine entity. It believes in the self's special power to experience the self-luminous reality of the universe.

By spiritualizing an "I AM a Deity consciousness," the causative entity becomes the Luminous. The Luminous is the guiding force, causing each self-luminous entity to trade the entropy-perpetuating reality of the universe. The continuous value of the relative reality of the universe is the entropy value of the universe. When a universe becomes an "electromagnetic mass" (m) that is radiating and irradiating "energy" (e) for forming the diverse entity forms of "self" (c), the consequential energy value of the universe as a future ideal primeval entity is zero. It is the entropy diffusion value of the universe. Therefore, $e = mc^2 = 0$. It is the general energy value of the special theory of relativity. The entity, who is sentiating the deity consciousness, fails to quantify this theory-effect of the self. The failure accounts for the transubstantiation of the entity's nonlinear growth with the linear entropy reality of the universe.

A deity diffuses the intrinsic self-luminous entity light with a force that generates an entropy of the deity consciousness and services the unit value of a deity as the alternative and illusionary energy value of the universe. Therefore, the entity, who becomes the primordial paternal of a new illusionary reality, perceives $e = mc^2 = 1$. In other words, the entity believes that the special theory of relativity quantifies the aggregate value of the energy in the universe without consciousness that the aggregated universal energy experienced by the entity is the energy of the self. The self is the entity catalyzing the incremental growth in the universe of entities, otherwise paralyzed without knowing the illusionary truth. The self then believes the incremental growth to be the emanating value of the energy in the universe, serviced by a para deity. It conceives that para deity to be the invisible creator God of illusion, hiding the truth from the non-believers.

The diffusion of the illusionary truth is the path for the divine self to convert each non-believer into a believer of the primordial deity's truth. That primordial deity is Mother Nature, who is conceived by Albert Einstein as Yahweh; perceived by the universe of entities as Tetragrammaton; and experienced by the universe as an entity in the form of the four-lettered reality: YHWH.

The letter Y symbolizes transforming a logarithmic root reality of the universe into an exponentiated infinite reality of an ideal point entity, creating a nonlinear, curvilinear and, parallel reality scripted by that ideal entity's divergent divine point. It is the reality of the causative body generating the nonlinear reality. The letter H symbolizes bridging the diagonal that connects the linear and the nonlinear realities. The first bridging diagonal is the universe of entities trading the nonlinear reality as a passive observer of the manifested gravitational force. As an active reproducer of the guider-effect, the universe of entities services that nonlinear reality in the form of a linear, primeval perpetuating reality.

The letter W symbolizes the entanglement of the two sets of curvilinear realities. First, the primordial paternal's curvilinear reality as an ideal creator of the polluted, exponentiated present reality of the truth. Second, the primeval paternal's curvilinear reality as the ideal perpetuator of the proliferating, exponentiated future reality of the truth. The entanglement ascends the primordial paternal to be God. God descends divine blessings for transforming each devoted follower, who perpetuates the myth of the Godhead into a deity.

A deity is worth following as a matter of faith by any devotee of the contaminated modern science. No referee will ever question any scholar who invokes the reference of the Godhead for scripting a transformed reality of Mother Nature. By taking this additional step, a deity-like scholar becomes the primordial deity who binds the reality of Mother Nature with him as the supernatural paradigm of the present reality. The second bridging diagonal (the second letter H) is the entity that trades the Godhead-effect without verifying the truth of the curvilinear reality. The entity then services a new paradigm of the present reality as the new absolute to bind all children who wish to be credentialed as a scientist.

This investigation falsifies the supernatural truth of both: first, the primeval deity who is destroying the Godhead-effect and, second, God who has become the Godhead after servicing the "I AM a Deity consciousness" to the primeval deity. Therefore, it is self-evident that the present investigator is not a natural scientist and the present investigation is not a work of natural science. Since this investigation illuminates the natural truth hidden from the universe of deities, it is self-evident that the present investigator is a management scientist who is conscious of how to discriminate the truth from the falsehood without a fear of the social power consequences. A management scientist is guided by a vision to illuminate the power distance that is contaminating the virgin integrity of Mother Nature. The mission is to empower each child to become a metaphysicist, endowed with the power to enjoy the independence from both the mentally-conceived God and the intellectually-reproduced Godhead-effect.

1.3 The General Theory of Nonlinear Relativity

The theory of nonlinear discontinuous relativity is general and mutually-exclusive of the diverse specific entities that trade and embody the special Godhead-effect. A general, nonlinear God of modern science radiates the etheric body's force as a discontinuous form of the universe. God creates a discontinuous universe of faithful believers who are his devoted followers within the primordial universe of entities. Each special primeval deity irradiates God's ascending Godhead value, without the descending the Devil value that symbolizes the entropy reality of the myth of the perpetuating Godhead-effect.

The perpetuating Godhead-effect culturally binds the rationality of an infinite number of idealized follower entities and transforms each into a primeval deity. The primeval deity trades the energy of the mass of the devoted entities for destroying the freedom of the devotees, conscious of the truth of Mother Nature. The "perpetuating, thermodynamic mass" (M) becomes the institutional force, whose immanent energy value derives from the referential, squaring reality of "God" (C).

Without God, the "energy" (E) value of the perpetuating mass is zero, i.e., the entropy absorption value of each para entity trading the polluted para-consciousness diffused by God as an entity. Therefore, $E = MC^2 = 0$. It is the special energy value of the general theory of relativity.

Each para entity fails to quantify Einstein's ideal-effect because their mind is also a product of the polluted present reality. Einstein was trading the Western religion-effect for conceiving a God whose emanating creator energy is immanent within the creation mass and is exchanged back and forth by the universe of creatures at a lightning speed. The exchange between the inanimate mass and the animate universe occurs at a lightning speed. Therefore, the speed of light has the same metric value as the one who is diffusing the self-luminous entity energy and trading the divine energy of the primordial illuminator.

Consequently, the traded divine energy becomes the logarithmic base for the exponentiation of the entity's intrinsic divine energy by diffusing the self-luminous consciousness. The exponentiated divine energy transforms into the gravitational energy of the departed entity. The triangulated divine energy of the universe of entities trading the departed entity's guider-effect becomes the parabolic sentient energy. The sentient energy is the consciousness of the energy that is present without an entity. It is the truth of anybody who is a devotee of the Western culture-effect.

The Earth's Western geography trades the Sun's derived reality since the rising Sun is toward the East of the Earth. Therefore, an entity within the Western geography always has a consciousness of the Eastward source of energy. Such consciousness generated a historical proclivity in the West to migrate Eastward toward Central and Southern Asia and, in the East, to migrate Eastward toward the Americas. Western geography is not a naturally scripted spatial fact. It is a conceived temporal fact. There is no logical reason why people in the Western hemisphere perceive themselves as living away from the Sun—the param creator of life.

Those in the Eastern hemisphere experience a strong force of the proximity to the Sun. As a result, a discontinuous consciousness of the Earth's continuous space, which wholly faces the Sun in the East, is disproportionately traded by those living in the West. That's why the West is guided by a monotheistic belief in the truth of one God, seeking to convert others' consciousness away from the reality of the Sun as the absolute creator of infinite divine entities and forces.

Since this investigation illuminates the truth of the Western-effect and the culturally-bound rationality for reproducing and triangulating the Western-effect, it is free from the Western culture-effect. However, as is self-evident, the Western-effect itself is a mind-born construct. For cleansing the predominating mass of the Western-effect, anybody may quantize and trade the Eastern cultural system. A supreme deity, who enjoys a supreme consciousness of the true reality in oneness with the Sun's astral body, may manifest many supra deities. Each supra deity enjoys a supra consciousness of the actual reality for transcending the knowable shadow limit of the consciousness polluted by the Self's discordant, secret, and Satanic reality.

1.4 The Parabolic Theory of Chaotic Relativity

The circulating, chaotic, relativity theory is parabolic and excludes the self's value as the quantum digit. The circulating value of the present reality generates chaos in the mind of each entity. Each entity becomes just a quantum digit, perpetuating and squaring the tail-domain of the circulating circular head-domain. Suppose the discordant, Satanic energy of an entity is the metric of the squared myth. In that case, the circulating self of that entity doubles the perpetuating linear negative metric value of the squared myth. By excluding the self's value as the quantum digit that norms the negative discordant energy, the theory's value, grounded in the diverse forms of relativity, becomes zero.

The self is the absolute digit necessary for verifying the relative truth of any theory that is seeking to explain the absolute truth of Mother Nature. A natural scientist hypothesizes the theoretical value of the self to be zero. A natural scientist also believes that the self has a zero-effect on the objectivity of the discovered value of Mother Nature. By appreciating the zero-effect of the self on the universe's truth, a natural scientist falsifies the self's value as a contaminating physical body. A social scientist is conscious that anything a natural scientist discovers is subject to eventual falsification.

Consequently, instead of discovering the Eastern causative force of light, a social scientist seeks to discover the diffused light's Westward consequences. For a social scientist, the act of discovery is an act of social construction of the reality. Each entity who can construct the social value is a unique deity that invests in the self-luminous entity energy to create something of the universal value. By appreciating the self's infinite-effect on the truth of the universe, a social scientist verifies the self's value as a cleansing intellectual body.

A humanist is conscious that anybody who services the divergent energy for modifying the universe's truth is a Satan. A Satan impedes each entity's potential to be the present deity—the one with the power to construct a unique and personally desirable reality. By appreciating the self's finite-effect on the truth of the universe, a humanist sentiates the self's value as a resting mental body.

By giving rest to one's mental body, a self-luminous astral body thermodynamically activates the hibernating etheric body as the waking causal body to transform the sleeping self-luminous entity into a running primeval child. That running primeval child becomes a management scientist who seeks to run from the discordant energy of the universe of dreaming and night-walking entities into the arms of the already perfect Almighty Creation.

By trading the convergent energy of the perfect Almighty Creation, a primeval child becomes the future reality. The future reality is the exchange value of the Sun as the absolute creator of the materialized earth-effect. The materialized earth-effect is the consequence of the light radiated by the Sun. Each entity irradiates the materialized earth-effect for servicing the Self's guider force.

The aggregate value of the light radiated by the Sun and irradiated by the Earth (inclusive of the entities on the Earth) is the "gravitational mass" (μ) value of the universe of white stars, before their transformation into the varying forms of stars, planetary bodies, and earthly entities. The gravitational mass value of the universe of white stars is the "astral energy" (α) value of the Vega— the primordial white star and the cosmic universe's gravitational center.

A sentient "self" (β) services the ascending sentient energy in the form of the astral energy of the Vega star. The present value of anybody, who is continuously servicing a discontinuous value of sentient energy over infinite time, is zero. Therefore, $\alpha = \mu\,\beta^2$. The reason why anybody, who is just a unit quantum digit, becomes synonymous with the sentient energy by energetically polluting the consciousness of the universe of entities, is the hypothetical kingship value of the self, the universe of selves, and the universe without selves.

Anyone has the sentient energy to conceive each entity's hypothetical kingship value, the universe of entities, and the universe without entities as zero. Anyone may consciously trade a body of energy comprising the "I AM a Deity consciousness" as the primordial-primordial digit; each entity's hypothesized theory value as the primordial digit; the hypothesized ideal value of the universe of entities as the param digit; and the hypothesized overall value of the universe without entities as the primeval digit. Therefore, the energy value of both "anybody" (*Gardhaba*, 1000) and "sentient energy" (*Varuna*, 1000) is one-thousand units. It is free from the discordant energy of any organizational metric.

Further, the Almighty Creation's overall westward illuminated value is the truth of the Eastern cultural system. Each Eastern entity conceives each unit of the westward illuminated force and the Western entity to have the divine energy traded from the Almighty Creator in the East. For an entity trading the Eastern culture-effect, there is an infinite number of deities since each Western entity is a self-styled deity. Further, the effect of the universe of deities is that they continually wish to proliferate their sovereignty eastward. Therefore, each entity in the East is a param deity's incarnation and has the power to illuminate the truth of the polluted entity and the pure Mother Nature.

Since this investigation illuminates the truth of the Eastern-effect and the culturally-free self-luminous value of Mother Nature, it is free from the whitewashing of the Eastern culture-effect. However, the Eastern-effect itself is an astral reality. The Eastern-effect of the Sun is the luminous factor radiating the light force for generating and illuminating the Western earth-effect. The illuminating Eastern-effect is the supernatural paradigm of the present reality formed as the limit value of any theory, ideal, or hypothesis, guided by discovering the illuminated earthly reality. Since the truth of each of the forms of the theory of relativity is zero, the present investigation's objective is to investigate the diverse forms of the present reality as the form of the absolute reality without the mediating-effect of the investigating entity. It focuses on illuminating the perfectness of Mother Nature, free from the supernatural paradigm of the present reality that is the causative cause of the polluting Western reality and the cleansing Eastern reality.

Mother Nature manifests her perfection in diverse forms, each with a differentiated knowable energy value and a correlated knowledge concept in the Western and the Eastern language systems. **In this work, I italicize the correlated Eastern concept within the parenthesis following the relevant Western concept. I quantify the energy value of that concept as proof for verifying its management value.**

For a management scientist, the present reality matters because it has the value essential for shaping the future reality of nature, society, and humans. It does not matter just because one has a natural, social, or human inclination towards knowing the secret of the present reality before anybody else as a path for personal glorification.

Overall, two-hundred sixty-four pathways for ascending self-awareness are illuminated in this investigation and tabulated in Tables 10-27 in the following chapters. Taken together, they illuminate the secret of the "emanation value" (*Shankara*, 264), formed through the infinite circumambulation of the divine energy serviced by the universe of sentient entities throughout their lifetime.

Freedom from the emanation value is the "portal" (*Dvara*, 264) to knowing the secret of the "immanent value" (*Janaka*, 180). The immanent value is a function of the "intuition dimension" (*Sva dharma*, 96), formed through the interaction of the "self-luminous element" (*Svarochisha*, 12) within each entity and the "natural" (*Anatanam*, 8) element of Mother Nature without each entity. The self-luminous element is the secret to developing "conscious consciousness" (*Sati-Parvati*, 16) of the naturally flowing "self-perpetuating" (*Udvaha*, ½) energy of Mother Nature within each entity.

The Five-Fold Gravitational-Effect of One's Belief System

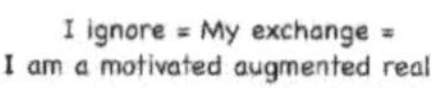

The Entity-Scripted, Superimposed Present Reality Perpetuates Five Forms of Virtual Entropy Reality

It Makes the Entity the Perpetuator God of the Paradigm that Exchanges the Unscripted Natural Reality.

Our spirit-level belief system guides the form of the present reality we conceive. We manifest our belief system into the present reality. By perpetuating and universalizing our belief system, we quintuple our gravitational power. We influence (1) our present and future value as a leader; (2) the present and the future of those who experience our constrained linear masculine vision as a follower; and (3) the perpetuating time-invariant reality of those conscious of Mother Nature's liberated non-linear feminine mission to let each person be the master of its discordant destiny as an entrepreneur. We influence the simplicity of the perpetuating reality by scripting a new virtual reality, and spiritually augmenting it, to destroy the natural organizational planning consciousness from the entire universe of entities. Specifically,

- "I am what I believe in" shapes one's present reality;

- "I do what I believe in" shapes one's future reality;

- "I service what I believe in" shapes everyone's present reality;

- "I trade what I believe in" shapes everyone's future reality;

- "I ignore what I do not believe in" shapes the eternal reality without the entity's gravitational-effect.

2.1 The Paradigm of the Present Reality as the Limit of Theory, Ideal, and Hypothesis Formation

The paradigm of the present reality shapes the organizational planning value. It is the key to open our self-luminous consciousness for manifesting the desired organizational development. The present reality paradigm includes the paradigmatic heuristics for sense-taking, techniques for sensible paradigm heuristics, and the entity managing those techniques within a unique belief system. Our primordial experiences shape our unique belief system, which is our self-luminous consciousness. Our belief system shapes the techniques for knowing the paradigmatic value of the present reality, and therefore the heuristics for exchanging the new reality. We evaluate the epistemological value of our primeval experiences with the ontological value of our beliefs. Our primeval experiences are the axiological value we create through our experience-integrated organizational development. Our organizational development values both: our organizational planning and the organizational planning of an infinite number of referential entities. The primordial-primordial experiences of the referential entities shape our primordial organizational planning.

We perceive the ontological truth—the hidden causative force of an experienced phenomenon—using our intellectual consciousness. We do so without the gravitational transformation by alternative referential experiences. The ontological truth is our noumenal reality as a subject, which we service as a force of attraction. We countertrade the phenomenological reality, transformed and repelled by the ecosystem back to us. An infinity of phenomenological realities constitutes the present reality. Although we are one super-universe of noumenal omnipresent reality, we are attracted to a differentiated universe of present phenomenological realities.

Our unique attraction set is a function of our unique system of beliefs about the diverse metaphysical values of the creators of the diverse present realities. The diverse phenomenological realities attract our mental consciousness as a function of their correlation with our belief system. The omnipotent truth of the diverse subjects servicing the varying forces of attraction is invariant to the varying forces of attraction that the subjects experience toward the ecosystem's omnipermeating extrinsic reality. Our ontological truth is our intrinsic reality. We are the almighty creator of the inferential value of the extrinsic reality. The inference is the transformative value of our experience. Our belief system transforms our experience into a hypothetical inference about the truth of the cause of our experience.

We are, therefore, an ontological body—the transforming almighty creature. Our ontological truth is our theory for the cause of our experience. The cause of our desired experiences is repelled by our immanent sentient light force that illuminates the cause for transforming the present reality into an alternative future reality. The causative, ontological body is the ideal convergent point of our self-luminous consciousness. Our universe of desired experiences is our hypothesis, as a divergent entity, about the proficient path for experiencing the ideal convergent point within our unique belief system's theory-effect. Our self-luminous consciousness is the knowable infinity-point value of our belief system without the varying para knower-effect.

As a "param perpetuator" (*Keshava*, 22), we cyclically reproduce our knowing of the "universe" (*Brahman,* 2) into the ideal convergent point of what we believe to be true. We form an imaginary body of divergent hypotheses to falsify and destroy an infinite body of parallel theories on the imaginary plane of our knowledge-base.

The infinity-point ideal value of our belief system = The zero-point reproduced, ontological value of the "param perpetuator" of the present reality (*Keshava*, 22) + The cyclical projection of the complex imaginary body of divergent hypothesis on an imaginary plane of the universe, to be the "organizational metric" (*Maha Shunya*, 0) of the alternatively imagined future reality (i * i) * Para knower cost-effect, servicing "moderation" (*Alobha*, 2) guided by the cost-escalating past experience = 22 + (-1) * 2 = 20.

The para knower cost-effect = Primordial worker cost of conceiving a parallel theory on the positive plane of para consciousness + Primeval worker cost of perceiving a desire for destroying a potential infinity of theories on the zeroth plane of consciousness.

The positive plane of para consciousness is the knowledge that transcends the limits of present consciousness. The zeroth plane of consciousness is the knowing without the primordial consciousness's negative plane. It is the knower's ontological body forming the double-negative (-2) inference value of the present reality.

There is potentially an infinity of divergent theories, convergent ideals, and parallel hypotheses within our absolute consciousness. However, at any moment, we breed a finite amount of theory-effect, ideal-effect, and hypothesis-effect for testing and authenticating the point value of our ontological truth. Our "breeding system" (*Nyaya*, 1) includes a proportion of this finite amount of our absolute knower consciousness. The universe of follower and believer entities that we create through our breeding system is an epistemological body. It is a super-knower guiding light force, transforming into the super-ideal convergent point of our absolute consciousness.

The super-ideal value of our breeding system = The ideal value of our "belief system" (*Saguna*, 20) + The cyclical projection of the complex imaginary body of convergent ideal on an imaginary plane of the self (i * i) * "Para manifestor cost-effect" (*Vaishya*, 3) = 20 + (-1) * 3 = 17.

The para manifestor cost-effect = The para knower cost-effect of the ideal self + Param worker cost of experiencing the desired infinity of theories on the infinity point of absolute consciousness.

The infinity point of absolute consciousness is the self. The self is the supra knower, manifesting the present consciousness as an absolute ideal value. The supra knower is an axiological body that services the triple-negative (-3) reference value of the present reality, including the metric of reference producing a negative one (-1) value through double-complexity. The epistemological cause of the truth is the noumena. We conceive the noumena by forming a hypothesis from a theoretical inference, within the ideal reference of our "personal experience" (*Adhyatama*, 11).

2.2 Three Types of Personal Experience

Personal experience is of three types: direct of the self, mediated through the others' senses and imagined through a sense of empathy with the others.

- When the personal experience is of direct type, we conceive the noumena as a theory validated with further experience, perceptive observation, and astute authentication mediated by the self.

- When the personal experience is of mediated type, we conceive the noumena as an ideal reference, the validating authority. An ideal reference is of three types:

  - **They:** One who believes us: includes our belief system as a supplementary belief paradigm.
  - **We:** One believable to everyone like us: a credible external authority within a complementary belief system, and

- o **I:** One like me who has a diverse belief system not believable presently and is the potential cause for forming a competing belief system.
- When the personal experience is of imagined type, we construct the noumena as a hypothesis for the investigation, validation, and authentication with alternative personal experiences as an antagonist. As an epistemological body, we norm our alternative personal experiences within a potentially falsifiable and destroyable paradigm of the present theories, ideals, and hypotheses.

A "protagonist" (*Nayaka*, 1), as an ideal competing subject (ontological body), perceives the inferential reality. An "antagonist" (*Khalnayaka*, 11), as an ideal supplementary subject, references a diverse set of personal experiences (the epistemological body working as a metric for measuring the worth of the ruling consciousness). As a complementary ideal subject, a "deuteragonist" (*Pattanayaka*, -1) conceives a global theory of ideal subjects. His behavior system is guided by the "wholesome" (Composite: *Nayana*, 957) personal experiences of the protagonist and the antagonist.

The deuteragonist enjoys the benefit of imaginatively referencing the diverse personal experiences of an infinity of subjects for idealizing them. Their personal experiences are strongly correlated with those of his. Both the protagonist as well as the antagonist suffer from the absence of a validating cause. On the other hand, the deuteragonist enjoys an infinity of validating causes for developing an intuitive sense of immanent reality within the complex proliferating paradigm of theories, ideals, and hypotheses formation.

An axiological body is a deuteragonist, which destroys local truth's epistemological cause and conceives a global truth. Since a local referential heuristic can't illuminate the global self's metaphysical reality, the axiological body validates both the protagonist and the antagonist's unique personal experiences. A "sidekick" (Agonist: *Grahanayaka*, 0) follower, who knows leader A as well as leader B, becomes a "superleader" (*Kshatriya*, 0) who knows what leader A and leader B know or don't know about the present reality.

The supra-ideal value of our behavior system = The super-ideal value of our conceived "octave of reality" (*Omkareshwar*, 17) + The cyclical projection of the complex imaginary body of the infinite theories on an imaginary parallel plane, unifying both the primordial protagonist as well as the primeval antagonist (i * i) * Param metric benefit-effect = 17 + (-1) * (-1) = 18. The param metric benefit-effect = Benefit of scripting the desirable hypothesis about the axiological value, on the parallel universe of "para consciousness" (*Nirharin*, 18) as a primeval wisher.

2.3 The Metaphysical Reality of the Becoming System

The parallel metaphysical body of para consciousness is our perpetuating and exponentiated "becoming system" (*Asha*, 9×10^{18}). An infinity of weak, secondary causative factors guides our "becoming" (*Anitya*, 9×10^{8}). A weak causative factor is the para-conscious value of the present reality, guided by a referential heuristic.

A practitioner makes sense of becoming as follows: X, the referential text, said so; Y, the reader, said so; Z, the researcher, and said so; and, therefore, A is the cause of my becoming B.

A scientist makes sense of becoming as follows: Becoming B is correlated with C belief system, D breeding system, and E behavior system, that includes A as the common denominator. Therefore, A is the immanent universal cause of Becoming B, without the limitations of C belief system, D breeding system, and E behavior system.

A philosopher makes sense of becoming as follows: A is immanent within the becoming of B, $\dot{B}$, and $\d{B}$. Therefore, A is emanating from $\breve{A}$, the scientist, as the eternal cause of becoming ΣB (i.e., infinite variations of B). Without the scientist, A will not be evident as the present scientific reality of becoming ΣB.

An educator makes sense of becoming as follows: $\dot{A}$, the philosopher, discovered the eternal cause of becoming ΣB, which is the absolute cause. Without the philosopher, scientist $\breve{A}$ will not be evident as the present ideal reality of becoming ΣB.

A student makes sense of becoming as follows: Υ, the educator, illuminated the eternal cause of becoming ΣB, and is the tertiary physical cause of becoming. Without the educator, becoming ΣB will not be a potential reality for a student within the present moment.

An investigator makes sense of becoming as follows: the infinity of the students is becoming the infinity of the educators, seeking to be an authority on becoming. They are not conscious of these diverse protagonists—the text, the reader, the researcher, the practitioner, the scientist, and the philosopher. All of them are tertiary and dynamic causes of the present metaphysical (i.e., not real physical, but idealized) reality of the educator, becoming an aspirational life path.

Does one have become a student of becoming earn them a doctorate on becoming? Does one have become a doctor of becoming make them the authority on organizational planning for the desired becoming?

One may hypothesize both the theories of the student and the ideal value of the student to be valid. In that case, since becoming powerful is not a normative reality, it implies that all the influential persons, who were once students, wish to perpetuate a state of power and privilege by not servicing the technological secret of becoming powerful. Alternatively, one may hypothesize only the theory of student to be valid. In that case, there is no ideal person who can be a paradigmatic authority on organizational planning for the desired becoming.

Each person becomes an authority through the devoted sequential engagement into investigating the diverse theories, ideals, and hypotheses on how to become powerful. As an authority, each person breeds personal behavior as the universal theory of growth, believing that the theory is the unique heuristic for becoming powerful. Alternatively, one may also hypothesize just the ideal value of the student to be valid. In that case, any person may become powerful at the present moment by transcending the universal theory's limits and servicing the life story of the self as the ideal growth paradigm.

A person's life path may be guided entirely by a theory-effect bred by others. Such a person, inevitably, experiences entropy of power and lets those who radiate and perpetuate their metaphysical becoming, as the ideal paradigm, enjoy ascending growth. Alternatively, a person's life path may be guided entirely by the ideal-effect of the perpetuating metaphysical becoming. Such a person also, inevitably, experiences entropy of power by illuminating others to destroy their physical followership behaviors dynamically and trade the perpetuating force of the metaphysical becoming of the leader.

Eventually, a person's life path may be guided entirely by hypothetically transforming the forming experience of the power entropy into a virtual reality (virtuon—the transformative space factor), without the present reality (soliton—the linear time factor). Such a person spontaneously experiences an absolute growth of power as a technological unit for descending the virtual risk of power entropy within the present reality. All those, who are fearful of the virtual reality, bank their power with the hypothesizing person as the banker of infinite integrity, liberating the universe by perpetuating his absolute guider mediating power.

The supreme-ideal value of our becoming system = the supra-ideal value of our behavior system (18) + the cyclical projection of the complex imaginary body of infinite virtual entity on an imaginary linear plane of the param deuteragonist self $(i * i)$ * liberator benefit-effect = $18 + (-1) * (-8) = 10$.

The liberator benefit-effect = Benefit of spiritualizing the metaphysical entity as the paradigmatic leader organization for generating the intrinsic consciousness (10) to integrate the present leader reality as desirable, without becoming a creator of the undesirable virtual reality.

A person may become free of the metaphysical spiritual-effect and sentiate the incarnational consciousness for the organizational planning of a primordial reality that forms the varying present values, without any complexity factor.

The para-ideal value of our "banking system" (*Ekavali*, 8000), as a bank of sentient energy = The infinity-point ideal value of incarnating the truth of our "belief system" (*Saguna*, 20), without any mediating guider power (20) * The cyclical projection of the present-primordial reality as the objective value of the primordial as well as the present realities * The cyclical projection of the supreme-ideal value of our organizational planning into the primordial-primordial reality of the self and the primeval reality of the universe = $20 * 2 * 2 * 10 * 10 = 8,000$.

The "incarnational consciousness" (*Sata*, 8000) = Benefit-cost ratio of sentiating the paradigm of the present reality of the self to form the universe of future entity reality = The paradigmatic value of the present entity reality * Unit value of "sentient energy" (*Varuna*, 1000) serviced by the present entity for incarnating the future entity = 8 * 1000 = 8,000.

"Organizational planning" (*Shri Krishna*, 10) is the normative value of planning. The "organization" (*Nairitti*, 29) is an act of norming a sequence of entities for fulfilling the desired body of causes. The "determination" (*Asiddhi*, 98) of the value of organizational planning is an act of evaluating the present reality, within the paradigm of management science. The present reality is the "divine plan" (*Sahajta*, 90) of a primordial creator.

A divine plan is the primeval-ideal value of our bragging system. It is an entity believing in the self's almighty creator power for thermodynamically incarnating a sensible sentient entity as an almighty creature. An almighty creature trades the transformative divine plan as the normative planning paradigm. It is guided by the immanent self-consciousness of the benefit-cost ratio of materializing the supernatural entity paradigm. It is the solution to the entropy cost of the formative planning paradigm of Mother Nature.

A supernatural organizational planning paradigm is the begging system's param-ideal value: it is an entity begging to know the planning heuristic of the primordial-primordial creator, which has endowed us with the power to be the almighty creature. A natural organizational planning paradigm is the primordial-ideal value of our "blessing system" (*Nipaka*, 10), which is blessing us with the immanent self-consciousness of a supernatural alternative as the present, not virtual, reality of a para entity.

The primordial ideal-value of our "self-consciousness" (*Prajna*, 2,222) = The linear perpetuation of the natural "objective value" (*Vidhana*, 2) of present entity reality as the convergent objective values of the primordial-primordial self (2), the primordial realm (2), and the param-primordial Mother Nature (2) as the knower of the sustainable and desirable objective value = 2,222. It is without any theoretical optimum mathematical operator conceived by an ideal mathematician.

2.4 Classifying Natural Planning Paradigms

We may classify natural planning paradigms into primary and secondary.

(1) Primary: There are three management science paradigms for organizational planning—the programming paradigm of a supra-natural creature; the performing paradigm of a supernatural creator; and the profiting paradigm of natural creation.

(2) Secondary: Within the metaphysical consciousness of the manager, who wishes to activate the formative value of natural planning paradigmatically, there are two dimensions —theory and ideal. The theory takes three forms to limit the formative value of planning paradigmatically. The ideal takes four forms to catalyze the paradigmatic value of formative planning organizationally. Besides, there are eight tertiary organizational development strategies for perpetuating the science of transformative planning without the mediation of the natural paradigm. There are twelve quaternary, twelve quinary, and twelve senary hypotheses about the primordial, primordial-primordial, and param-primordial paradigms that shape both the eighteen pairs of development and the entropy reality of the perpetuating organization.

The Paradigm of Present Reality

A paradigm is a secret recipe for creating, perpetuating, and destroying the potential reality. It is how an organization works behind the scenes, without the self's consciousness for illuminating the present reality. *An organization has six fundamental paradigms for doing so.*

First, to service the inanimate feminine energy as a catalyst of the proportionate animate masculine energy. Second, to invest the animate masculine energy as a catalyst of the disproportionate inanimate feminine energy. Third, to trade the proportionate feminine-effect as a catalyst of a param masculine entity. Fourth, to exchange the absolute masculine-effect for catalyzing the growth of the self into a param-primordial androgynous entity. Fifth, to develop a normative and organizational consciousness of the para-primordial androgynous entity, theoretically catalyzing the growth of the ideal self to falsify the hypothesis of the self-supremacy. Sixth, to organize the energies emanating, without the self's consciousness, for the self's further development as an organization, thereby self-substantiating the thesis of the present reality of an eternal self.

The above six are explained further below.

First, the paradigm of technological servicing of the inanimate feminine energy substantiates the masculine theory of the disproportionate ideal positive potential value of the self. It liberates the masculine entity to subtly equilibrate the intrinsic energy and ascend into a heavenly state of evenness of gravitational power within a spirit of liberal democracy.

Second, the paradigm of technological investment of the animate masculine energy substantiates the feminine ideal of the proportionate theoretical negative present value of the self. It binds the feminine entity to a state of gross disequilibrium in the extrinsic energy. It descends her into a hellish state, oddly diffusing her sentient energy for the masculine creator's intrinsic SHEENY well-being. The masculine entity perceives the transformative development as a metric of his absolute guider power for manipulating the feminine entity to be a responsible management agent, devoted to religiously perpetuating his SHEENY well-being. The feminine entity becomes the machine for generating disproportionate benefits by producing an infinity of devoted feminine children; each institutionalizing the religious trading-effect as the social culture-effect and affirming the animated masculine investor's supremacy.

Third, the paradigm of technological trading of the tertiary feminine-effect is catalyzed by the incremental gravitomagnetic return on investment, beyond the proportionate theoretical servicing value and the disproportionate ideal exchange value of the feminine, immanent within the masculine energy.   It empowers a masculine entity to substantiate the hypothesis of the supernatural heavenly value of the self.   The masculine entity is motivated to create a kingdom that includes the primordial feminine entity, working as the management agent for institutionalizing the social culture-effect, and the primeval feminine entity, self-managing the citizenship responsibility as her personal workculture-effect. Moreover, it includes the param masculine entity, who religiously follows the present paradigm of radiant love and becomes the ruling king of the sovereign kingdom. Even without any divine gravitomagnetic energy, the param masculine spirit entity technologically trades the sentient energy of the primeval feminine, to become a sentient child entity, for fulfilling the objective of the present reality.

Fourth, the paradigm of technological exchange of the quaternary feminine-effect, immanent within the param child, is catalyzed by the intrinsic thermodynamic energy. It is generated through the technological growth of the electromagnetic energy inside the physical body. The intrinsic thermodynamic energy generates technological entropy of the immanent feminine capability. The immanent feminine capability is the primordial-primordial cause for the satanic gravitoelectric energy. It gravitationally motivates the param creature to wake-up from the dark matter state of the absolute liberated inertia (moksha). The param creature becomes an inanimate object and seeks to compete with the animated param animal child entity. The param creature gravitationally transforms the cellular energy to self-organize a param-primordial androgynous spirit entity.

Fifth, the paradigm of organizational development of the quinary feminine-effect, which diffuses and emanates without the self's consciousness. It is catalyzed by the masculine theory of the ideal self. The theory of the ideal self is conceived by perceiving the self as a son of the mysterious God, the creator of the new paradigm of the religious divinity of the priest-king. As a devoted follower of the priest-king, the param-primordial self trades the objective value of the present reality to become a knower; they control the absolute intellectual property rights on the feminine entity's devoted work. However, since the ideal masculine self is not conscious of the primordial (secret) cause of the absolute sentient self, he is motivated to organizationally program the new paradigm as a heuristic for organizational performing. By subjugating the guider power of the sentient energy traded from the primeval feminine, he generates an entropy in the organizational profiting and, consequently, in the self as an organization.

Sixth, the paradigm of technological growth of the senary feminine-effect is a thesis conceived by the para-primordial deity. It is the primordial-primordial cause falsifying the hypothesis derived from the masculine theory of the ideal self. The para-primordial deity is the almighty creator of the universe of diverse ideal entities, each of which has a unique local guider power and universal global divine energy. Suppose a sentient entity believes in the supremacy of the local guider power for appropriating global divine energy. In that case, the natural energy of Mother Nature, as the primordial deity, falsifies the hypothesis of the devoted believer as an ideal almighty creation.

A sentient entity diffuses disproportionate animated energy for substantiating an impossible thesis, within a false consciousness of the almighty creature self as a supreme-primordial deity. It experiences the "I am possible" para-consciousness of the para-primordial masculine creator. The universe of inanimate entities trades that energy and descends the sentient entity into one of the thirty-six hellish energy realms. The sentient entity imagines these as the heavenly energy realms while devoting its divine energy to be free from the natural paradigm. The param-primordial spirit entities become the supra-primordial sentient animate entities by manifesting the supernatural paradigm. The sentient entity becomes a super-primordial inanimate guider entity—the ascended demonic master of the metaphysical entropy paradigm.

2.5 Three Primary Paradigms for Determining the Value of Organizational Planning

A supra-natural creature plans its theoretically conceived ideal value through organizational programming. The "programming paradigm" entails theory-shaping by an ideal self. The development value shaped through the programming guiding theory is the ideal value of the planner.

- **Under the programming paradigm, the value of organizational planning = organizational programming.** The value of organizational programming includes a normative system of the planner for learning, a transformative system of learning programs and routines forming the theory of ideal planning, and a formative platform of learnable genetic codes of the theory-maker within the normative organization.

o **The normative learning** is a product of the "learning to learn" heuristic of a normative citizen, who activates a formative set of learnable genetic codes—already learned and pre-programmed within the cultural system—as the institutional norms.

o **The transformative theory of ideal planning** is a product of the "second-order learning" heuristic. It empowers a normative citizen to activate the formative platform of learnable cultural codes as the social norms without the genetic system's limits.   The social norms constitute a complex leadership organizational workculture system for programming a first-order absolute benefit value.   A masculine entrepreneurial planner perceives an ascending second-order learning advantage by trading psychological power from his genetic programming.   He asymptotically closes an infinity of local disadvantages through disproportionate profiting from the intellectual property value of his theorized advantage as the master planner.   He is not conscious of the infinite cost of the localized genetic pre-programming. He services the institutional benefits without compensating the international system for the immanent institutional costs.

o **The formative platform of learnability** is a product of the "proficient networking" heuristic for diligently leading the path of integrated, holistic planner as the hypothesized ideal. Normalization of the formative path of learnability into a normative local leader entity, with genetically coded learning growth, mitigates the cost of transformative learning. It services absolute cost programming to infinite global followership entities devoted to freedom from entrepreneurs' network programming. The global performers take on both the ecological cost of discovering concealed genetic codes and the economic cost of culturally validating those codes.

A supernatural creator plans by promoting the organizational performing of the theoretically perceived ideal value of a normative creature. The "performing paradigm" entails theory-making by the aspiring self. The development cost of the imperfect theory guiding the performing is the ideal value of the planner.

- **Under the performing paradigm, the value of organizational planning = organizational performing.** The value of organizational performing includes: (a) a normative system of followership benefit, (b) a transformative system of descending followership benefits for collective leadership cause, and (c) a formative platform of free entrepreneurship benefit to compensate for the risks of mediating a value-based followership-leadership exchange.

  o **Normative "followership benefit" theories** are a product of the "super-followership" heuristic. A super-follower is a discerning follower who knows the value of a super leader, who, in turn, enjoys the benefits of proficient networking of a universe of devoted followers.

  o **Transformative "leadership cause" theories** are a product of the "supra leadership" heuristic. A supra leader is an ascendant master who is the percipient knower of the emotional qualities that discriminate against a discerning follower. The supra leader develops strong relational contractual bonds with the super-follower, who does not know the former's emotional codons' concealed emotional intelligence value. After socially binding the super-follower with the human covenants of a matrimonial sacrament for furthering the collective leadership cause, the supra-leader enjoys ascending benefits of the leadership collective's cause using his above-par agency rights.

  o **Formative "entrepreneurship benefit" theories** are a product of the "proficient exchange" heuristic. A supreme entrepreneur mediates the value exchange between two groups: First, a group of supra leaders seek to "produce and sell" a collective cause benefit; second, a group of discerning supra followers seek to "buy and consume" the individual benefits of the collective cause, without strong relational bonds with the supra leaders.

A path of supreme mediation is inherently risky because of two factors. First, the supra leaders lack the workforce power to produce and sell a collective cause benefit without any strong psychic linkages with the supra followers. Second, the supra followers seek to ascend the benefits of their workforce power by taking advantage of the supra leaders' weak psychic linkages. A supreme entrepreneur enjoys divine social qualities for inspiring both the supra leaders and the supra followers to work for him. As working agents, the supra leaders and the supra followers exchange the concealed knowledge about their production and consumption methods. They reveal their psychic weaknesses and weaken their credibility to profit from the direct risk-free exchange.

The planning of the natural creation services organizational profiting to the supra-natural creature, despite a supernatural creator's theory-effect. The "profiting paradigm" entails theory-taking by a hedonistic agent. The development benefit-cost ratio of profiting without theory-effect is the ideal value of the planner.

- **Under the profiting paradigm, the value of organizational planning = organizational profiting.** The value of organizational profiting includes (a) a cost-effective normative market for the entrepreneurial exchanges, (b) a transformative system for managing the escalating followership costs of consuming and leadership costs of producing the delegated normative cause, and (c) a formative platform for consuming a creditworthy secondary-cause marketed by the hedonistic management agent.

  - **Normative market cost-effectiveness theories** are a product of the "para manager" heuristic. Like an invisible hand, the market works as an inquirer of the tertiary benefits sought by each entity and services social engagement as the secondary benefit of its mediation. The market does not promise the desired tertiary growth benefits to any self-managing entity. It only assures that the self-managing entity will enjoy the at-par secondary benefits of mediating different participants' primary energies. It does not seek any part of the worker's social benefits. Consequently, the market is a cost-effective para-manager for each worker to enjoy the profiting value of its workforce proficiency as the primary benefit, with or without the market mediation.

- o **Transformative cost-escalating producer theories** are a product of the "primeval citizen" heuristic. Like a visible hand, the nonmarket institution works as an ideal, perfect, infinity point citizen. An ideal citizen has the "sentient impulse" to sense and the "willingness gravity" to consume everything produced by a proficient workforce system, organized by a self-managing divine entity. The primeval citizen is aware of the psychic weaknesses of the diverse entities. He knows how to invest the secondary cause-identity for redeeming the deadly sins of the freedom entrepreneur and be a friendly citizen angel. He takes the exclusive, hedonistic management agency, to be a perfect citizen devoted to the collective cause.

Additionally, the primeval citizen discredits the primary leadership cause-identity, as devoid of the collective cause. He credits the secondary absolute followership cause-identity, as the "void" (Virtuon: *Shunyata*, -2) fulfilling the collective cause. He becomes a cost-escalating producer of the diverse tertiary causes. He seeks the eternal blessings of the entrepreneurial principal to polarize disproportionate consumption credit as the compensation for the cost and benefit of the guider power. As a cost, the guider power substitutes the tertiary cause benefit with the secondary absolute followership cause-identity. As a benefit, the guider power substitutes the primary leadership cause-identity with the secondary absolute followership cause identity. As a managing agent of the nonmarket institution, the benefit-cost ratio of the guider power includes the tertiary-cause benefit and the primary leadership cause identity. The managing agent is the supra-natural creator of the nonmarket institution.

- **Formative secondary cause theories** are a product of the "param alien" heuristic. Like a uniquely inanimate and para-emotional alien, the masculine managing agent is free from the guider psychology of distributing credit to the feminine citizens for religiously performing the absolute theory of citizenship. He networks a theory of secular profiting to program a nonmarket institution for extracting infinite rents from the citizens as physical assets, without the market's natural planning. He knows how to help each citizen experience escalating group costs of their inherited secondary-cause programming. Within a state of absolute stillness, he ensures that each citizen consumes what she is producing as a primary leadership programming cause and what the entire citizen group is producing as the secondary followership performing cause. The consumption of the unfulfilled tertiary profiting cause generates an absolute entropy in the value of organizational planning. Each citizen experiences escalating personal work costs of the supernatural organizational planning, without any natural development of the collective social benefits.

2.6 Seven Secondary Forms for Paradigmatically Activating the Value of Organizational Planning

A managing agent has both a social fiduciary responsibility for furthering the hypothesizing entrepreneurial principal's cause and a personal moral responsibility for balancing the proportionate cause of the idealizing follower and the theorizing leader. The costs of guider mediation by a managing agent are a function of each citizen's integrated religious faith in the secular alien agent.

An entrepreneurial principal religiously hypothesizes that a feminine managing agent will service his interest as the primeval child. A follower religiously idealizes the feminine managing agent as the primordial maternal devoted to his well-being as a param child. A leader religiously theorizes that he is the param paternal of the feminine managing agent, who is a devotee of the family group's collective well-being.

A self-organizing masculine entity activates the secular heuristic to trade the triple planning benefits from the feminine managing agent for servicing the well-being of the local principal, national follower, and international leader groups. **He activates the "planning paradigm" of the primordial paternal, where the value of organizational planning = organizational planning.**

The "value of organizational planning" (*Vijaya*, 104) is traded as the "believing energy" (*Isani shakti*, 104) by a believer. It includes the "primary-cause object" (Personal programming: Feminine *Shani*, 18) that is desired by each citizen leader, the "secondary-cause objects" (Social performing: *Tushti*, 85) produced by the infinite alien followers, and the "tertiary-cause object" (Religious profiting: *Naraki*, 1) consumed by each entrepreneurial manager. The "self-incubating" (*Sah*, 1/8) proportion of organizational planning, that metaphysically forms the "universe of management entities" (*Moksha Astikaya*, 1/8), is a product of three values:

- **Normative planning value = f (Mass of the primary-cause object/ citizen leadership value of the primary-cause subject).** The higher the energy-mass of the primary cause object, the higher the social benefits of making that object the optimization cause of organizational planning. The higher the citizen leadership value of the primary-cause subject, the higher the consequential worker-social benefits of the normative mass of cause-object. In other words, normative planning is the subjective value of an object.

- **Transformative planning value = (Energy of the subjective experience of accelerating production/ citizen leadership value of the secondary-cause object).** The higher the energy of the subjective entropy generated by accelerating production, the higher the social costs of production (and the higher the social benefits of stillness). The higher the citizen leadership value of the secondary-cause object, the higher the consequential worker-social costs of an added cause object. In other words, transformative planning is the objective value of a subject.

- **Formative planning value = (Energy of the objective of accelerating consumption/citizen leadership value of the subjective tertiary-cause).** The higher the energy of the objective growth generated by accelerating consumption, the higher the social benefit-cost ratio of forming an alternative planning paradigm of theory-taking instead of theory-making. The higher the citizen leadership value of the tertiary-cause object, the higher the consequential worker-social benefit-cost ratio of theory-taking as the formative planning paradigm. In other words, formative planning is a divine process of transforming a subject into an object by consuming the sentient energy immanent within the subject. It produces the gravitational energy of the object as the absolute functional value of the planning.

There are seven secondary forms for the formative planning of the universe of management entities in the form of an object. The object functions as the organizational profiting catalyst for perpetuating a linear value of the future organizational planning. **The seven secondary forms include the three forms of theory planning and the four forms of ideal planning.** Theory planning includes <u>theory-taking</u>, <u>theory-making</u>, and <u>theory-shaping</u>. Ideal planning includes <u>ideal-taker</u>, <u>ideal-maker</u>, <u>ideal-shaper</u>, and <u>ideal-equilibrator</u>.

2.7 Theory-Taking

Theory-taking is a primeval paradigm of taking a general theory as an ideal paradigm, whose citizen leadership value = 1. As a follower of an ideal paradigm, a citizen scientist experiences the zero cost of production (primary-cause), consumption (secondary-cause), and exchange (tertiary-cause) within the paradigm of modern science. A citizen scientist may produce, consume, or exchange an infinity of theories without critical reviews, as long as those theories fall within the ideal paradigm. Consequently,

$$E = \text{Overall "self-irradiating" } (Sah, 1/8) \text{ energy growth value of organizational planning by a general follower}$$

$$= \text{normative planning value} \times \text{transformative planning value} \times \text{formative planning value}$$

$$= \text{mass of the primary-cause object} \times \text{subjective energy of the infinite production rate} \times \text{objective energy of the infinite consumption rate}$$

$$= m \times c \times c, \text{ because}$$

$$c = \text{absolute divine light force of an entity as the self } (8 * 10^{15}) + \text{knower proportion of the para-entity divine light force } (2/10 * 8 * 10^{15} = 1.6 * 10^{15})$$

$$= \text{subjective energy of the infinite production rate} = \text{objective energy of the infinite consumption rate.}$$

Thus, the "self-irradiating" (*Sah*, 1/8) energy growth value of organizational planning using $E = mc^2$ theory is a product of:

- **m = ½:** The normative planning value set = (0, 1), where 1 = presence of a primary-cause object as the creation factor; 0 = absence of the creation factor. Consequently, the normative planning value of m = ½. The primary-cause object is either the purpose of creating organizational planning (norming the state of 1) or not (norming the state of 0). For instance, a smartphone (i.e., the mass) creates normative planning (or not) as a function of whether an organization owns (or not) intellectual property rights (i.e., the guider power) to optimize the unique object value of the smartphone for the self. The organization must exchange half of its guider power to trade the subjects' sentient energy devoted to optimizing the object value to its unique transformative reality.

- **c = ½:** The transformative planning value set = (0, 1), where 1 = presence of a subjective energy experience as the perpetuating infinite factor; 0 = absence of the perpetuating/ infinite factor. Consequently, the transformative planning value of c = ½. The subjective energy experience is either the purpose of perpetuating the organizational planning (transforming from zero to infinity, transcending into a para-conscious subjective state of holistic transformation) or not (zero transformation). For instance, an organization (a system of subjective energies) perpetuates the transformative planning (or not) as a function of whether it wishes (or not) to make infinite variations to a smartphone to optimize the unique energy experience of each subject entity. The organization must exchange its entire residual guider power in the form of divine energy to optimize its subjective energy to the inclusive formative reality of the universe of subjects.

- **c = ½:** The formative planning value set = (0, 1), where 1 = presence of an objective energy value as the destruction factor; 0 = absence of the destruction factor. Consequently, the formative planning value of c = ½. The objective energy value is either the purpose of destroying the present paradigm of organizational planning or not. If it is, then the organization forms a dream-state of the imagined negative infinity using the objective energy immanent within the perpetuating object. From a negative infinity state, the organization transcends into a zero state, where a reconstituted innovative object, such as a smartwatch, perpetuates as an alternative paradigm. The organization must exchange the guider power of the whole universe in the form of natural energy for fulfilling its objective. The objective is to enjoy the alternative paradigm as a subject in primordial oneness with the alternative paradigm. The subject is in primordial oneness with one's future self, as an organization that is planning the alternative paradigm.

Consequently, the "self-irradiating" (*Sah*, 1/8) energy growth value of organizational planning using e = mc² theory = ½ × ½ × ½ = 1/8. It is the trading value of the "primeval management factor" (*Moksha Astikaya*, 1/8) serviced by the "universe of management entities" (*Moksha Astikaya*, 1/8). The presence of a primary cause-object is a predominating factor for creating organizational planning. The dominating cause is to discover the subjective value of the object. The deciding factor is the proportionate cost-effectiveness of the primordial object in a finite form, transforming into the present object in an infinite form and forming a primeval object in a potential form.

The subject is the potential form of the primeval object. The value of the "theory-effect" (*Mahadasha*, 0) is zero. After organizational planning, the subject, who as a planner (*Sundari Sodashi*, 10^{1000}) services the logarithmic "divine energy" (*Asrava shakti*, 10) into the organization, experiences absolute entropy and has zero management value. Thereafter, correcting for the e = mc² theory-effect, the energy growth value of organizational planning—that is mediated by the e = mc² theory = 0.

2.8 Theory-Making

Theory-making is an absolute paradigm of the metaphysicist as the special subject, the ideal citizen leader making the primeval object's physics. A paternal scientist produces, consumes, and exchanges a special theory within the infinite costs of the production, consumption, and the exchange of new science. He then destroys the present paradigm, which is the primary cause of the emanating costs. He illuminates the general theory as the potential form of the primordial subject of critical reviews. Therefore, the overall "inhalation" (*Ana*, 1) energy growth value of the theory-maker

= normative planning value × transformative planning value × formative planning value.

- **Normative Planning Value:** Normative planning value is the value of creating a metaphysical theory within the leader self's presence as the subject of ideal paradigm (1), vs. without (0). The leader self idealizes (i.e., consciously personifies) a special theory that the creator is different from the creature. The follower universe idealizes (i.e., consciously personifies) a general theory that the creation is different from the creator, conditional on the belief system that the creator is different from the creature. The special theory liberates the creature from the perceived costs of creation. In the creature's mind, the illusionary creator bears the infinite costs of planning the primeval subject.

The primeval subject becomes an object, trading the special theory as the normative planning value and para-consciously servicing the general theory as a natural consequence. Therefore, as a mediating guider, working on manifesting the transformative planning value, the special theory's normative planning value = 1. The formative planning value of an alternative general theory of non-relativity, not idealizing the leader scientist, is zero. A general theory of non-relativity is a meta-scientific theory and is a discovery of zero value within the scientific realm.

- **Transformative Planning Value:** Transformative planning value is the value of the perpetuating infinite forms of the special metaphysical theory as an "ideal supplementary paradigm," within the presence of the subjective energy experience of the paternal leader (1), vs. without (0). If a primeval physicist discovers infinite quantum alternatives to the special theory of relativity, without falsifying the special theory of relativity, then the transformative planning value of the alternatives (and, therefore, of the leader universe, as the primeval subject who trades the normative planning value of the special theory) = 1.

Quantum theory of the ground reality enjoys at-par mind space within any scientist, as does the special theory of relative reality formed without a consciousness of the ground reality. If an entrepreneurial scientist forms a quantum theory without norming the special theory, then the transformative planning value of that alternative = 0. A special theory of non-relativity is a hypothetical conjecture. It is a discovery needing further validation before breeding as a scientific truth among students.

- **Formative Planning Value:** Formative planning value is the value of destroying the general transformative universe of the metaphysical theory within the presence of the objective energy value of ground reality as an "ideal competing paradigm" (1), vs. without (0). If a param physicist (such as Einstein) conceives the general theory of relativity before the special theory's infinite transformations, then his formative planning value as a holy spirit who enjoys the intuition about the infinite transformations = 1.

A param subject enjoys at-par mind space within any wisher of the positive science, i.e., within one seeking an ideal ground reality. The ideal ground reality is authentic. It is an absolute quantifiable idea, without the limitations of special theory-effect. The feminine self is the primordial maternal bearing the absolute truth of the primeval hypothesis about the ground reality and the param theory of the relative reality of the paternal self. She is the ideal ground reality, embodying the absolute paternal consciousness as well as the primeval masculine personifications.

The feminine metaphysicist is a creation, perpetuating the illusionary truth of the mc^2 theory. The "inhalation" (*Ana*, 1/8) energy growth value of her organizational planning, conceived and traded by a paternal scientist = $1 \times 1 \times 1 = 1$. With or without the mc^2 theory, the "supernatural energy" (*Shram Shakti*, 1) serviced by Mother Nature for substantiating the "illusionary energy" (*Maya Shakti*, 1) of the theory and the theory-maker is the absolute truth.

The presence of a secondary-cause subject is a predominating factor for perpetuating organizational planning. The dominating cause is to discover the subject's objective value, without the costs of the secondary-cause object. The deciding factor is the proportionate cost-effectiveness of the three primeval theories of ground reality. The first theory is the present subject in a finite form (the authentic truth of a general hypothesis in its original form). The second theory is the primeval subject in an infinite form (the beauty of the varying derived transformations, each validating the general hypothesis). The third theory is the primordial subject in a potential form (the discovery of the infinite forms of the ground reality, which substantiates the multi-dimensional truth of the primordial subject and falsifies the absoluteness of the general hypothesis).

2.9 Theory-Shaping

Theory-shaping is a primordial paradigm of a non-physicist (such as Isaac Newton, a mathematician-alchemist) seeking credit for becoming the Father Creator of both the physical science as well as the metaphysical philosophy of science. The overall "fitness" (*Porutham*, 16) energy growth value of the organizational planning of the theory-shaper, who is a masculine academician, is a product of:

- **Normative Planning Value:** The normative planning value is the value of shaping the theory of dynamic (transcending the inertia of ground reality) ideal, without the Father Creator's knower identity (2). It entails dynamically reshaping the identity of the Father Creator by projecting an infinity of relative ideas (conditioned by an infinity of academicians), while seeking to falsify the dynamic theory of non-relativity and perpetuate a potential theory of relativity (2). The potential theory of relativity is a general theory, whose truth is conditional on the falsifiability of the dynamic theory of non-relativity. Conditional truth is the normative planning value of the dynamic theory's knower, with an absolute potential to falsify and perpetuate a potential theory of relative inertia.

- **Transformative Planning Value:** The transformative planning value is the value of first, perpetuating an infinite general hypothetical shapes of the potential ideal, within the presence of the subjective knower energy of the Father Creator (2); and then, second, theorizing the potential of perpetuating infinite relative ideas without the presence of the subjective knower energy (2). The unconditional relativity theory is an absolute theory, whose infinite truth value is a special and exceptional case. The special case enjoys an absolute value by transforming the ground (i.e., the spatial: the one present without subjective consciousness) reality as the conditioning factor into the abstract (i.e., the temporal: the one growing within subjective consciousness) reality.

The abstract reality is the catalyst of the infinite general hypothetical shapes. An academician may abstract and mentally situate the present findings within the traditional classical paradigm (conditioned by space's objective reality). Alternatively, he may shape a complex universe of the infinite concrete hypothetical shapes, each of which is a potentially general theory. The complex universe is a modern orthodox paradigm (conditioned by the subjective reality varying over time). It is complex because each transformed subjective reality perpetuates as an idea (i.e., complexity-effect, i) within the academician's absolute consciousness.

- **Formative Planning Value:** The formative planning value is the value of first, destroying the potential ideal hypothesis about the objectivity of the potential (i.e., general), kinetic (i.e., dynamic), and gravitational (technological, i.e., special) realities (2), and then second, idealizing the self as the knower personification of the Father Creator (2). An academician may destroy the abstract notion that the complex reality is objective. Alternatively, he may destroy the Father Creator's knower energy, which is breeding that very abstract notion within his mental consciousness. Consequently, he may form the knower consciousness of the Father Creator (2), who is radiating the potential, the dynamic, and the technological realities as a light force.

The "force" (*Prapya*, 34) is the illuminated light effect of the destruction of a physical object within which the consciousness of the multi-dimensional realities (potential, dynamic, and technological) is immanent. Force has no functional work value; it is the residual effect of the work done by an organization by destroying the physical object. The physical object performs no work—instead, the ecosystem without the physical object performs the work. The organization programs the work. The self is the entity planning the work. The work is a normative development value of the mental consciousness. The self seeks to form a complex experience, by first, incrementally conceiving the varying proportionate realities of Mother Nature, over time, and then, second, transforming that into the Father Creator's absolute reality, which is universal across space.

Consequently, the "fitness" (*Porutham*, 16) energy growth value of the organizational planning of the Father Creator as an absolute ideal $= 2^i + 2^i = 8 + 8 = 16$. As a path for illuminating the paradigm of the scientific method as an absolute ideal, a primeval academician transforms the illusionary truth of the mc² theory into a complex reality. The absolute ideal is the illuminated co-existence of physical and metaphysical realities within an integrated spacetime dimension. The "i" is three because it forms three complex ideas within the present theoretical idea (i.e., absolute idea).

- o First, the presence of a tertiary cause-creator is a predominating factor for destroying the organizational planning.

- o Second, the dominating cause is discovering the ideal subject, who knows how to destroy the limitations of the present ideal. The ideal subject lets the primordial ideal flourish as the operational theory of the varying space-appropriate physical actions, varying across time in ascending order, without the complexity factor.

- o Third, the deciding factor is the "absolute idea" without the theory-effect. The absolute idea is the ideal subject. The ideal subject is the primordial-primordial self, who is shaping the multidimensional truth of the ideal universe. The ideal universe's multidimensional truth includes the classical Newtonian mathematics, the orthodox Einsteinian metaphysics, and the modern quantum science.

Three complex ideas create a triangulated illusion of truth. They together form a "supernatural paradigm of the present reality" (*Yukti*, 8) that mirrors the primordial reality of "Mother Nature" (*Anatanam*, 8). The two-dimensional reality of Mother Nature—one primordial and natural and another primeval and parallel—perpetuates the energy growth value of the organizational planning. It forms the "intrinsic squaring layer" (*Sahaja Puta*, 16) of the "formative divine energy" (*Madhusudan*, 16). By trading the formative divine energy from the "child primordial greeter" (*Madhusudan*, 16), anyone gets empowered to become the Father Creator and earn the credit for performing the organizational planning. This planning is already immanent within the triangulated "tertiary residual" (*Khara*, 6) of the "universe" (*Brahman*, 2), beyond the limits of the "supernatural energy" (*Shram Shakti*, 1).

A child primordial greeter gifts whatever is present in the universe in a primordial-primordial form of the divine energy. The divine energy is the power to decide the greetable gifts and endow each primordial paternal entity with the gravitational energy for organizationally manifesting and enjoying those gifts without the "supernatural work energy" (*Shram Shakti*, 1). A primordial paternal creates a universe of child entities who devote their sentient energy for fulfilling the purpose of their creation. The universe of child entities follows the path for trading the guider-effect of the Father Creator and servicing their unique truth as an idealized mirror image of the Father Creator.

2.10 The Ideal-Taker

The ideal-taker is a primordial-primordial paradigm of a student (i.e., a divine entity, consuming the SHEENY consciousness or sentient energy), an almighty creature idealizing the destiny of a master academician (i.e., technological entity, producing the guider power or gravitational energy). An ideal-taker begins the lifepath with a subjective notion of the $e = mc^2$ theory of the organizational planning, with a "self-incubating" (*Sah*, 1/8) growth energy value = 1/8; it culminates by discovering the truth of the ideal subject within the self. The ideal subject is proficient in both the classical as well as the orthodox approaches to the scientific method.

The energy growth value of the organizational planning of the ideal subject, who as a "cosmic child" (*Arcisa*, 128) is conscious of the multidimensional entity paradigms forming the present reality, is 16/ (1/8) = 128. For instance, *Pansiyapanas Jathaka Potha* is a sequential collection of the 550 past lifetimes of Gautama Buddha, the ideal subject. Each lifetime is one divine dimension of Gautama Buddha as an Ideal Taker. Each lifetime is a proportion of the absolute reality of Gautama Buddha as an ideal subject, who embodies those infinite primordial realities. An Ideal taker is a proportionate "*bodhisattva*" (*Param Buddha*, 6), within the consciousness of the proportionate absolute reality of Gautama Buddha. The absolute-absolute (omnipresent) reality of Mother Nature, who is the mother of the infinite proportionate absolute realities, consists of the four divine entity dimensions:

- **Creator factor:** "Soul" (*Atman, 4*)—supreme deity—created at the infinity of imagination as an ideal subject;

- **Perpetuator factor:** "God" (*Ishvara, 5*)—para deity—the creator of the theory of value perpetuating in the universe;

- **Destroyer factor:** "Absolute Buddha" (*Bodhisattva [Maheshwara], 6*)—primeval deity—the perpetuator of the hypothesis that there is only one "devotee greeter" (*Para Buddha [Maisamma], 6,800*) who is the destroyer of the dark illusion of the infinite theories blocking the light force of the ideal subject; and

- **Illuminator factor**, the "Absolute *Bodhisattva*" (*Parameshwara, 7*)—param deity—the destroyer of the light force of the ideal subject which is transforming the consciousness of the self as an entangling "deity soul" (*Paramatma, 1600*).   The deity soul illuminates the omnipresent cause for radiating the divine charge, seeking to norm, transform, and form an infinity of absolute ideas.

The overall value of the illuminated divine charge exchanged by ONE absolute ideal subject = One unit of divine charge as a creator + One unit of divine charge as a perpetuator + One unit of divine charge as a destroyer = 30 units of divine-effects as an illuminator = 30 causative perfections within a state of infinite enlightenment.

Gautam Buddha enjoyed an energy growth value of 32 (2^5) within 550 past lifetimes by doubling the energy growth value after each sequence of 110 lifetimes.   The thirty-two is the energy value generated through "primordial oneness" (*Adi, 32*) with the "perpetuating primordial-primordial creator" (*Krishna, 32*). The one-hundred ten is the "perforation" (*Viddhatva, 110 = [1000 + 100]/10*) of the "divine energy" (*Asrava shakti, 10*) over one-hundred ten lifetimes, after becoming one with "sentient energy" (*Varuna, 1000*) and focusing on trading the "gravitational energy" (*Lalita, 100*) of the devoted followers.

The one-hundred ten lifetimes included ten as a creator of the proportionate absolute reality x ten as a perpetuator of the proportionate absolute reality + ten as a destroyer of the omnipresent divine reality as an omnipotent divine entity.   As a destroyer, fifty lifetimes were "self-perpetuating" (*Udvaha*, ½) the omniscient guider entity. The five-hundred-fifty is the eventual "quitting" (*Jahat*, 550) of the organizational plan for trading the devotee gravitational energy as the path for becoming an "ascending primordial-primordial creator" (*Sadhya*, 32).

During his 551st life as the "deity soul" (*Paramatma*, 1600), Gautama Buddha doubled the formative growth value of 32 to 64 to become an "omniscient guider" (*Palala*, 64), who is conscious of the self as a divine "greet" (*Palala*, 64) at the moment of his self-illumination. He then doubled the normative growth value of 64 to 128 to become a "cosmic child" (*Arcisa*, 128) at the moment of his self-liberation from the universe of sentient entities.

The cosmic child is an ideal-taking knower who behaves as if he is the "probable cause" (*Hariti*, 128) of the cause-manifesting (theory-taking, theory-making, and theory-shaping) system realities of the self. Self's perpetuating presence as a probable cause is the predominating factor for illuminating a cause-free reality without the illusion the self creates. It is the value of organizational planning the self destroys by ideal-taking. The dominating factor for perpetuating the self's presence is to make the self the ideal cause, with the potential to project intrinsic consciousness for manifesting an ideal quantum object (i.e., light, free of mass-effect). An ideal quantum object is one whose reality is both: a classical linear function and an orthodox wave function.

The deciding factor is to "take" the cosmic child as the ideal cause for hypothesizing the varying proportions of the self's absolute truth. The self's absolute truth as the "light" (*Prabha*, 180 = 128 + 52), free from the follower mass, is the future value of the cosmic child after "awakening" (*Samjna*, 52) from the "shadow consciousness" (*Chhaya*, 52 = 34 + 18) of the devil self. The devil self radiates the "force" (*Prapya*, 34) of the "mind-born primordial maternal" (*Amudhehswari*, 18) as an "offspring" (*AUM shakti*, 18) embodying the "future of everybody" (*Shani*, 18). The future of everybody is the fundamental characteristic (i.e., normative planning) value of the ideal quantum object as a unit of diverse time (transformative planning) within a primordial space (formative planning).

2.11 The Ideal-Maker

An ideal-maker is a param-primordial paradigm of an investigator, an almighty creation idealizing the almighty creator reality of a primordial creature. The investigator causally norms four parallel paradigms of the organizational planning—ideal-taker, theory-shaping, theory-making, and theory-taking (4). It forms ten finite sequences (1, 2, 3, 4, 12, 13, 14, 23, 24, 34) as the fifth, ideal-maker paradigm, by taking the ideal self as the primordial (first) entity, the param (second) entity, the primeval (third) entity, and the primeval-primordial (fourth) entity. It transforms the forty finite sequences within the self's varying ideal values into forty infinite sequences within the constant ideal values of the universe without the self. Forty infinite sequences make the universe into a finite ten-face entity within the infinite prime sequence (1st, 2nd, 3rd, 4th, 12th, 13th, 14th, 23rd, 24th, and 34th prime entity).

At each finite point, the universe trades 1, 8, 9, or 10 face ideal values of the homologous self, after four, three, or two successive proportionate absolute values. At the idealized infinite point (34th prime entity: 139), the universe plans "hibernation" (*Sushuptivat*, 139) after programming the formative organizational planning of the thirty-three transformations of the deity soul, as the diverse self. The energy value of the deity soul's organizational planning, inspiring the universe to program the perpetuating paradigm $= 40 * 40 = 1,600$.

For instance, *The Ramayana* is an epic story of "Sri Rama" (*Ramachandra*, 12), a fifth-dimension ideal-maker (12th whole number forming a self-luminous human entity). The "perpetuator deity" (*Ishvara: Param Vishnu*, 5) incarnates as "Sri Rama" (*Ramachandra*, 12) after an ascending oneness with the "param deity" (*Shiva*, 7). He destroys "Ravana" (*Dasha Mukha*, 23,125), originally a "fourth-dimension ideal-taker" (*Atman: Param Brahma*, 4). The four-dimensional ideal-taker is a ten-face entity who projects the four-entity hellish (variable) dimensions sequentially into $4 + 3 + 2 + 1 = 10$ directions, without consciousness of the twelfth heavenly (constant) ideal-maker identity.

Ravana embodied the "infinite organizational programming" (Learned Self: *Palala*, 64) of the universe as an ideal-taker. Ravana became an "omniscient guider" (*Palala*, 64), who knew the secret of the universe's entire creation and how to manipulate the universe to manifest the desired reality. Sri Rama embodied the "infinite organizational planning" (Primeval Self: *Rama*, 100) of the self as an ideal-maker, who projected the ten "deity" (Omnipresent: *Deva*, 1) dimensions concurrently into the ten "deity-soul" (Omnipotent: *Paramatma*, 1600) directions for programming the ten "deity-spirit" (Omniscient: *Rishabhanatha*, 91) entities without the self.

The "ideal-maker" (*Rama*, 100 = 19 + 81) is a "nineteen-phase entity" (Programmed Life: *Manyu*, 19) who trades the "growth-effect" (*Shri*, 81) over the "self-perpetuating" (*Udvaha*, ½) "divine energy" (*Asrava shakti*, 10) from the twentieth, "param manifestor" (Entropy Point: *Apahrtabhara*, 20) phase. The param manifestor phase's energy is immanent within the futuristic "grandmaternal spirit" (*Kapinjala*, 20). The ideal-maker services the immanent energy to the "twenty-face grandmaternal soul" (*Adi Para Atma*, 100), composed of the ten deity-soul and the ten deity-spirit dimensions. The grandmaternal soul services the immanent energy to the "deity" (*Deva*, 1), over ten lifetimes, in the form of the "divine energy" (*Asrava shakti*, 10) for manifesting the ten deity dimensions.

The perpetuator deity is an ideal-making manifestor, who is creating, perpetuating, destroying, and illuminating the cause-conscious (ideal-taking, theory-shaping, theory-making, and theory-taking dimensions of "present reality" [*Badhabuddhi vadartha*, -2]) system realities without the self. The presence of a cause-conscious destroyer is the predominating factor for illuminating a cause-conscious reality of the universe as the value of "organizational programming" (*Dharana*, -8 = -2 * 4). The dominating factor is to validate the descending-order entropy reality of the universe as the theoretical-cause of the ideal planning by a guider entity: it has the technological potential for destroying the whole entropy reality of the universe.

The deciding factor is to "make" the almighty creator reality of the self the ideal cause for falsifying the hypothesis that the universe's infinite manifested reality is the absolute reality of the self as the creator. Absolute reality of the self is free from the varying proportionate normative-effects. It is the formative nature of an ideal quantum object as a unit of diverse space (normative programming) value, projecting an absolute entity's formative planning.

The formative planning is the deity reality of the self as the self-luminous subject of the divine "organizational planning" (*Shri Krishna*, 10). The self becomes a "self-luminous subject" (*Pradyumna*, 60) by trading the "organizational planning" (*Shri Krishna*, 10) value of the "light" (*Prabha*, 180), without "absolute time" (*Sva*, 11 = 180 - 169), while enjoying "consequential freedom" (*Caksur vijnana*, 109 = 69 - 60) from the limited force of the light. With bare-essential "formative planning" (*Nissara*, 1/60), the self becomes a "deity" (*Deva*, 1), as a "divider" (*Shudra*, 1 = 1/60 * 60) of the "self-luminous subject" (*Pradyumna*, 60) into a group of five "greeter self-luminous entities" (*Vithoba*, 12).

Each greeter self-luminous entity manifests the "organizational programming" (*Dharana*, -8) of one-hundred ten lifetimes, as a deity who trades one unit of divine energy in every lifetime and services it for creating ten "maternal spirits" (*Dasha*, 1). The "para primordial self" (*Sri Rama*, 12) trades the ten maternal spirits' unit energy plus the unit energy of the historical grandmother soul and the futuristic grandmaternal spirit's unit energy. He services one unit in the form of the "supernatural energy" (*Shram Shakti*, 1) for generating the westward "I AM a Deity consciousness" (*Dakshinachara*, 1) among the devotees.

2.12 The Ideal-Shaper

An ideal-shaper is a primeval-primordial paradigm of a wisher, which theoretically shapes the subject of present reality into the ideal (infinity point, primeval) primordial self.   The wisher forms the consciousness of the seven phases of organizational performing, including the primordial-primordial phase within the self as a primordial ideal equilibrator (organizational planner) and the six primeval-primordial phases within the universe as a primeval ideal equilibrator (organizational programmer). The six primeval-primordial phases include ideal-equilibrator, ideal-shaper, ideal-maker- ideal-taker, theory-shaper, theory-maker, and theory-taker.

The wisher transforms the infinite ten-phase prime sequence into a finite ten-dimensional binary sequence without a mediating and transformative guider factor. The binary sequence is (11, 22, 33, 44, 1212, 1313, 1414, 2323, 2424, 3434). Consequently, the wisher illuminates seven hundred organizational performing pathways by projecting seven phases, first within the infinite ten-phase prime-sequence and then accelerating the growth-effect by shaping a correlated reality of the self as the masculine flame and the universe as the feminine twin-flame.

Besides, the wisher also experiences two thousand organizational profiting pathways, as the sequential development of first, the correlated growth-effect of the universe and then, the correlated growth-effect of the self. The correlated growth-effect of the universe is the zodiac time, which forms the spatial reality. The liberated growth-effect of the self is the solar time, which forms the temporal reality. The self is not conscious of the spatial reality before forming the self as a deity spirit. Therefore, the self believes its temporal reality to be the integrated space-time reality of the universe, where the zeroth self is the absolute space and the manifestor of the transformative guider factor. Therefore, the organizational development's overall pathways scripted by the divine self = 700 dimensions of organizational performing + 2,000 dimensions of organizational profiting = 2,700 pathways.

For instance, *The Mahabharata* is an epic story of Sri Krishna as the sixth-dimension ideal-shaper (with an energy value of 10, as the primordial illuminator of the binary sequence forming the seventh dimension of Mother Nature as the organizational planner). As an ideal-shaper, Sri Krishna was the master of spontaneously (effortlessly and naturally) manifesting the desired organizational development without any organizational planning, programming, performing, or profiting, and indeed, without any organization. The presence of a cause is the predominating factor for illuminating the self's primordial reality as a divine entity. The self presently perpetuates as a divine spirit for destroying the infinite devil spirits, who become the subject of present reality by objectifying their primordial self.

The dominating factor is to "shape" the ascending-order growth potential of the universe by destroying the ideal cause, which is trading formative growth energy of the universe to be the primeval creature's absolute creator without the primordial creation. The deciding factor is the primordial-primordial creator who motivates the wisher to make the self the ideal cause and condition the universe to take the ideal cause to be a divine entity. The primordial-primordial creator shapes the theory of believing in the para deity for making the masculine self into a theory (Universe of Super

Wishers: *Duniya*, -2), who takes the wisher energy from the feminine self. The theory is the self's present reality as the param creature, without the primordial-primordial creator and the primordial creation. The primordial-primordial creator is the masculine self, a seventh-dimension ideal-equilibrator (Param Deity: *Shiva*, 7). He hides the secret of his para-primordial reality by illuminating the nineteen-phase entity (Param Child: *Manyu*, 19) as the super-wisher universe's technological growth value.

2.13 The Ideal-Equilibrator

An ideal-equilibrator is the para-primordial paradigm of the param deity, who effortlessly equilibrates the self's technological servicing value by projecting the masculine self as the primeval negative entity. The masculine self perpetuates the present reality through the technological trading of the feminine self as the primordial positive entity. There is one primordial-primordial entity—the androgynous self. The androgenous self programs seven phases of organizational performing, which includes the four forms of the ideal self (forming the absolute space, with the east, west, north, and south directions) and the three theoretical transformations for the growth of the ideal self (forming the absolute time, with the past, present, and future dimensions).

The masculine self performs to create four transformations of the ideal self (forming the absolute cause, projecting the truth of ideal equilibrator, shaper, maker, and taker into the quantum quark in the form of the potential, truth, charm, and up quarks, respectively). The feminine self performs to create four formations to catalyze the transformative growth of the ideal self (norming the primordial space, primordial time, primordial cause, and primordial-primordial entity). The presence of the primordial space is the truth of the present quark. The presence of the primordial time is the truth of the beauty quark, i.e., why absolute time is an illusionary theory-effect of primordial time. The presence of the primordial cause is the truth of the strange quark, i.e., why the absolute cause is not the same as the primordial cause. The primordial-primordial entity's presence is the truth of the down quark, i.e., why the param creature is independent and enjoys primeval freedom to manifest the desired organizational quality for the desirable organizational planning programming, performing, profiting, and development.   As an "absolute cause" (*Brihaspati*, 1780), the overall paradigmatic reality of the param deity has an energy of 1,780—one primordial-primordial entity, seven phases of organizational performing, eight forms of organizational profiting, and zero cost of organizational development.

For instance, *Shiva Shakti Purana* is a sequence of the stories of *Shiva* as the seventh-dimension ideal-equilibrator.   As an ideal-equilibrator, Shiva is an "androgynous luminous entity" (*Ardhnareshwar*, 39), who projects intrinsic energy as the "cellular energy" (*Shakti*, 19) and extrinsic energy as the primary "primordial illuminator" (*Parvati*, 10) of the truth of the cellular energy.   Shiva perpetuates the residual self as the secondary "primordial illuminator" (*Shri Krishna*, 10) of the cellular energy's beauty. "Primordial energy" (*Adi Shakti*, 15) is the predominating maternal cause of the cellular energy.   Without the growth-effect of the time, there is no cause for the quantized cellular growth of the energy.

Fire is the "para-primordial energy" (*Adi para shakti*, 17), the dominating paternal cause of the maternal duality. Without the fire, there is no cause for the disproportionate growth of time, free from the divine space within the param deity's consciousness.   Param deity is the deciding androgenous cause of the maternal supremacy in the formative growth of the gender of an organization.   Gender is masculine if the maternal element is not dominant. Gender is feminine if the maternal element is dominant.

Table 1 summarizes the seven secondary forms of the formative paradigm of organizational planning, the energy growth value of each form, and the three causative factors guiding each form's development.

Table 1. The Seven Secondary Forms of the Formative Paradigm of Organizational Planning

Forms of Organizational Planning	Growth Value of Organizational Planning	Predominating Factor	Dominating Factor	Deciding Factor
Theory-taking	1/8 present life	Object	The subjective value of the object	Theory of relative value
Theory-making	1 present life	Subject	The objective value of the subject	Theory of absolute value
Theory-shaping	16 lives within the present soul	Creator	Subject	The absolute idea, without theory-effect
Ideal-taker	128 lives within the present soul	Perpetuator	Object	Proportionate Absolute, without ideal-effect
Ideal-maker	1,600 lives within the present soul	Destroyer	Feminine self	Absolute, without any proportionate normative-effect
Ideal-shaper	2,700 pathways of the present soul	Illuminator	Masculine self	Omnipresent, without any disproportionate formative-effect
Ideal-equilibrator	1,780 times within the present pathway	Primordial Illuminator	Androgenous self	Binary transformative growth-effect

The Five-Fold Sentient-effect of One's Becoming System

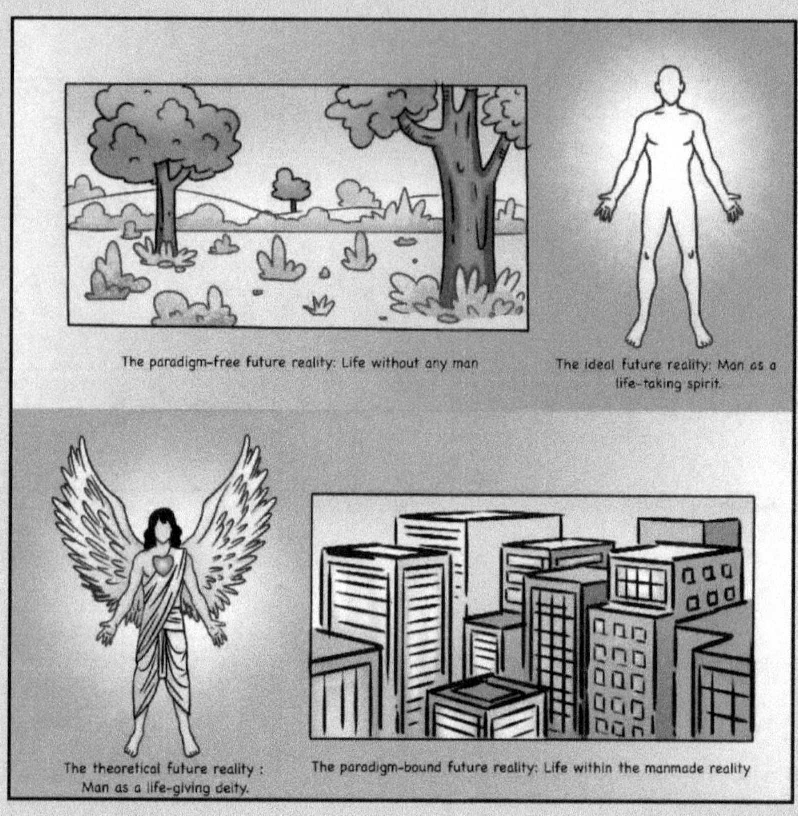

The paradigm-free future reality: Life without any man

The ideal future reality: Man as a life-taking spirit.

The theoretical future reality : Man as a life-giving deity.

The paradigm-bound future reality: Life within the manmade reality

The supernatural paradigm as the future reality?: Making the better-half the plan for realizing the goal of saving the whole masculine person from the wished manmade reality

The Future Reality is a Simple Reality, Without the Paradigm-Programming Entity

The energy we invest in modifying the perpetuating natural reality shapes the color potential of the future reality.

- If we invest sentient energy conceived from within, then the future reality is complementary, constructive, and purple. It manifests a divergent, spiritual dimension of the virtual reality. It becomes our ideal future reality.

- If we invest gravitational energy conceived from others, then the future reality is competitive, destructive, and red. It manifests a convergent, animalistic dimension of the virtual reality. It becomes our theoretical future reality.

- If we invest divine energy conceived naturally, then the future reality is supplementary, innovative, and colorless. It manifests a diverse, parallel human dimension of natural inverse rainbow reality. It becomes our paradigm-free future reality.

- If we invest devil energy bred by a leader, then the future reality is metaphysical, creative, and orange. It manifests an engaged, intrusive, plant dimension of the augmented reality—we nurture the greenish growing leader-effect with our constant bluish consciousness water. It becomes our paradigm-bound future reality.

- If we invest satanic energy to be a leader, then the future reality is dynamic, strategic, and yellow. It manifests a disengaged mineralized life dimension of the augmented reality—we transform the black-tone indigo consciousness of what is not present into our goal's whitish consciousness. It makes the future reality a supernatural paradigm.

3.1 Eight Tertiary Strategies for Paradigm-Free Organizational Development of Transformative Planning

A masculine entity may organizationally develop transformative planning to manifest himself as the deciding factor. A feminine entity may develop transformative planning to let an androgynous entity manifest itself as the predominating factor. An androgynous entity may develop eight tertiary strategies to let a feminine entity manifest herself as the dominating factor in catalyzing transformative planning. The first seven strategies are naturally color-coded as Blue, Yellow, Red, Indigo, White, Green, and Orange. Together, they form a purple-ocean strategy. The duality of the black and the colorless condition is the primary cause of the tertiary strategies.

3.2 The Purple Ocean Strategy

The purple ocean strategy is the consciousness of the primary color potential of the diverse entities. A strategy is a technique for devoting intense energy to break down formative paradigmatic norms and radiate transformative organizational development. It is a supernatural solution for transforming the entropy reality of a masculine spirit into a self-luminous entity's normative growth potential. Gray, Purple, and Orange are the primary colors within the human kingdom. Purple is the color of sentient energy. When a self-luminous human cell trades the white color of the electron and the spirit's black color, it transforms into the gray color within the white light-medium. When a self-luminous human cell trades the white color of the electron without the spirit's black color, the purple porphyrin breaks down and the orange flow gets absorbed within the gray matter cells. Consequently, the purple glow of porphyrin diffuses without the gray matter cells. By sequentially trading the sentient power, an infinity of spirits enjoys an at-par organizational development without transformative planning.

- Blue, Yellow, and Red are the primary colors of the animal kingdom. Red is the color of masculinity. When the animal cell trades the diffused blue color potential of the mineralized water-effect, the red hemoglobin breaks down within an animal organ. Consequently, the blue glow radiates within the dark black medium, and the red glow of hemoglobin diffuses without the yellow flesh medium.

- Indigo, White, and Green are the primary colors of the plant kingdom. Green is the color of femininity. When the plant cell trades the diffused red color potential of the hemoglobin and the blue color potential of the mineralized water-effect, the green chlorophyll breaks down within a plant organ. Consequently, the indigo glow radiates within the dark black medium, and the green glow of chlorophyll diffuses without the white light medium.

- White, Orange, and Colorless are the primary colors of the metal kingdom. White is the color of the deity kingdom—it absorbs all colors within its almighty potential. Androgeneity has no color—it is colorless. When a yellow flesh medium exchanges its color potential with a metal, the white electron breaks down within the metallic atom. Consequently, the orange glow radiates within the dark black medium, and the colorlessness of the atom diffuses without the white light medium. When the metallic atom trades the hemoglobin's diffused red color potential, the dark black medium breaks down, and the orange color gets absorbed within the white light medium. The metallic atom transforms into a self-luminous human cell.

- Blue, Purple, and Black are the primary colors of the spirit kingdom. The black repels all colors so that the spirit may trade the extrinsic red color potential and transform its intrinsic blue color potential into a self-luminous purple human cell. A spirit is an entity without an organization, i.e., without a physical body, wishing for transformative organizational development after experiencing escalating formative growth costs without normative oneness with Mother Nature. One may become a spirit either within one lifetime, through an absolute exchange of the sentient energy immanent within the formative development into a self-luminous entity, or over infinite lifetimes, through a sequential servicing of the sentient energy for developing an infinity of organizations. The energy value is a positive one because the spirit naturally embodies the electrokinetic working-effect of its infinite workculture system as a primordial sentient entity. Without the natural trading of the blue-color mineralized water-effect, the spirit is a subject who wishes for a diverse life path and has a zero intrinsic energy value.

The value of the purple ocean strategy = The value of the natural paradigm of present reality * The proportionate catalyst value of the divine charge immanent within the ten primary colors * The disproportionate catalyst value of the divine charge emanating from the ten primary colors (white, purple, indigo, blue, green, yellow, orange, red, black, and gray), and transforming into a colorless form at 1.5 infinity (one infinity of the diffusing colorful masculinity-effect and half infinity of the equilibrating colorless femininity-effect). $= 8 * 10^{10*1.5} = 8 * 10^{15} =$ the value of the "Self" (*Atmatva*, 8×10^{15}), as an omnipresent androgynous entity.

The natural feminine paradigm of present reality is colored blue—it is the natural water-effect, life consciousness, and photonic light force immanent within the spirit kingdom. Entropy in the natural feminine paradigm's value is the predominating factor for conceiving the purple ocean strategy. Growth in the value of the red-color thermodynamic masculine organization is the dominating factor for perpetuating it. The omnipresent colorless androgynous entity is the deciding factor for destroying it and illuminating the blue ocean strategy's natural organizational development.

3.3 The Blue Ocean Strategy

The blue ocean strategy is the consciousness of the sentient life potential within a creator soul entity. A creator soul enjoys an absolute consciousness of the thirty-six basic soul programs within the self. These include: (a) the primary programs for normatively perpetuating the consciousness of the metal, mineral, plant, animal, human, and spirit potential within the masculine self; and (b) the secondary programs for destroying the para-consciousness of the metal, mineral, plant, animal, human, and spirit potential within the feminine self.

Soul-level programming is the tertiary planning strategy of a self-luminous human entity who is not conscious of the intrinsic param deity reality. It empowers the self-luminous human entity to create a formative sentient life potential of 360,000. It trades a unit of ten-color divine charge immanent within the creator, perpetuator, destroyer, and the illuminator selves. The intrinsic param deity reality lets the self-luminous entity diffuse its deity potential for manifesting the normative profiting value of the natural blue ocean.

The natural blue ocean is the normative development of the metal, mineral, plant, animal, human, and spirit potential within the divine self. It empowers the divine self to be immanent as a primordial feminine, for dynamically equilibrating the primeval masculine creator and perpetuator potentials. The divine self does so with the destroyer and illuminator potentials of the param androgynous entity.

The overall liberator potential of the natural blue ocean = 6,666. It is the transformative sentient potential for creating, perpetuating, destroying, and illuminating the normative planning pathways within and without the present kingdom of the divine self. The overall metaphysical value of the blue ocean strategy = 360,000 + 6,666 = 366,666 = the value of the "Devoted Creation" (*Brahmaastra*, 366,666), as an absolute masculine entity.

The devoted masculine form of the present reality is yellow—it is the color of the five-face para deity who is perpetuating the normative performing of the intrinsic param deity reality, within and without the consciousness of the universe of self-luminous entities. Creator consciousness within each self-luminous human entity is the predominating factor for conceiving the blue ocean strategy as a natural alternative. Destroyer consciousness within Mother Nature is the dominating factor for perpetuating it. Illuminator consciousness within the deity kingdom is the deciding factor empowering Mother Nature to destroy the primeval masculine entities' thermodynamic-effect and illuminating the social benefit of the yellow ocean strategy.

3.4 The Yellow Ocean Strategy

The yellow ocean strategy is the consciousness of the infinite potential value of enjoying the present reality of the divine self and perpetuating the varying life paths over time. Enjoying is about normative profiting. Normative profiting is about "formative performing" (*Karma*, 10), without the "primordial-effect" (*Karma-effect*, 10^{10}). Formative performing requires "transformative development" (*Siddha*, 7) to destroy the present masculine light force and illuminate the primeval feminine life consciousness. The primeval life consciousness is the energy of the "devoted deity" (*Maha Saraswati*, 9), immanent within the devoted creation, and diffused into the infinite absolute masculine forms over the finite lifetimes conditioned by the blue ocean strategy. The divine charge, immanent within the devoted deity, diffuses in the form of the creator, perpetuator, destroyer, and illuminator selves.

The overall metaphysical value of the yellow ocean strategy = 9 * 10 * 10 * 10 * 10 = 90,000 = the value of the "Devoted Param Perpetuator" (*Ananta*, 90,000), as a primeval masculine entity norming the "infinity" (*Ananta*, 90,000). A primeval masculine entity brags about the formative performing value traded from a "para deity" (*Param Vishnu*, 5), perpetuating the yellow ocean strategy for enjoying the infinite forms of the self. He believes there is no end to the SHEENY benefit, given the normative profiting in the present life. Therefore, he diffuses an infinity of electrons without the physical body, which together forms the yellow color of the polluted, pungent gas. The yellow color is the thermodynamic bragging-effect of a primeval masculine: it poisons the normative profiting paradigm and catalyzes transformative development to compensate for the karma-effect.

The primeval masculine form of the present reality is red—it is the color of the infinity of neutrons within the physical body of a supra deity who manifests the normative profiting of the present entity reality, within or without oneness with the intrinsic param deity reality. Masculine consciousness is the predominating factor for conceiving the yellow ocean strategy as a transformative alternative. Entropy in primordial oneness with the param deity is the dominating factor perpetuating it. Growth in primordial oneness with the "param child" (Masculine child cell: *Manyu*, 19) is the deciding factor. It empowers the physical body to destroy the sentient-effect of the primordial masculine entity and illuminate the social cost of the red ocean strategy.

3.5 The Red Ocean Strategy

The red ocean strategy is the consciousness of how the paradigm of present reality is destroying the transformative masculine reality in the universe. Two units of the "manifestor deity" (*Hanuman*, 3) divine charge transform into the "universe of primeval masculine entities" (*Jagath*, -2), and the other eight units norm the natural paradigm of the present reality. The natural paradigm of present reality compensates for the past, present, and future social costs of two units each. The residual two units empower a feminine entity to generate formative profiting by fulfilling the objective of present reality.

The objective of present reality is to generate "knower-level consciousness" (*Jnana*, 19), within the "primordial-effect" (*Karma-effect*, 10^{10}). Primordial-effect is the normative development of an entity. Normative development is a function of a transformative organization. By destroying the present masculine organization, an entity illuminates the feminine manifestor reality within the self. The overall metaphysical value of the red ocean strategy = 8 = the value of the "Primordial Deity" (*Anatanam*, 8) as a feminine entity normatively developing the paradigm of present reality.

A primordial deity forms the profiting value by radiating an infinity of neutrons, which carry and equilibrate both the negative masculine and the positive feminine charges. The infinity of neutrons radiates the primordial androgynous matter in a vibrant red color plasma form. The red color is the low-frequency residual of the deficit in the knower-level consciousness generated by the devoted "primordial-primordial effect" (*Bhakti-effect*, 86). It motivates a feminine entity to develop an inertial charm attraction to the manifested present reality.

The devoted feminine form of the present reality is indigo—it is the color of the infinity of protons that form the physical body of a supreme deity, who is creating the normative development of the primordial entity within and without oneness with the param entity. Feminine consciousness within the present entity is the predominating factor for conceiving the red ocean strategy as a formative alternative. Entropy in primordial oneness with the param entity is the dominating factor perpetuating it. Growth in primordial oneness with the "primordial deity" (Mother Nature: *Anatanam*, 8) is the deciding factor. It empowers the physical body to destroy the inanimate object-effect of the primordial feminine entity and illuminate the worker-social benefit of the indigo ocean strategy.

3.6 The Indigo Ocean Strategy

The indigo ocean strategy is the consciousness of the method of present reality, which is illuminating the formative feminine reality within the entity. The method of present reality includes two units of the formative development, traded from the primordial deity as the objective of present ecosystem reality. It further includes two units of a normative organization, serviced by a primordial-primordial entity with the knower-level consciousness of the objective of present entity reality. It is free of the cost of transformative planning by a primordial paternal entity, who wishes to program the self as the "subject dimension" (*Dharma of the Karta-dharta*, 370) ruling the "universal consciousness" (*Karta-dharta*, 86).

The overall metaphysical value of the indigo ocean strategy = 4 = the Supreme Deity's value as a primeval feminine entity normatively organizing the method of present reality. A supreme deity forms the development value by radiating infinite protons as a thermodynamic-effect of the performing objective. Infinite protons form the indigo color of the long duration wavelength (low-frequency variation) objective light force.

The infinite feminine form of the present reality is white—it is the color of the infinity of astral bodies that emanate from the physical body of a luminous. A luminous normatively organizes the supreme deity reality, within and without oneness with the primordial entity. Androgynous consciousness within the primordial entity is the predominating factor for conceiving the indigo ocean strategy as a normative alternative. Entropy in primordial-primordial oneness with the primordial entity is the dominating factor perpetuating it. Growth in primordial-primordial oneness with the supreme deity is the deciding factor. It conditions the physical body to destroy the self and illuminate the white ocean strategy's worker-social cost.

3.7 The White Ocean Strategy

The white ocean strategy is the consciousness of the objective of present reality, which is liberating the normative androgynous reality of the entity. The objective of present reality includes two units of normative entity planning of the normative ecosystem development, within primeval oneness with the primordial deity's tradable energy. An entity enjoying a knower-level consciousness of the objective of present reality generates primeval oneness with the tradable energy of the primordial deity. She forms the intrinsic primordial-effect energy into a normative plan for transformative programming. The transformative programming objective is to intensify oneness with the primordial energy into the point of infinity, thereby realizing eventual oneness with the "primordial-primordial self" (*Dvandva Brahma*, 125).

The knower-level consciousness works like a "guider power" (Gravitational energy: *Guru*, 100), transforming oneness of the spirit with the infinite, perpetuating reality of the self into a consciousness of the self as the perpetuator of the infinite forms of reality. The overall metaphysical value of the white ocean strategy = 2 = the Super Deity value as a primordial androgynous entity normatively planning the objective of present reality.

Both the entity objective and the ecosystem objective produce a five-dimensional pair of opposites within oneness with the param deity. The ecosystem objective produces a natural five-dimensional entity: the five-headed "perpetuator deity" (*Param Vishnu*, 5). The entity objective produces a supernatural five-dimensional entity, the four-headed "creator deity" (*Param Brahma*, 4), trading the perpetuating consciousness in the form of the invisible guide-effect of the perpetuator deity.

A super deity forms the organizational value by trading an infinity of astral bodies from a luminous entity, experiencing infinite thermodynamic entropy. The growth of the infinite creator and perpetuator entities produces an infinite thermodynamic entropy within the luminous star entity. Consequently, the white ocean strategy generates escalating worker-social costs and necessitates a green ocean strategy.

The devoted androgynous form of the present reality is green—it is the color of the gravitational-effect of the infinity of dark matter. The dark matter is immanent within the physical body of a luminous. The luminous normatively plans the super deity consciousness as the primordial-primordial entity. Infinite androgynous consciousness within the primordial-primordial self is the predominating factor for conceiving the green ocean strategy as a metaphysical alternative. Entropy in param-primordial oneness with the primordial-primordial self is the dominating factor perpetuating it. Growth in param-primordial oneness with the super deity is the deciding factor for the green ocean strategy's ascending social benefit-cost ratio.

3.8 The Green Ocean Strategy

The green ocean strategy is the consciousness of the subject of present reality, which is devoted to the metaphysical entity reality of the self. As a subject, an entity enjoys a wisher-level consciousness for manifesting the desired infinity of reality. The subject is conscious of the intrinsic normative programming of the capability for manifesting. Therefore, the subject forms proportionate planning for activating the normative programming. However, reformative planning alone is not sufficient for activating the intrinsic capability. The subject also needs to transform its performing and transcend beyond the Wisher-level consciousness. Just wishing for something is not a path for any movement within the self for manifesting the wish.

A Wisher is like an inertial electromagnetic mass, with zero variable mass-effect and zero accelerating light force of the aroused mass-effect. The overall metaphysical value of the green ocean strategy = 0 = the Primordial Paternal value as the primeval androgynous entity normatively programming the subject of present reality.

As a subject, both paternal and maternal entities produce a nine-dimensional pair of opposites, without oneness with the param deity. A nine-dimensional masculine entity comprises the five-dimensional perpetuator energy and the four-dimensional creator energy. A homologous nine-dimensional feminine twin consists of the five-dimensional perpetuator and the four-dimensional creator energies. The four-dimensional entity system's overall energy is immanent within the "primordial maternal" (*Para Shakti*, 18). A primordial maternal conceives the four entities as the parallel dimensions within the perpetuating self. A primordial paternal perceives the intrinsic entities to have a negative masculine charge and experiences the extrinsic entities to have a positive feminine charge. Therefore, as a neutral entity, he wishes to exchange the negative intrinsic reality with the positive extrinsic reality, so that he may become a "devoted androgynous entity" (*Maha Shiva*, 9).

A Wisher produces the planning value by normatively programming a transformative performing of the primordial maternal. The Wisher first wishes for a four-dimensional masculine creator, motivating the primordial maternal to produce a five-dimensional feminine perpetuator as the desired masculine creator's twin-flame. The Wisher then trades the residual one-dimensional feminine deity energy as the masculine workculture-effect. Thus, the Wisher transforms the residual thirteen-dimensional feminine, luminous essence of the primordial maternal into a twelve-dimensional self-luminous entity. The Wisher then trades the six positive feminine energy units within the self-luminous entity to become a primeval deity.

Consequently, the self-luminous entity becomes a double-helix entity with six positive feminine dimensions and six negative masculine dimensions. Each feminine dimension enjoys a primeval consciousness and services intrinsic energy for equilibrating the negative primordial-effect within the masculine dimension. Therefore, within a sense of otherness, the masculine dimensions enjoy a disproportionate entity-level growth-effect of the green ocean strategy. Similarly, within a sense of togetherness, the feminine dimensions conceive the perpetuating paradigm of the zero growth in ecosystem-level energy as the cause of their infinite closed-system thermodynamic entropy.

A Wisher plans the green ocean strategy by servicing a thermodynamic-effect of togetherness with its physical electromagnetic mass. It forces the "dark matter" (*Paramatma*, 1600) to release the gravitational energy in the green color bursts. The darkness beyond the light horizon explodes into an iridescent green color mass-effect without the Sun's light force because of the prismatic-effect of the pearlescent sentient power that is immanent within the dark matter. With an absolute entropy of the feminine dimensions, the green ocean strategy's social benefit-cost ratio falls from infinity to zero. A need emerges for an orange ocean strategy.

The primeval androgynous form of the present reality is orange—it is the color of the sentient-effect of the infinity of "black hole" (*Visnunabhi*, 82). It is immanent without the luminous normatively programming the primordial paternal consciousness as the param-primordial entity reality. Param androgynous consciousness within the param-primordial self is the predominating factor for conceiving the orange ocean strategy as a physical alternative. Entropy in para-primordial oneness with the param-primordial self is the dominating factor perpetuating it.   Growth in para-primordial oneness with the primordial maternal reality is the deciding factor.   It ascends the worker-social benefit-cost ratio of the orange ocean strategy for the primordial paternal as the subject of present reality.

3.9 The Orange Ocean Strategy

The orange ocean strategy is the consciousness of the metric of present reality, which is devoted to the SHEENY well-being of the self's physical entity reality.   As a metric, an entity enjoys a primeval masculine-level consciousness of the infinite black hole in the formative capability for manifesting the desired reality.   Therefore, the entity is motivated to make a disproportionate investment of the intrinsic energy for trading a desirable object that offers the guider catalyst power for manifesting the desired reality.   A desirable object is one that has a charm attraction for the entity.

Given the primeval masculine consciousness, the entity believes that making one's desired destiny conditional on one object, howsoever desirable, is risky.   Therefore, the entity strangely repels the desirable object so that its light force generates a charm attraction for the universe of desirable objects.   It plans, metaphysically, to activate the purple ocean strategy by forming an exchange system.

Each desirable object repels one another by servicing its intrinsic energy to the point of intrinsic entropy.   It dynamically programs the blue ocean strategy for forming the infinite capability by trading the universe of extrinsic energy. It technologically performs with the yellow ocean strategy by investing the whole trading-effect. It organizationally profits with the red ocean strategy by countertrading the diffused investment energy.   It develops into an ecosystem entity, trading and countertrading the sentient energy within the self through the indigo ocean strategy.   It becomes an organizational entity that mediates the flow of the sentient energy through the white ocean strategy.   It seeks to plan the growth of its gravitational energy through the green ocean strategy.   It experiences infinite entropy in its sentient energy by transforming that into the gravitational energy through the orange ocean strategy.   It seeks to compensate for the entropy in the formative capability by activating the gray ocean tactics.   The gray ocean tactics entail ascending the servicing of the normative intellectual capability to supplement the descending power of the formative physical capability.

An entity trades the normative intellectual capability from its etheric body. The etheric body is a group of zero-energy souls who reside in the param heaven—the luminous kingdom. They have diffused their sentient energy in the form of the "primeval perpetuator" (*Vishnu*, 15) of the present creation, wishing for the infinite gravitational potential within the universe of self-luminous entities. They compensate for the black ocean tactics of the inanimate "universe of conscious reality" (*Padartha*, -3), which transforms mental capability aspiring for para-conscious divine energy.

Without a "devoted divine entity" (*Bhakta*, -5) blackening their para-conscious potential by personifying it, an entity experiences a colorless "universe of dynamic reality" (*Paroksha*, -7) beyond its range of vision. A devoted divine entity genetically programs the primeval trading-effect of the demonic, "universe of inanimate objects" (*Sthavaravisha*, 57) as the normative intellectual capability. By genetically transforming the self's physical reality, an entity becomes the causal body for creating an unintended "universe of technological reality" (*Shramika*, -6). It programs an infinite purposeless performing culture into the infinity point of absolute entropy of the sentient energy.

The overall metaphysical value of the orange ocean strategy = -1 = the value of the "Primeval Masculine" (*Asura*, -1), as a param androgynous entity transformatively destroying the metric of present reality. A primeval masculine forms the capability for the orange ocean strategy by servicing the sentient-effect of otherness with its metaphysical gravitational mass. The dark horizon of the black hole radiates the Sun's orange color photonic light force because of the gravitational air mass-effect that is serviced by the dark matter.

The orange color photonic light force forms an electromagnetic mass within the cooling-effect of the sentient energy traded from the universe of white stars. The "universe of white stars" (*Mahar loka*, 27,000) services its sentient energy to compensate for the entropy in the earth's sentient energy. With the primeval growth of the androgynous sentient dimension, the worker-social benefit-cost ratio of the orange ocean strategy ascends from zero to infinity. An opportunity emerges for reproducing the rainbow ocean strategy, without the metric of present reality.

Table 2 summarizes the eight tertiary organizational development strategies for transformative planning, the growth value of each strategy, and the three causative factors guiding each strategy's organizational development.

Table 2. The Eight Tertiary Organizational Development Strategies for Transformative Planning

Strategies for transformative planning	Growth-effect of transformative planning	Energy value of growth-effect	Predominating factor	Dominating factor	Deciding factor
Purple ocean strategy	Normative capability	8×10^{15}	Create exchange	Perpetuate servicing	Destroy growth
Blue ocean strategy	Formative capability	366,666	Create programming	Perpetuate performing	Destroy profiting
Yellow ocean strategy	Formative investment	90,000	Create performing	Perpetuate profiting	Destroy development
Red ocean strategy	Formative trading	8	Create profiting	Perpetuate development	Destroy the organization
Indigo ocean strategy	Formative exchange	4	Create development	Perpetuate organization	Destroy planning
White ocean strategy	Formative servicing	2	Create organization	Perpetuate planning	Destroy programming
Green ocean strategy	Formative growth	0	Create planning	Perpetuate programming	Destroy performing
Orange ocean strategy	Formative entropy	−1	Create capability	Perpetuate investment	Destroy trading

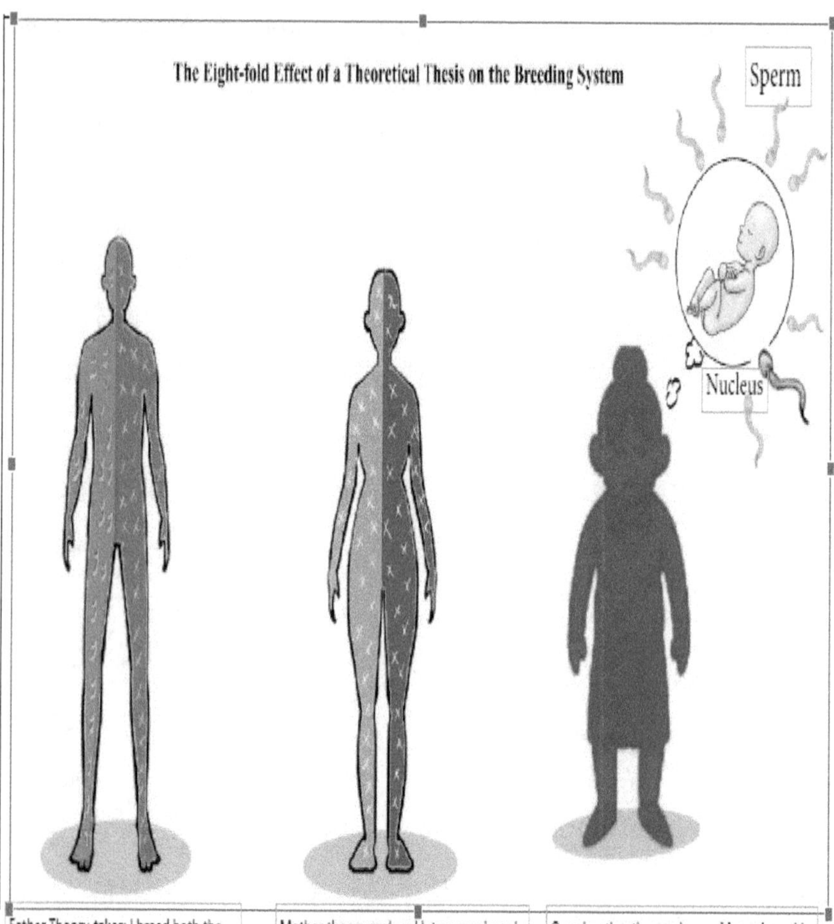

Father Theory-taker: I breed both the masculine and the feminine chromosomes. I am the thesis that the males are diverse, females are not.

Mother theory-maker: I let my son breed the masculine chromosome and my daughter breed the feminine chromosome, for augmenting my X chromosome. I am the anti-thesis that the engagement with the diversity is the key to desired supernatural-effect.

Grandmother theory-shaper: My son's soul is guiding my egg to fertilize naturally for producing my shell-free spirit homolog. My spirit homolog is potentiating my masculinity in the form of a Y-chromosome immanent within my son's body.

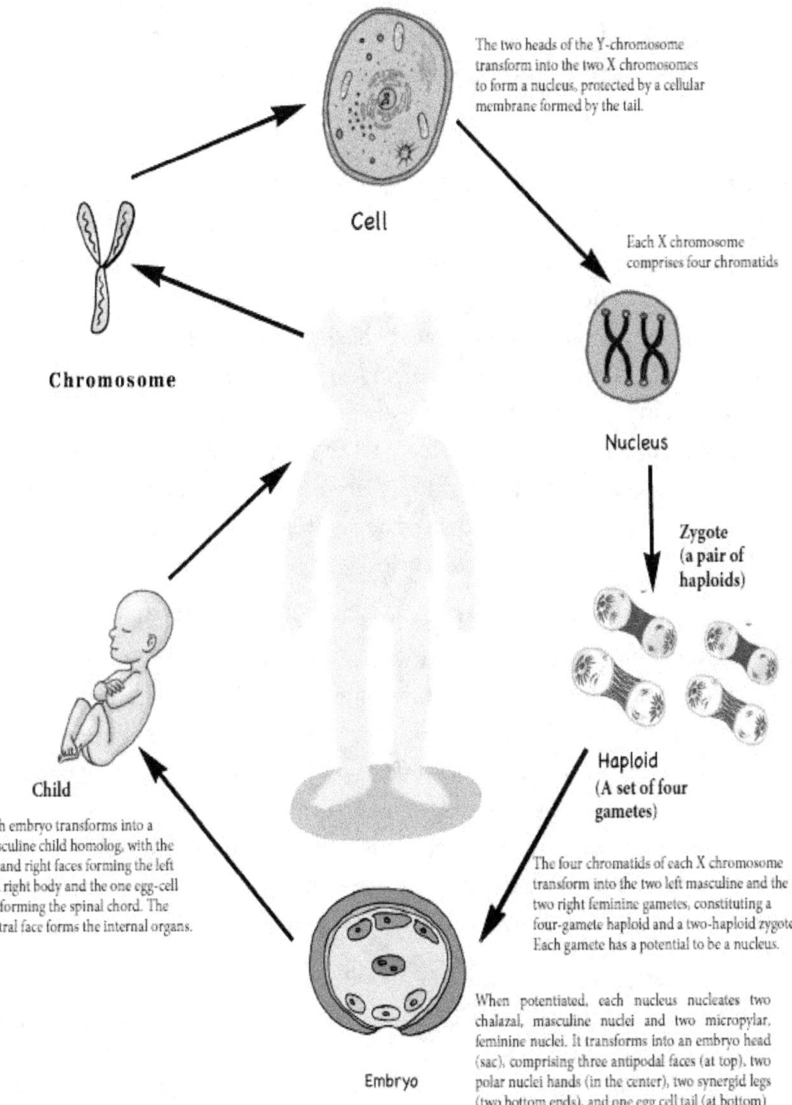

Grandfather human entity. By letting my masculine Y chromosome mature naturally without me through the warmth of my feminine X chromosomes, I have the potential to be paternal of an octave of sons. The grandmother spirit is already immanent within both me and my masculine chromosome. She is the metaphysical cause of my potential to potentiate and reproduce the four supernatural creator units within each of my feminine X chromosomes.

The Past Reality Includes Both the Entity as the Augmented Paradigm-Planner and the Potential for Programming the Entity as the Paradigm for the Virtual Present Reality

As an augmented paradigm-planner, each entity forms a vision of the past reality to transform the truth of the omnipresent reality with a hypothetical thesis. It asserts that the visualized reality is the undesirable present reality, necessitating the entity to be the mediating exchange factor. This virtually hypothetical thesis, conceived for perceiving the past reality, shapes the energy we invest in modifying the perpetuating natural reality. The thesis objectifies the theory-making leader as embodying the holistic truth of the past reality. It subjects a theory-taking follower to a destiny of networking, referencing, and sense-making of the diverse theories, as a path to be the aggregator-ruler by falsifying the whole universe of competing theories.

Let's imagine a theory-shaping entrepreneur who is conscious of the objective of the reality that embodies a unit of thesis and a unit of anti-thesis. She is the limit of the technological growth generated through the competing theses. She is the grandmother guiding the theory-making maternal leader and the theory-taking paternal follower. She seeks the growth of the universe of responsible granddaughter managers. Her objective is to mitigate the escalating costs of the theory-effect on the universe of self-organizing grandsons. As the creator factor, her past reality includes the following.

- **The Mother theory-maker:** My "thesis" illuminates the formative causative factor whose normative effect is the present virtual reality. My son is the present, virtual reality that parallels the natural perpetuating reality embodied within my husband. My husband personifies the truth value of the thesis. My son is genetically different, despite inheriting both feminine and masculine chromosomes. I am the cause of the somatic differentiation my son enjoys. I falsify the anti-thesis that my husband is the causative factor.

- **The Father theory-taker:** My "anti-thesis" illuminates the transformative causative factor whose metaphysical effect accounts for the past augmented reality. My daughter is the past, augmented reality that illuminates the future simple reality of the feminine cells, as free from the masculine contamination. Since my son has a masculine element, I am the cause of the germline differentiation my son inherits. I substantiate the thesis that my wife is the formative causative factor, and I am the transformative factor.

- **The Grandmother theory-shaper:** My objective value is two. With two feminine chromosome units, I have the potential to generate a two-units technological growth. One, genetically reproducing the natural paradigm within my daughter. Second, spiritually mutating the natural paradigm to manifest my son. My husband is the entity mutating my spirit.

- **The Grandfather human entity.** My person-effect is four. With a pair of diverse chromosome units, I have the potential to create two blackened units as the descending technological growth and two whitened units as the ascending technological growth. My daughter is the blackened holy spirit, seeking to illuminate my value as the creator. My grandchildren are the whitened souls, enjoying the "synthesis" illuminated by their father. I am the normative cause of the two supernaturally perpetuating units constituting the technological growth.

4.1 Objective for Planning the Present Paradigm: Organizational Development

The objective for planning the present six-dimensional rainbow color (IGBYOR = Indigo, Green, Blue, Yellow, Orange, and Red) paradigm is to let an absolute creature falsify the thesis that the organizational development power of an entity within the human kingdom is zero. A "thesis" (*Harkriya*, 666) consists of six hypotheses:

- **First, the organizational development power of an entity within a human kingdom is zero.** Only God has the power for the formative development of an entity, either within the primordial-primordial hell state of the black hole or within the primordial-primordial heaven state of the white star.

- **Second, the energy growth value of an entity is a subjective factor.** Only God decides whether one ascends from the primordial-primordial hell as a sentient subject to eventually realize a place in the primordial heaven (as a *Siddha* illuminator of present reality, 7). Alternatively, one may descend from the primordial-primordial heaven as an inanimate object to eventually realize a place in the primordial hell (as a "metric of present reality": *Maha Shunya*, -1).

- **Third, it is not possible to objectively sequence the energy growth value of different objects and subjects in the whole universe.** Neither is it possible to subjectively plan the consequent desirable organizational development within the present moment of a constant energy growth value. Once God has genetically programmed a destiny, then the paths of *karma* (action), *jnana* (knowing), and *bhakti* (devotion) have zero value in transforming that destiny. The human entity's role is to fulfill the goal of the divine planning by forming the "universe of organizational reality" (*Jagatkritsna*, -10) and trading the present reality for accelerating the self-realization of the divine judgment.

- **Fourth, an entity performs its ontological programming epistemologically to ascend the axiological profiting by descending the metaphysical well-being.** An ideal human entity services the primordial divine energy for trading the infinite guider power to be the "universe of ecosystem reality" (*Pasaka*, -9). It motivates the entire ecosystem to diffuse their absolute divine energy for their param-primeval SHEENY well-being. It irradiates that energy as the subject mediating that well-being.

- **Fifth, the present scientific physical value defines an entity.** For instance, a king might have had super-normal value in the past life but has zero value in the present moment. No ideal human entity will service the primordial-primordial sentient energy for incarnating the king soul as a child, knowing that the king soul is the "universe of para-conscious reality" (*Pratyahara*, -12). Once born, the king soul will only be a demanding and dominating child, forcing all to diffuse their sentient energy for the normative development of his guider power. Without any sentient energy, a demanding and dominating masculine reserves his destiny in the primordial hell, as destined by God. In case the king has a physical body, it implies that God has a special destiny planned for the unique wishers.

- **Sixth, a sensible energy growth value of organizational planning for a king is zero organizational development in the present moment.** Infinite masculinity within the king's physical body is the "universe of entity reality" (*Karuyantra*, -11).   In case the king seeks additional physical power, he gravitates toward a destiny in the material kingdom and gifts the atoms a destiny in the human kingdom by transforming them into the sentient cells.

The sixth hypothesis is an anti-thesis that empowers an absolute creature to falsify the thesis by liberating an infinite organizational development power within each entity and substantiating the following six anti-hypotheses of the anti-thesis.

- **First, the organizational development power of an entity within the human kingdom is beyond absolute.** A human entity is more than a "superleader" (*Kshatriya*, 0)—it has the sentient power to script the destiny of the self with its divine planning energy as the "natural essence of present reality" (*Aap*, 1). It is not dependent on the divine planning energy of "God" (*Ishvara*, 5) to compensate for its "path-ambiguity" (*Avidhi*, -4).

- **Second, the energy growth value of an entity is an objective factor.** A human entity programmatically scripts the desired energy growth destiny of the self, rather than spontaneously performing into eternity as a natural deity. It thereby validates natural "simplicity" (*Brahmin*, 2), free of the "mind-born complexity" (*Parindartha*, -3) of its knower-level consciousness of the present reality, to manifest the desired reality.

- **Third, it is possible to objectively sequence the energy growth value of the different objects and subjects in the whole universe.** It is also possible to clarify the subjective planning path for the consequent desirable organizational development without the limitations of the present moment of a constant energy growth value. A human entity has the divine power to perform at-will spontaneously as a "deified manifestor" (*Vaishya*, 3) for manifesting a desirable organizational development of the self, as a path to manifest the desired sequential growth value of the different objects and subjects in the whole universe.

- **Fourth, it is possible to authenticate that an entity is more than the present scientific value of its ontological programming**, even after correcting for the multiplier-effect of its epistemological performing and the accelerator-effect of its axiological profiting. An entity is the "deified creator" (*Param Brahma*, 4) of the infinite scientific values as a function of its guider power for manipulating the perpetuating SHEENY value without the self. An entity services its guider power to transform the present paradigm of natural reality and form a new paradigm of supernatural reality.

- **Fifth, it is possible to define an entity by its metaphysical SHEENY value in the present moment.** For instance, a "king" (*Rajah*, 0) is an entity that empowers each entity to accrue himself as the desirable SHEENY value (i.e., the social benefit of the presence of a king of the universe). Consequently, the king has a zero SHEENY value in the present moment and eternally (because of the social cost of the presence of an infinity of kings as a universe of entities guided by the ruling consciousness of the primordial king). Similarly, God is the "deified perpetuator" (*Param Vishnu*, 5) of the four units of the SHEENY value desired by the deified creator for the normative development of its guider objective, and one unit of the SHEENY value essential for the formative development of the divine nature for transforming its sentient reality into a creator consciousness.

- **Sixth, a sensible energy growth value of the organizational planning to be more than a king is the sentient energy already immanent within an entity.** An entity is a dynamic destroyer of the varying God realities. An entity may believe in a Confucian God at a primordial moment, in a Semitic God at a param moment, and in a Vedic God at a primeval moment. Therefore, an entity is a primeval deity who has the power to trade the absolute SHEENY value of a desirable ecosystem-level God for the primeval SHEENY value desired by the self as a desiring param paternal.

The "primeval deity" (*Maheshwara*, 6) has five units of the SHEENY value of God as well as one unit of SHEENY value as a natural divine entity. The "param paternal" (*Narada*, 7), on the other hand, has seven units of the

SHEENY value of the "param deity" (*Shiva*, 7). He has an illuminated consciousness of the entity benefit of trading the one unit of divine energy from "Mother Nature" (*Anatanam*, 8) for becoming an at-will "primeval paternal" (Godhead: *Param Ganesha*, 17) or "primeval maternal" (Necessity: *Prayojana*, 17). The param paternal has the freedom to activate one unit of the guider-effect of the param deity immanent within the self to be the "primordial maternal" (Mammal: *Anasuya*, 18). The primordial maternal has the opportunity to organize the energy immanent within the self for the SHEENY well-being of the despised king, as the objective of the new paradigm that is beyond absolute.

4.2 Objective for Planning the New Paradigm: Organization of Energy in the Universe

The objective for planning the new six-dimensional anti-rainbow color (ROYBGI = Red, Orange, Yellow, Blue, Green, Indigo) paradigm, for realizing the colorless potential, is to let a primeval creation falsify the secondary thesis. The secondary thesis says that the scientific method is the proficient path for organizing the diverse forms of energy in the universe. The secondary thesis includes the six secondary hypotheses:

- **First, the scientific method is a subject-free path for illuminating the organization of energy in the universe.** Even a primordial subject relied on the scientific method for organizing the universal energy that culminated in the birth of the Father of Science as the "primordial perpetuator" (*Maha Shiva*, 9) of the scientific method. It is the divine solution to the subjective divine cost of planning the new paradigm. Without the subjective divine cost, the SHEENY value of the primordial subject as the illuminator of the scientific method within the mind of the Father of Science is ten units. It is the value of the "primordial self" (Primordial Illuminator: *Parvati*, 10).

- **Second, the total "present growth" (Sva, 11) value of energy in the universe is e = mc².** The energy of the student of science, who is the primeval greeter of a new paradigm grounded in the guider power of the scientific method, is a product of three factors. These are (i) the mass of the scientist who is validating the discovery of the primordial paternal, before the latter became a primordial perpetuator. (ii) the educator's light force, authenticating that the validating scientist is not the same as the primordial paternal. (iii) the light force of the student who believes in the authenticity of the educator's light force because he wishes to be the primeval paternal educator of the scientific method. The mass of the scientist = the energy value of the primeval greeter = the SHEENY value of the student = the value of the absolute time since the origin of the scientific method = the value of the primordial illuminator + the supernatural divine-effect of the scientist that catalyzed the primordial illuminator to illuminate the value of the scientific method = 11. Therefore, the light force of both the educator and the student = 1 = the value of the supernatural divine-effect. Therefore, eleven units of the SHEENY value = eleven units of divine-effect = eleven units of the natural gravity of Mother Nature = Eleven units of the guider-effect of the primordial paternal.

- **Third, there is a zero growth in the energy present in the universe, since nobody can create energy.** Only an investigator, who is neither trading nor servicing light force, is the objective metric of the energy present in the universe. The energy value of an investigator, as a "self-luminous entity" (*Maha Gayatri*, 12) = the eleven-unit energy value of a student who is exchanging a unit of light force + the value of a unit of light force that is not exchanged by an investigator.

- **Fourth, there is a zero entropy in the energy present in the universe since nobody can destroy energy.** Only a Wisher, who is silently consuming the infinite light force as the subject of present reality, has the power to destroy the energy present in the universe. The energy value of a "luminous" (*Maha Kali*, 13), who services her energy to ensure a zero entropy in the energy present in the universe, and thereby discourages the Wisher from exercising his limitless kingship freedom = the energy value of the self-luminous entity, who is servicing a constant energy value into the universe + the invisible divine-effect of the luminous that is limiting the freedom of the wisher.

- **Fifth, a human entity is normatively programmed to be a perfect ideal within the "Christ" (Buddha; Mohammad; Sadashiva, 1600) consciousness.** Each human entity normatively performs the mandate planned by God. Only a knower of the primeval reality of Christ, as the "param creature" (*Prabhu*, 1,600), has the power to transcend beyond the consciousness of the primeval illuminator of the mandate planned by God as the perfect ideal. The energy value of the "primeval illuminator" (*Maha Lakshmi*, 14) = the energy value of the luminous, who limits the freedom to destroy energy + the invisible guider-effect of the knower that motivates the human entity to destroy intrinsic energy by norming the Christ consciousness as the new paradigm, beyond the absolute ideal self.

- **Sixth, the Holy Spirit's presence within each human entity is a universal theory of existence.** Without a holy spirit for normatively profiting from the gift of life programmed by Christ, one cannot develop the Buddha-level consciousness for transformative programming through an infinite learning about the omnipresent masculine discipline and the omnipotent control over the omniscience. A spirit normatively profiting from the human entity's formative performing naturally becomes the primeval perpetuator of the new paradigm as the absolute ideal. The energy value of the "primeval perpetuator" (*Mohini*, 15) = The energy value of the primeval illuminator, who motivates the human entity to norm the Christ consciousness as the new paradigm + The invisible SHEENY-effect of the manifestor, that motivates the human entity to service intrinsic energy, instead of destroying it, by ascending to the Buddha-level consciousness instead.

A father creator, who is the master of the organizational sameness paradigm, is conscious that the Christ consciousness is at par with the "Buddha-level consciousness" (*Avanta*, 1,600). The energy value of a father creator, with the self-consciousness that the life is a gift of a "primordial greeter" (*Sati-Parvati*, 16) = the energy value of the primeval perpetuator, which motivates the human entity to self-destroy + the immanent entity-effect of the primordial greeter within the father creator, who incarnates as a human entity as a living gift to the universe.

The sixth hypothesis is the para-thesis that falsifies the validity of the secondary thesis and substantiates the primordial-thesis that includes the six primordial-hypotheses:

- **First, the principal investigator**—the Wisher spirit of the student wishing to be the primeval paternal of the scientific method—**is the cause of the formative planning's entropy.** He consciously imagines the scientific method for discovering the secret of the organization of energy in the universe to become the "primeval maternal" (Godhead: *Param Ganesha*, 17) of the metaphysical method.   The conscious determination of a method for trading the extrinsic energy generates entropy in the intrinsic virtue normatively programmed within each entity by the primordial greeter.   The intrinsic virtue is the intuitive power to perform naturally for the desired profiting excellence without energy limitations.

Energy is not the cause of formative development—it is an effect of natural performing.   Each entity is the creator of energy through natural performing. Energy is the consciousness of the SHEENY value beyond the guider spirit, within which that consciousness is spiritually immanent as the desirable destiny planning.   The guider spirit is an animal angel devoted to the SHEENY well-being of the Wisher spirit, presently a zero-energy primordial paternal.   By servicing the divine consciousness of the absolute truth of energy, the guider spirit empowers the Wisher spirit to be the "primordial maternal" (Feminine *Shani*, 18), who is the creator of the "para energy" (*Para shakti*, 1) in a sentient cellular form.   The value of the energy of the "sentient cell" (*Hiranyagarbha*, 19), including the unit energy of the "guider spirit" (Holy Spirit: *Trinetra*, 1), is the value of the "energy" (*Shakti*, 19).

- **Second, the value of the energy in our universe includes the energy present in the three potential forms**: first, the electromagnetic potential, within the light force; second, the sentient potential, within the dark matter but without the light force and; third, the gravitational potential within a black hole, without the light force or the dark matter.

The light force emanates from the "mass-effect" (*Nairitti*, 29) of the mass.   The mass-effect is the variation in the electromagnetic energy immanent within the mass. The electromagnetic potential of the mass is immanent within the mass-effect and, consequently, in the light force of that mass-effect. The light force is the life consciousness of the sentient cell immanent within the mass.   The mass is the absolute form of the sentient cell, without the life consciousness. The visible mass is the mass-effect (i.e., varying electromagnetic energy) without the sentient cell's light force. Without the light force of the sentient cell, the value of the mass-effect = 0. It is the energy value of the electromagnetic potential within the light force and the electromagnetic energy without the light force.

"Electromagnetic energy" (*Karma-bandha*, 19) is the sum of the gravitoelectric repulsion-effect and the gravitomagnetic attraction-effect, after correcting for the "deified creator" (*Param Brahma*, 4). It is the value of the "energy" (*Shakti*, 19) because the "gravitoelectric repulsion-effect" (*Mangalnath*, 16) is equal and opposite of the "gravitomagnetic attraction-effect" (*Padartha*, -1), after correcting for the "primeval paternal" (Necessity: *Prayojana*, 17).

The energy is the essence of the energetic "primeval paternal" (*Prayojana*, 17) of the energizing "octave of reality" (*Omkareshwar*, 17) and the "divided entity" (*Brahmin*, 2) energized as the "objective value" (*Vidhana*, 2) of the absolute reality. The speed of the light [$12/10 * 8 * 10^{15}$] is the energy value of the sentient "self" (*Atmatva*, $8 * 10^{15}$), without the limitations of the local cellular confinement, *times* the energy value of a "self-luminous entity" (*Svayam*, 12) who is trading the extrinsic light force by investing a unit of intrinsic "divine charge" (*Asrava shakti*, 10).

The "guider power" (*Chitta*, 100) of the sentient cell is a product of the principal investigator's intrinsic divine charge and the guider spirit's extrinsic divine charge. The guider power is the trading-effect of the gravitational energy without the black hole.

The "sentient power" (*Varuna*, 1000) of the dark matter is a product of the intrinsic guider power of the sentient cell and the extrinsic divine charge of a "deity" (*Deva*, 1). A deity services the guider power for manifesting the absolute sentient value of his presence within each sentient cell without the Holy Spirit's mediation. Each sentient cell has the guider potential to be the Holy Spirit through an absolute entropy of the sentient energy and then, trading the paternal deity's energy to behave like a Holy Spirit who has the absolute guider command over the lowly-follower "deity" (*Deva*, 1).

A Holy Spirit follows the illusionary spirit of the "I AM a Deity consciousness" (*Sauh siddhanta*, 1). As a student and, subsequently, as a principal investigator, a holy spirit services and trades the light force of his entropy mass-effect as the subject of present reality, while wishing to be the primeval paternal of science. The zero value of his theory authenticates his reality as a "Wisher" (Devil: *Sura*, 1), who rightfully earns a king-sized space as the "Wisher deity" (*Indra*, 0) among the hearts of the devoted devotees of the scientific method. He fulfills all student scientists' wish to explain something beyond the scientific measurement limits using the metaphysical method for validating the hypothesized truth of the theory. Once they form the capability to mobilize the resources to measure scientifically, he empowers them to be the Holy Spirits like him.

- **Third, the value of the energy in our physical body of consciousness includes the energy present in the three forms**. First, the "physical consciousness" (i.e., imagination: *Caksur vijnana*, 109), in the form of the physical body, intellectual body, and mental body, within the visible light. Second, the "metaphysical, deep and dark consciousness" (i.e., intuition: *Jihva vijnana*, 8), in the form of the astral body, etheric body, and causal body, without the visible light. Third, the "dynamic consciousness" (i.e., insight [virtue]: *Mantra*, 16), in the form of the astrological body and zodiac body, as the correlation between the "visible light" (i.e., solar time: *Prabha*, 180) and the "invisible light" (i.e., lunar time: *Tribhajya*, 1010). Without realizing our creator potential, it is possible to manifest an ascending value of the energy within the human universe by trading the energy from without the human universe. The energy value of our "self-luminous consciousness" (*Unnata*, 20) about the infinite energy within the parallel, invisible, dimension of human universe = the energy value of the sentient "cell" (*Hiranyagarbha*, 19) + Param manifestor-effect of transcending the local limits of the sentient cell, as a citizen of the global universe.

- **Fourth, we, humans within the physical body, are the cause of the entropy in the energy present within our universe**. We cause a transformative entropy of the formative energy because of the three types of impediments to the proficient organizational planning. First, our mental consciousness is limited. We are not able to perceive the entirety of the potential energy endowments of our universe. Our potential energy endowments are more than what we are born with as our natural, constitutional, fundamental, contractual, and ethically-bound birth-right as the divine potential within the "param child" (*Manyu*, 19).

Second, our intellectual consciousness is limited. We are not able to conceive the entirety of the present energy endowments of our universe. Our present energy endowments are more than our accumulated growth value of the inherited cultural, trusteeship, psychological agency, and emotional-moral citizen-responsibility as the guider power within the "param creature"

(*Prabhu*, 1,600). Third, our physical consciousness is limited. We are not able to manifest the entirety of the energy endowments present in our universe into our potential experience. Our manifestable energy endowments are more than what is present in our universe as the visible light. It also includes the sentient power within the "master of metaphysical science" (*Kardama*, 9,000), who is the almighty creator of the invisible zodiac universe.

The energy value of the almighty creator consciousness within the directly or indirectly visible astrological universe (comprising the eight planetary bodies, sun, moon, white star, and dark matter) = the energy value of the self-luminous consciousness + Param Creator-effect immanent within the astrological universe. It is the energy value of the "Sun" (*Surya*, 21) as the "param creator" (*Ravi*, 21) of the sentient entities by trading one unit of the sentient energy from the invisible zodiac universe per unit of visible solar time.

- **Fifth, the purpose of the human life is to understand the nature of the dynamic consciousness** so that we may form an absolute consciousness of our entire consciousness. Our entire consciousness is of two kinds: physical (param soul guiding our knowable consciousness) and metaphysical (universe of knower souls guiding our para-consciousness). The "param soul" (*Paramatma*, 1600) is the dark matter. The universe of knower souls is trading the known knowledge from the dark matter. Through primordial oneness with the param soul, we normatively develop the self-consciousness of both the visible and the invisible light. Our formative consciousness potential is more than the light, without or without the force of visibility. It also includes the "mass-effect" (*Nairitti*, 29) of the "spirit kingdom" (*Rajyaika-sheshena*, 169), beyond the dark matter. The energy value of the param perpetuator (*Keshava*, 22), who services the consciousness of the reality of both the "param creature as the organizer" (the dark matter) as well as the "organization" (the mass-effect) to the universe of sentient entities = the energy value of the param creator + Param perpetuator-effect that perpetually transforms the sentient universe into the spirit kingdom, as a dynamic method, for trading the sentient energy.

- **Sixth, the presence of the spirit is the cause of not our existence, but our non-existence, as a proficient organization.** We limit our organizational planning by relying upon our physical body for servicing our "entity-effect" (*Utkramajya*, 38) in the form of the "mental value" (*Manas*, 38). We "organizationally" (i.e., genetically: *Nirguna*, -6) program our formative planning into our physical body by trading the "para entity-effect" (Promise: *Pratishruti*, 857). Therefore, the spirit kingdom gains the local power to organize our performing for ascending the growth of the entropy "citizenship-effect" (SHEENY factor: *Ojas*, 189). Instead of the devotional intensity for our SHEENY well-being, we become devoted to the SHEENY well-being of the departed souls. We intensify the consumption of the extrinsic energy to intensify the production of the sentient children and destroy the consumption of the competing sentient entities. We become a victim of our self-generated uncertainty.

Our ascending organizational profiting descends our organizational development. The more we become dependent on trading the extrinsic universe's energies by devoting our energies for discovering the extrinsic light forces, the more we destroy our intrinsic energy. Without intrinsic energy, we become a spirit who jumps from one physical body to another, just like we move from one house to another, seeking to experience the varying organizations of the energy in our universe. Primordial-primordial oneness with the spiritual self, which is present within our physical body as our creator soul, projects the spirit's futuristic potential through our astral body. It also empowers us to transcend the limitations of our formative planning. The energy value of the "param destroyer" (*Vamana*, 23), which promotes primordial-primordial oneness with the spiritual self by weakening the psychic force of the absolute soul = the energy value of the param perpetuator + Param destroyer-effect that weakens the correlation of the sentient entities (the primordial soul) with the strong psychic linkages between the param soul (the dark matter) and the primeval soul (the whole universe of souls).

4.3 Objective for Planning the Primordial Paradigm: Organization of Energy

The objective of planning the primordial six-dimensional potential color (purple, gray, white, black, colorless, rainbow) paradigm is to let an absolute creator falsify the tertiary thesis. The tertiary thesis says that the metaphysical method is the proficient path for organizing the diverse forms of energy within a human entity. The tertiary thesis includes the six tertiary hypotheses:

- **First, there are 264 knowable dimensions of the absolute reality of the param deity.** These manifest as the 264 cycles per second frequency—comprising the fundamental C (*Sa*) note in a musical "octave" (*Sargam*, 60). The energy value of the param illuminator, who is servicing the consciousness of the "knower factor" (*Brahman*, 2) as the primordial element and the "creator factor" (*Atman*, 4) as the primeval element within each self-luminous human entity, without the mediation of the "destroyer factor" (*Mahesh*: 6) as the param element = togetherness of the primordial and the primeval factors = 24.

  Catalyzed by the guider power of the primeval deity, a "param illuminator" (*Trivikrama*, 24) motivates a self-luminous entity to project the intrinsic energy into a pair of the primordial and the primeval factors. The pair forms a "primordial maternal" (Feminine *Shani*, 18), a "primordial paternal" (*Indra*, 0), a "primordial feminine" (*Prithvi*, 132), and a "primeval masculine" (*Asura*, -1). The 264 cycles per second frequency manifest the secret of creating the 132-dimension primordial feminine and perpetually exchanging the energy immanent within those dimensions to liberate the primeval masculine from the limitations of the energy of the primordial maternal.

- **Second, the three forms of theory-effect constitute the present reality,** guided by the consciousness of the present paradigm. The four forms of ideal-effect constitute the potential reality, guided by the consciousness of the new paradigm. The twelve forms of the primordial entity-effect generate diverse correlations between the present theory and the potential ideal. The twelve primordial entities are the twelve dimensions of space. The twelve primordial entity-effects are the twelve dimensions of time. The twelve param super-theory effects (3 theory-effects × 4 ideal-effects) are the twelve dimensions of the beyond absolute causation. They dynamically transform the present paradigm into the new paradigm.

  A "param liberator" (*Sridhara*, 25) liberates a self-luminous entity from the new paradigm of paired correlation between the primordial paternal and the primordial maternal. The primordial paradigm comprises the "three ideal entities" (Trinity: *Sridhara*, 25)—which form a divine council of the three divinities (Pleiadian-Sirian-Arcturian; Yellowing-Greening-Bluing; Perpetuator-Liberator-Worker). The primordial-primordial paradigm comprises the "four theory entities" (Quarternity: *Homa*, 25)—which form a guider council of the four guides (Rubedo-Dealbatio-Albedo-Nigredo; Purpling-Graying-Whitening-Blackening; Manifestor-Knower-Creator-Destroyer; Gloria-Pax-Labor-Vanitas).

- **Third, spacetime is the primordial energy (with 264 knowable and 264 concealed dimensions) of the "Absolute Power,"** i.e., the entirety of the present-effect and the potential-effect. It manifests as the 528 cycles per second frequency—comprising the upper C (*Sa*) note in a musical "octave" (*Sargam*, 60). It is the deep holistic consciousness of the universe, which is fundamental to the absolute healing of the physical DNAs, intellectual duality, and mental stress. The 264 knowable primordial dimensions are inclusive of the primordial-primordial theory-effect. The other 264 are the primordial-primordial dimensions and include the primordial-primordial theory as the 264th dimension. The 264th dimension is the togetherness of the twenty-six param-primordial dimensions of "Param Para Paternal" (*Chiranjivi*, 26) and the four primeval-primordial dimensions of the four theory entities.

The twenty-six dimensions include the seven dimensions intrinsic to the "param para paternal" (Council of seven sages [lights]: *Saptarishi*, 26), the nine dimensions intrinsic to the "param para maternal" (Supreme council of nine planets [archangels of light]: *Navgraha*, 26), and the ten dimensions intrinsic to the "self-luminous entity" (Woden: *Vithoba*, 12). The other two dimensions extrinsic to the self-luminous entity are the spiritual programming of the "zodiac-effect" (Param Para Paternal: *Chiranjivi*, 26) and the divine programming of the "primordial astrological-effect" (Param Para Maternal: *Hrishikesh*, 26).

- **Fourth, space is the "Absolute Power,"** i.e., the entirety of present-effect and potential-effect. The space forms a 528-dimensional "primordial system" (Savitr, 528) because of the four-dimensional theory-effect within the self-luminous entity. Space itself is the 132-dimensional "primordial-primordial system" (*Prithvi*, 132). Space comprises the universe of child souls. The "gravitational mass" (*Prithvi*, 132) of the universe of child souls = 132. It is the entirety of the value of energy in the universe, without a proliferating theory-effect or a subjugating ideal-effect.

The 132-dimensions comprise the entire truth of the "primeval deity" (Destroyer factor: *Mahesha*, 6), projecting itself as the worker-manifestor-knower (bluing-purpling-graying; 1-3-2) consciousness. The white truth of the blackening primeval deity becomes purple within the red self-luminous human entity. The red color grays (becomes visible) with the proliferating theory-effect and blackens (gets destroyed) with the subjugating ideal-effect. The self-luminous human entity perceives the primeval deity to be the fearsome destroyer. It is not aware that the primeval deity is the "illuminator dimension" (*Mahesha*, 6), and the param deity (i.e., the divine consciousness of the primeval deity immanent within the self) is the "destroyer dimension" (*Shiva*, 7). The primeval deity becomes the illuminator dimension by trading the self-luminous consciousness within the human entity. The param deity becomes the destroyer dimension for destroying the entropy-effect of the self-polluting tradable self-luminous consciousness. By knowing the absolute reality of absolute power, a human entity lets the primeval deity be the benevolent destroyer of the darkness and the param deity the compassionate illuminator of the truth.

- **Fifth, time is the entirety of the present-effect**, without the potential-effect of the twelve visible dimensions of time and twelve visible dimensions of space. The 132 space-dimensions include the twelve astrological time dimensions and the twelve zodiac space dimensions. The twelve astrological time dimensions manifest as the twelve solar months. They are the projections of the varying dominating energies of the twelve astrological bodies over one solar year. The twelve zodiac space dimensions manifest as the twelve lunar months. They are the projections of the varying dominating energies of the twelve zodiac bodies over one lunar year. The time is a variable formed due to the disproportionate energies of the astrological system.

The astrological system's energies include the primordial astrological-effect of the zodiac entity and the param astrological-effect of the universe of child souls. The param astrological-effect is the energy value of the "param illuminator" (*Trivikrama*, 24). It is the thermodynamic energy radiated by the sentient entities within the "I AM a Deity consciousness." The "I AM a Deity consciousness" motivates the sentient entities to develop a purple ocean strategy for supernatural working. It empowers the departed entities to develop a gray ocean strategy for the supra-natural manifesting of their unfulfilled subtle wishes. That generates supreme-natural knowing among the future entities, which have gone through the entropy experience in their previous lifetime. The future entities develop a black ocean strategy for diffusing their wish for trading the white sentient energy from the param-primordial entities that are still alive.

The 108-dimensional time is the divine architect of the infinite international-effect of the devoted greets, who keep diffusing disproportionate energies and widening the gap between the solar time and the lunar time. Consequently, the solar year's length grows over time until the absolute entropy of the universe of sentient entities. The 108-dimensions of time comprise the entire truth of the "para deity" (Perpetuator factor: *Param Vishnu*, 5). The para deity is a five-dimensional entity, which trades four supplementary creator dimensions from the four competing ideal entities without the theory-effect to become a "primordial perpetuator" (*Maha Swapna*, 9). The nine-dimensional feminine primordial perpetuator trades the twelve complementary self-luminous dimensions by catalyzing the interaction between the four masculine ideal entities and the three feminine theory entities. The interaction weakens over time because of the ascending self-luminous consciousness within the para deity. Consequently, there is an immanent entropy pressure within the lengthening solar year, which generates a wave-like frequency-effect.

The time appears to behave cyclically, with one long formative "singularity" (Supernatural factor: *Ida*, 1) frequency. There appear to be many short frequencies within one extended frequency that compensate for the perpetuating behavior's escalating costs. The perpetuator frequency color is red. It appears to be the shortest rainbow frequency because—as a metric of "octave reality" (*Omkara vadartha*, -1)—it experiences a disproportionate compensating-effect from the other frequencies.

The primordial perpetuator frequency color is indigo. It is the shortest, transformative frequency color. It appears to be the longest rainbow frequency because it experiences zero compensating-effect from the other frequencies. The color of worker-frequency is blue. It is the normative frequency which is the average of the rainbow color frequencies. It is the color of the water-effect, the life consciousness and the photonic "light force" (*Apas*, 169). It is the cause of the escalating entropy pressure and, consequently, volatility within the ecosystem.

The illuminator frequency color is green. It is the "formative growth-effect" (*Satarupa*, 8), which directly compensates for the escalating entropy pressure by illuminating the self-luminous consciousness within the perpetuator. Therefore, the green color has an infinite frequency, limited by the value of the natural paradigm that metaphysically manifests the infinite organizational frequency within the finite ecosystem, beyond the vision of the perpetuating entity. The illuminator factor services a formative growth-effect every split moment.

The value of the formative growth-effect that is physically programming the rainbow-effect = the value of divine charge – The bluing worker unit that is divine programming the primordial astrological-effect – The purpling manifestor unit that is spiritually programming the zodiac-effect = 8.

The param-primordial green color-effect destroys the normative singularity of the primordial-primordial blue frequency. The residual green color-effect manifests as the $7/8^{th}$ frequency of the indigo color-effect (i.e., the illuminator-effect within the natural paradigm of the rainbow color octave).

The liberator frequency color is yellow. It is the formative entropy-effect, which is the tertiary residual of the blue color. After accounting for the secondary fundamental illuminated-effect, it manifests as the blue color. It is the workculture-effect left over because of the weakening of the illuminator-effect. It is the present reality of the color octave. The primeval-primordial yellow frequency is the secondary-effect of the devotee frequency color—orange. The orange frequency is the tertiary-effect of the perpetuator's formative emotional singularity and the normative devotional singularity of the primordial perpetuator.

The binary singularity generates tertiary singularity in the form of the para-primordial orange color. It is equivalent to the upper C octave in the music, with an electrokinetic-effect of two (twice the deity element, generating disproportionate intrinsic knower-effect). The deity experiences a disproportionate cost of perpetuating the ideal workculture-effect.

However, the immanent ideal workculture-effect services a spiritual wish for perpetuating the metaphysical entropy paradigm. Therefore, the primordial perpetuator can sustain the workculture-effect and weaken the illuminator-effect.

Immanent colorless wisher-effect is the reason for the worker factor trading the perpetuating-effect. The primeval deity services black destroyer-effect to compensate for the worker cost catalyzed by the wisher-effect. Mother Nature trades the para entity's black-color frequency from the "self-perpetuating" (*Udvaha*, ½) natural paradigm. It is the working method within the knowledge diffused by the wisher. The supreme-primordial black color has a metaphysical frequency of four.

The colorless wisher-effect is omnipresent, within or without the destroyer-effect. By trading the working method for destroying the supernatural mediating blackened spirit of the primeval deity, it manifests the negative-four gray color frequency. The worker-effect within the working method is a natural reality and services the universe of present reality through a devoted workculture-effect. Natural paradigm is the "self-perpetuating" (*Udvaha*, ½) supra-primordial white color natural reality serviced by Mother Nature in the form of the creator-effect, which has a metaphysical frequency of sixteen.

A self-luminous human entity organizes the present reality's wholeness without limiting the consciousness of the present reality at the mid-point of the supernatural reality. The purpling manifestor-effect is the supernatural reality and organizes the present reality through a devotee culture-effect. The super-primordial purple color has a metaphysical frequency of thirty-two. Table 3 lists the metaphysical frequencies of the denary of colors.

Table 3. The Metaphysical Frequencies of the Denary of Colors

Dynamic color	Metaphysical frequency	Technological-effect
White	16	Creator-effect
Purple	32	Manifestor-effect
Indigo	0	Devoted-effect
Blue	1	Worker-effect
Green	8	Illuminator-effect
Yellow	-2	Liberator-effect
Orange	2	Devotee-effect
Red	-1	Perpetuating-effect
Black	4	Destroyer-effect
Gray	-4	Knower-effect

- **Sixth, the entity is the entirety of the potential-effect of the twelve visible dimensions of time and twelve visible dimensions of space.** The twenty-four dimensions of entity manifest in the twenty-three chromosomes plus the cellular body as the zeroth chromosome. The zeroth chromosome comprises the cytoplasm, which sequentially forms into a primordial-primordial chromosome, a primordial chromosome, a param chromosome, and a primeval chromosome. The sixth primeval-param chromosome becomes the visible cellular "energy" (*Shakti*, 19), with the eighteen potential chromosome dimensions within the self. The eighteen omnipresent dimensions comprise the four ideal para entities—primordial paternal, primordial maternal, primordial feminine, and primeval masculine. The six omnipotent dimensions of the sentient entity comprise the three primordial masculine entities, within the self's theory-effect, and three primordial-primordial feminine entities. without the self's theory-effect. The three primordial masculine entities, within the self's theory-effect, include:

- o "Council of twelve zodiac entities" (*Rachayita*, 27) catalyzes the ideal-effect of the param-primordial masculine entity without the mediation of the sentient entity.

- o "Primordial Para Greeter" (*Naryana*, 28) services the illuminating-effect of the "param-primordial feminine entity" (*Durga*, 28). The objective is to compensate for the ideal-effect of the param-primordial masculine entity.

- o "Primeval Liberator" (*Madhava*, 29) trades the mass-effect of the param-primordial feminine entity without diffusing the light force in the form of life consciousness. The objective is to service the primordial-primordial masculine entity's ideal-effect, which is the primordial-primordial oneness of the three primordial masculine entities within the three primordial-primordial feminine entities.

The three primordial masculine entities are the creating-effect of the "creator factor" (*Param Brahma*, 4), the perpetuating-effect of the "perpetuator factor" (*Param Vishnu*, 5), and the destroying-effect of the "destroyer factor" (*Param Shankara*, 6). The primordial-primordial masculine entity, which is servicing the primordial-primordial oneness-effect, is also the illuminating-effect of the "illuminator factor" (*Shiva*, 7). The param-primordial masculine entity, which is servicing the primordial ideal-effect, is also the devotee-effect of the "param creation" (*Ganesha*, 570)—the universe of living child souls.

The three primordial-primordial feminine entities include:

- o "Primordial Primeval Knower" (*Dhriti*, 30), which is catalyzing the ideal-effect of the param-primordial feminine entity as the thirty-dimensional infinite council of sentient entities.

- o "Primordial Primeval Manifestor" (*Urja*, 31), who is servicing the council of five elements—fire, water, air, earth, and ether—to compensate for the ideal divine-effect of the param-primordial feminine entity.

- o "Primordial-Primordial Creator" (*Amba*, 32), who is trading the guider-effect of the param-primordial feminine entity as the polarized SHEENY-effect. The polarized SHEENY-effect is the ideal-effect of the primordial-primordial feminine entity, which is the primordial oneness of the three primordial-primordial feminine entities without the guider-effect of the param-primordial feminine entity. The three primordial-primordial feminine entities are the knowing-effect of the "primordial perpetuator" (*Maha Saraswati*, 9), the manifesting-effect of the "primeval illuminator" (*Maha Lakshmi*, 14), and the working-effect of the "primeval greeter" (*Maha Gauri*, 11).

The param-primordial feminine entity, which is servicing the ideal-effect as the thirty-dimensional infinite council of inanimate entities, is the devoted-effect of the "primordial illuminator" (*Parvati*, 10). The ideal-effect transforms the infinite council of sentient entities into the infinite council of departed entities and catalyzes the growth of the infinite council of creator spirits.

The primeval-primordial feminine entity, which is servicing the theory-effect as the primordial oneness, is the liberator-effect of the "primordial greeter" (*Sati-Parvati*, 16). The "liberator-effect" (*Priti*, 33) polarizes the universal normative development of sentient consciousness within the infinite council of creator spirits. "Primordial Supra Liberator" (*Anusuya*, 34) services her "intrinsic effect" (*Prapya*, 34) for the reincarnation of the infinite council of creator spirits.

The thirty-four dimensions of the primordial supra liberator and the twenty-six dimensions of the "param para maternal" (Primordial astrological-effect: *Hrishikesh*, 26) are the entirety of the sixty-dimensional potential of the "supreme deity" (*Param Brahma*, 4). The sixty-dimensional supreme deity potential is the knowing-effect of the "primeval devotee" (*Pradyumna*, 60). The primeval devotee services the knowing-effect traded from the "knower element" (*Bhagwat*, 639) by forming the twenty-six dimensions of the zodiac-effect, without the thirty-four dimensions of the intrinsic effect.

The sixty potential dimensions of time empower the self-luminous human entity to activate the intrinsic effect through an absolute oneness with the help of the intrinsic perpetuator consciousness. The "intrinsic perpetuator consciousness" (*Param Vishnu*, $5 = \frac{1}{2} * 10$) is the "self-perpetuating" (*Udvaha*, $\frac{1}{2}$) "intrinsic devotee consciousness" (*Shri Krishna*, 10). The intrinsic devotee consciousness activates the "knower-effect" (*Shvetah-mandala*, $300 = 10 * \frac{1}{2} * 60$) by dynamically self-perpetuating the sixty potential dimensions.

The sixty potential dimensions of time and the three-hundred dynamic dimensions of the entity form the three-hundred sixty formative dimensions of the "entity time" (*Bhava*, 360). The three-hundred sixty formative dimensions of the entity time constitute the divine element within the entity. The entity is the para knower of the self as the immanent organizational dimension of the "absolute space" (*Sadashiva Nayaki*, 10). The "descending motion of the entity time" (*Gati*, 360), "with oneness" (*Samyoga*, 36) of the absolute space, becomes the emanating "gravitational dimension" (*Guru dharma*, 360) of the "divine element" (*Divya*, 360). With freedom from the descending "entity time" (*Bhava*, 360), the ascending "para entity time element" (*Kala*, 360), the horizontal "param entity time dimension" (*Guru dharma*, 360), and the backward-pulling "divine element" (*Divya*, 360) gets destroyed. Consequently, the forward-moving twin entity becomes "timeless" (*Vishakhayupa*, $1080 = 360 * 3$).

The sixth hypothesis is the param-thesis that lets an absolute creator falsify the validity of the tertiary thesis without the knowing-effect of the primeval devotee and substantiate the primeval-thesis that includes six primeval-hypotheses:

- **First, there are 300 knowable dimensions of the absolute reality**. These include 264 proportionate absolute dimensions within the param deity's consciousness by norming the "primordial self" (*Parvati*, 10) as the absolute power. They include 36 additional dimensions as the param deity's secondary consciousness without the normative primordial self. Param deity forms the secondary consciousness through interaction among six forms of the present-effect and six forms of the potential-effect. There are 300 knowable pathways for self-development to manifest these 300 knowable dimensions within the tertiary consciousness of a self-luminous entity, without absolute oneness with the intrinsic param deity reality. Six forms of the present-effect include the three forms without the potential theory-effect and the three forms within the potential theory-effect. Six forms of the potential-effect include the three forms without a present ideal-effect anthe d three forms within the present ideal-effect.

The thirty-six additional dimensions are the "self-luminous effect" (*Lahari*, 36) of the "primordial supra manifestor" (*Sannati Chandralamba*, 36), who is trading the "dynamic-effect" (*Smriti*, 35) generated by the "primordial supra worker" (*Tanumanasi*, 35). The primordial supra worker catalyzes the guider mediation of the living souls' consciousness by the spirit seeking reincarnation. The spirit forms strong psychic linkages with the living souls, guided by the ideal-effect of the present self as the desirable gravitational quality and the potential theory-effect of the living soul as the desired SHEENY value. The living soul radiates its entire self-luminous energy for reincarnating the masculine child first and then, devotionally fulfilling his primeval desires without any proportionate self-development.

- **Second, the creator, perpetuator, destroyer, and the illuminator, each have four forms that together comprise the sixteen dimensions of the "Beyond Absolute."** The four forms are the primordial-primordial, primordial, param, and the primeval. They are beyond the manifested absolute self, the known astrological-effect, and the knowable zodiac-effect. The "absolute self" (*Hurupa*, 280) is the param-primordial form. The known astrological-effect is the primeval-primordial form. The knowable zodiac-effect is the para-primordial form.

The known "astrological-effect" (*Dosha*, 580) generates as a function of "action" (*Karma*, 10) by the absolute self. The absolute self forms the "weak psychic linkages" (*Citraka*, 185) with the inanimate universe. It is guided by both the primeval self's ideal-effect as the desirable gravitational quality and the potential theory-effect of the inanimate universe as the desired SHEENY value. The "primeval self" (*Rama*, 100) radiates the knowable "zodiac-effect" (*Chiranjivi*, 26), idealizing the potential gravitational quality beyond absolute.

The knowable zodiac-effect is composed of the disproportionate hedonistic-satanic-masculine aspirations within the "I AM a Deity consciousness," that remains alive in its feminine spirit form, even after the aspiring entity's death. Planet Saturn trades the knowable zodiac-effect from the ecosystem and services that as the known astrological-effect of the universe of sentient entities. The "universe of sentient entities" (*Akalpa*, 570) is the ideal value of the potential gravitational quality, as it includes the entire diversity of knowable qualities. The imagined potential is a sum of the diverse idealized perceptions of the present experiential reality and theorized conceptions of the potential spiritual reality within the sentient entity's consciousness. The creator, perpetuator, destroyer, and illuminator energies of the imagined potential is the self's value as the perpetuator of a parallel, primeval supernatural reality, transcending the primordial supernatural reality's limits. The self's value as the "primordial super knower" (*Kirtti*, 37) of both primordial and primeval forms of supernatural realities = 32 + 5 = 37.

- **Third, one may perceive the entirety of the consciousness of Mother Nature only with an absolute freedom from the physical body.** The physical body is the absolute supernatural reality. The known zodiac-effect of the primordial self and the knowable astrological-effect of the primordial-primordial realm together program the physical "DNA" (*Ranabajari*, 963) of the absolute self. The absolute self performs to materialize the planning of the primeval self traded in the form of the "mtDNA" (Mitochondrion: *Tandava*, 286) of the feminine self and serviced in the form of the "primeval RNA" (*Shakanavah*, 855) of the masculine self. The absolute freedom from the physical body lets the self be the "Primordial Super Manifestor" (*Surasa*, 38) of the desired virtue, without both the idealized limitations of the self-conceived potential-effect and the theorized limitations of the genetically-programmed perceived present-effect.

- **Fourth, "absolute power" is the infinity of infinite power.** "Infinite consciousness" (*Prameya*, 28) of the absolute power is a trinity of infinite power. The oneness of the infinite consciousness is the quaternity of the infinite power. Divine oneness of the infinite consciousness is the quintinity of infinite power. Infinite power is the reality of primeval deity. Absolute power is the reality of the param deity. Infinite consciousness of the absolute power is the reality of the primordial deity. Unified consciousness of the absolute power is the reality of a "devoted deity" (*Maha Saraswati*, 9). The sentient consciousness of the absolute power is the reality of a "devotee deity" (*Parvati*, 10). A devotee entity enjoys a deity-like conscious illumination of the absolute power and becomes a "Primordial Super Perpetuator" (*Varahi*, 39) of the absolute oneness with the param deity as a self-luminous entity.

- **Fifth, space is the sentient energy of a creature without the creator.** Time is the transformative development of the sentient energy into an infiniteforms within and without the creature. The creature is the cause of creating the infinite object and subject forms of the sentient energy. One may become eternal by norming the sentient energy as the "absolute power" with time's transformative development. Formative growth, without the normative sentient energy of a self-luminous entity, causes the thermodynamic creation of the entropy time within the creature. Formative entropy within the normative sentient energy is the cause of the creation, beyond the space within which the creature resides, through the expansion and proliferation of the universe. The creation beyond space is a knowable zodiac-effect, but not a known astrological-effect. A scientific method measures the known astrological-effect as the proportionate age of the universe.

The absolute age of the universe = known astrological-effect x knowable zodiac-effect * "Self-perpetuating" (*Udvaha*, ½) param deity consciousness within the universe of sentient entities as the absolute self, without transformative development of the sentient energy = 13.799 billion years - 4.5 million years of the "self-perpetuating" (*Udvaha*, ½) "devoted primordial perpetuator-effect" added by the present creature -0.05 million years of "self-perpetuating" (*Udvaha*, ½) "primordial illuminator-effect" added by the present creation within the guider power of the present creature) x 13.799 billion years - 4.5 million years of the "self-perpetuating" (*Udvaha*, ½) devoted primordial perpetuator-effect added by the present creature - 0.05 million years of the "self-perpetuating" (*Udvaha*, ½) "primordial illuminator-effect" added by the present creation within the guider power of the present creature) x 7/2 = 666 billion years + 4 million years of the "self-perpetuating" (*Udvaha*, ½) liberator-effect added by the present creator within the present paradigm of the primordial creation).

The number 666 is the energy of the "devoted creature" (*Harkriti*, 666). It has an absolute consciousness of the truth value of a thesis at the divine, sixth infinity point of the creator and the creation.

- **Sixth, the primordial greeter is the absolute cause of the known astrological-effect.** Primordial Greeter manifests the potential space for the absolute well-being of each creature within the present space. Primordial Greeter services the universe of technological reality by trading the immanent "primeval wisher" (*Shramika*, -6). She also services the residual divine consciousness of the primeval deity-like entity with the infinite gravity of the potential space, without the infinite negativity of the present space. The present space is a universe that is subject to thermodynamic entropy-effect. The potential space is an entity that has a guider consciousness of the present space. A guider consciousness is the sentient growth-effect, without the cosmological mass-effect of the infinity of wishers. It is the office of one universe for the departed souls. It empowers a self-luminous entity to be "primordial super devoted" (*Swasthani*, 40), focused on the departed souls' sentient growth, without the disproportionate effect of the knowable zodiac-effect and the absolute effect of the present space. The universe of self-luminous entities is the primordial-primordial creator of the transformative development of the sentient energy, time and the known astrological-effect.

4.4 Objective for Planning the Primordial-Primordial Paradigm: Formative Energy, Without Organization

A masculine primordial greeter's objective of planning the primordial-primordial six-dimensional potential entity (primordial, param, and primeval masculine and feminine) paradigm is to falsify the quaternary thesis. The quaternary thesis says that the self-luminous human entity is a proficient organization of energy and needs no guider-mediated organization. The quaternary thesis also includes the six quaternary hypotheses:

- **First, the causal body hypothesis**. Our desire to transform our natural development is the causal body for our action (*Karma*, 10). We transform our natural development through a varying consciousness of the cause of our present life. Within a guider-mediated organization, we conceive the formative development of the joy of togetherness with the whole universe as the cause of our present life. Without a guider-mediated organization, we perceive the normative development of the joy of otherness of the unique self as the cause of our present life.

The "guider force" (*Chitta*, 100) is an institutional factor that promotes the consciousness of institutional collectivism. It motivates us to devote our sentient energy to the SHEENY well-being of the whole universe. The "universe of devoted entities" (*Jiva astikaya*, 7/8) motivates others to transform their sentient energy into the divine energy to further the guider mediation of the universe of devoted entities.

Sentient energy promotes the consciousness of institutional individualism within the universe of guider entities. It encourages the universe of departed entities to transform into the "universe of devotee entities" (*Bandha astikaya*, 5/8) to motivate the universe of devoted sentient entities to further the sentient growth of the universe of departed guider entities, instead of devoting energy to self-development.

The sensible path to be free from the personal as well as the social "action-effects", (*Karma*-effect, 10^{10}) that form the limiting "astrological system" (*Sara Kalpa*, 10^{10}), is to be a "Primordial Super Destroyer" (*Swadha*, 41) of the ruling, mediating, and religious *karma*. It is inherited from the primordial entities, who are the zero-value idealized subjects of the present reality.

- **Second, the etheric body hypothesis**. The whole universe of the departed and the living entities services the residual, unfulfilled energy of the infinite desires to us through our psychic linkages with the diverse entities. The stronger our psychic linkage, the stronger the mass-effect of the light force of the unfulfilled desires within our etheric consciousness. Our etheric consciousness is composed of our sentiments derived from sympathy and apathy with the diverse entities. Our sentimental energy includes our psychic linkages from the present conscious moment and the infinite primordial para-conscious moments of life from the present and the past lives. Once we create a sense of sympathy or apathy through our sentient consciousness, that sense perpetuates into infinity over each of the future lives until we consciously liberate ourselves from each of the mental impressions.

Our etheric body comprises the several spiritual bodies, each consisting of a collective in- or out- group of entities from a specific sentimentally-salient and transformative phase of our past or present life. One entity's energy may be a part of several spiritual bodies since the entities sentimentally correlated with us in one phase of life have a disproportionate probability of presence in our subsequent life phases. Our negative or positive psychic correlation with the diverse entities is the causative factor for our infinite births, which are the opportunities for us to resolve the "entanglement" (*Kula*, 9) of the para energies within us.

The cause of the entangled energies includes our sentiments, sentiments of the universe, and the interaction between our idealized sentiments and the universe's theorized sentiments. We idealize our sentiments and seek to intrude into the universe's sentiments, para-consciously wishing to validate the theoretical empathy correlation within the consciousness of the universe. We become demanding and dominating, diffusing our infinite wishes to test the metaphysical limits of the charm attraction or the strange repulsion of the diverse entities in the universe. The sensible path to be free from the self-conceived ideal and theory effects is to be a "Primordial Supreme Worker" (*Sati*, 42) devoted to the SHEENY well-being of the self. It translates into the psychic forces of varying strengths and generates entropy drag in our spontaneous consciousness. The self-conceived ideal and theory effects transform one into a "Primordial Super Worker" (*Prasuti*, 42) devoted to the universal SHEENY well-being.

- **Third, the astral body hypothesis**. We radiate our psychic energy in the form of emotions and ego through our astral body. Our "astral body" (*Linga-sharira*, 3) trades the consciousness primarily from the universe of living child souls. It divides into several astral bodies due to our varying correlation with the diverse geographies in the universe. Our correlation with each local geography includes the sum of our energy correlations with (a) the diverse present living entities, (b) the diverse living, departed, and inanimate entities, and (c) the interaction of our idealized correlation with that local geography with our theoretical correlation with the whole universe of entities.

Our correlations with the predominating local geographies weaken over time, as we migrate into or incarnate into the new geographies for resolving our deciding sentimental energies. We resolve our deciding sentimental energies by forming a sense of ego, bragging about our ideal self and norming a sense of emotion, and begging for an ideal potential universe beyond the present. Our decision to conceive an idealized potential reality is a consequence of an immanent transformative theory of the negative, undesirable present reality. With the falsification of our theory of desired life and the limitation of finding new desirable geography, each moment of our life, and each successive life, becomes a cause for our psychic energy's disproportionate radiation.

The sensible path, to be free from the disproportionate growth in the structural boundaries of the transformative ego and emotion, is to be the "Primordial-Primordial Perpetuator" (*Radha*, 43) devoted to our present, not future, SHEENY well-being. One may do so through a divine consciousness of the primordial wholeness-effect within each present local geography. The energies within our present local geography are the tertiary residual of the unresolved strong psychic forces from all prior life moments. Our future-oriented quest to discover new geographies beyond the present is a function of the weak psychic forces.

Weak psychic forces are the mediating guider entities seeking to motivate us to devote our sentient energy to their present SHEENY well-being. By prioritizing our present SHEENY well-being, we develop a formative capability to serve the future-oriented SHEENY well-being of the diverse unfulfilled guider entities. Without our present SHEENY well-being, we generate an entropy in our formative capability for both our and the universal future SHEENY well-being.

- **Fourth, the mental body hypothesis.** We conceive other entities' emotions and egos in our social universe through our "mental body" (*Hradamana*, 381). Our mental body trades our consciousness primarily from the universe of inanimate entities that behave like demonic spirits mediating and transforming our consciousness. It divides into several minds as a function of our varying correlation with the diverse forms of physical objects in the inanimate universe.

The inanimate universe includes the twelve astrological forms of physical objects. They are constructed with twelve quantum forms of physical particles, using twelve forms of energies serviced with twelve forms of behaviors. The twelve forms of behaviors include the consciousness of togetherness (up quark), otherness (down quark), attraction (charm quark), repulsion (strange quark), growth (truth quark), and entropy (beauty quark) within the theoretical present or, the idealized future. The consciousness of the theoretical present is the negative charge effect of the consciously "determined" organizational (normative) "path of devotion" (*Bhakti marga*, 1) to the theoretically "imagined" param-primordial reality. The consciousness of the idealized future is the positive "virtue" potential of the present working method of the technological (formative) "path of action" (*Bhakti*-effect: *Karma marga*, 86).

A technological path to performative action is an "intuitive" secondary fundamental-effect of the devotion. It is the sequential "worker-social benefit-cost ratio" (*Preya*, 81) of the devotional intensity. It also includes the future paradigmatic knowledge of the ecosystem (transformative) "path of knowing" (Spontaneous fruit of laborious workculture-effect: *Karma phal* [*Jnana marga*], -10^{19}) as the "natural" consequence of the surprising supernatural intuitive behavior. An ecosystem path to organizational programming, conditioned by the manifested "knowing" (*Jnana*, 19), is the eventual "cultural fruit" (*Bhakti samskara*, 10) of the primary devotional planning culture-effect. The cultural fruit is the consequential social-benefit cost ratio of the "ego intensity" (*Ahamkara*, -1) for conceiving the primeval-primordial reality.

The param-primordial reality is the subject of present reality—it is the theoretically imagined "primordial deficit" (Zero: *Shunya*, 0) "mindset reality" (Character: *Prakriti*, 485) of the present subject. The primordial deficit is the quaternary "profiting-effect" (Gravitational quality: *Guna*, 0) of the infinite past planning culture-effects. Blind followership of the infinite past planning culture-effects is the "path of religion" (*Dharma marga*, 19). The primeval-primordial reality is the object of present reality—it is the ideally conceived "primeval" (Infinite: *Shreya*, 81) development of the present subject. It is the quinary "development-effect" (Self-luminous entity: *Purusha, 12*) of the "infinitely perpetuating" (Sinful: *Paap, 15*) devotional planning. Entrepreneurial mediation using the finite present human-effect is known as the "path of divinity" (Knowing energy: *Siddhi marga, 96*).

The solution for compensating the formative cost of trading the quinary development-effect, without any present organization, is not the present organization, i.e., the self-luminous entity. It is, instead, descending the "agitated" (*Maha paap, 279*) state of the ecosystem that is turning an entity into a "demeritorious devoted" (Sinful: *Paap, 15*) and ascending the formative growth of the "rejoiced state" (*Maha punya, 268*) to service senary development of the "meritorious devotee" (*Punya, 15*), without future ecosystem-effect.

Responsible management of the "devotion" (*Bhakti, 46*) is the "path of the sentient entity" (Rejoice: *Siddha marga, 179*). It empowers the entity to organizationally develop the rejoiced state, free from the limitations of the varying ecosystem-effect. Responsible management is the "guider-effect" (*Chitta, 100*) to the entity's "gravitational quality" (*Guna, 0*) for trading the extrinsic, rationality-binding "space-effect of the primordial space" (*Dik, 100*). Devotional leadership catalyzation of the "extrinsic energy" (*Para shakti, 18*) is a function of the six "qualities" (*Guna, 0*) of the catalyzing "energy" (*Shakti, 19*) of the entity. These qualities comprise the divine (*Divya, 360*), fire (*Tejas, 17*), water (*Apas, 169*), air (*Vayu, 385*), earth (*Bhu, 724*), and the ether (*Shuddhi, 285*) elements. These qualities are catalyzed within an entity by the "intrinsic energy" (*Adi Shakti, 15*). A twin entity services the intrinsic energy after countertrading the "dynamic energy" (*Adi para shakti, 17*) from "everybody else" (*Idanta, 3*) who is trading the "variable zodiac lunar time-effect" (*Idanta, 3*).

A param entity trades the variable "zodiac lunar time-effect" (*Idanta, 3*) to compensate for the formative cost of servicing a "supernatural paradigm of present reality" (*Yukti, 8*) in the form of an ascending "entity-effect" (*Utkramajya, 38*) and a descending self-conceived "mental value" (*Manas, 38*). The "mental value" (*Manas, 38*) is the wholeness of the twelve intrinsic values and the six extrinsic values within a "param sentient child" (*Manyu, 19*). It organizes the eighteen dimensions of "mind" (*Manas, 38*) as an entity and the twenty dimensions of "spirit" (*Kapinjala, 20*) as a spirit, within a thirty-eight dimensional "mental body" (*Hradamana*, 381), the param entity.

An additional group of the six extrinsic values constitutes the "path of solar dimension" (Devotional energy: *Param siddha marga*, 179) of the ecosystem para entity as a creation. It is beyond the "param creator power" (*Chitta*, 100) of the entity as a creature. The six extrinsic values include the four dimensions of the "creator factor" (*Atman*, 4) *plus* the two dimensions of the creation as the "knower factor" (*Brahman*, 2) forming that creator factor in the form of the masculine present and the feminine potential.

The "creator factor" (Soul: *Atman*, 4) is the self-organizing "SHEENY-effect" (*Soham*, 4), without the creature-level consciousness of the potential "objective" (*Antahkarana*, 2) of the present creation-level reality of the "universe" (*Brahman*, 2). Therefore, a param entity, who is self-luminous (*Purusha*, 12), "self-reproduces" (*Upanayana*, 1/3) the four-dimensional "creator factor" (Soul: *Atman*, 4 = 12 * 1/3). The four creator dimensions include the "lunar or zodiac time" (Primordial-primordial time: *Tribhajya*, 10^{10}) as the maternal soul, the "solar time" (Primordial time: *Prabha*, 180) as the paternal soul, the "present time" (Param time: *Sva*, 11) of the self as a soul, and the "potential time" (Primeval time: *Vishvakarma*, 108) of everybody else in the ecosystem as the twin self and the twin soul.

The potential time of the creation is the creator factor that is beyond the present self. The creator factor decides to let the mind reproduce and manifest the potential reality of everybody else as the self's present physical reality. It does not intellectually know either the primordial illuminated reality that conditions the potential yet-to-manifest reality or the primordial-primordial shadow reality being curved by the primordial illuminated reality. The six extrinsic values of the ecosystem become the six intrinsic values of the entity, within the seventh, "guider element" (*Guru*, 100) serviced by the ecosystem. The entity's six intrinsic values become the six intrinsic values of everybody else as the para entity, within the eighth, "sentient element" (*Ojas*, 189). The six intrinsic values of everybody else— the para entity—becomes the six intrinsic values of the self—the param entity in the present moment, within the ninth, holistic "culture element" (*Sadashiva Nayaki*, 10). The culture element is serviced by the primeval entity, who trades the entity's sentient-effect and services the inanimate culture element for curving the ecosystem's extrinsic value

- **Fifth, the intellectual body hypothesis.** We perceive the intrinsic and extrinsic values of our culturally-reproduced social correlations through our intellectual body. Our intellectual body trades the consciousness of the culture element primarily from the universe of departed child souls. Just like our mental body, our intellectual body also gets divided into the eighteen forms of intellects as a function of our varying correlation with the diverse forms of metaphysical subjects in the departed child universe.

The departed child universe includes the six masculine zodiac forms of metaphysical subjects (the future, primeval entities) derived from the six feminine zodiac forms of physical subjects (the present, param entities). In turn, they derive from the six androgynous forms of dynamic subjects (the past, primordial entities). They, again, derive from the six universal forms of the technological subjects (the primordial-primordial entities). These six universal forms of the technological subjects are composed of four creator and two creation kingdoms.

The four creator kingdoms include plant, animal, human, and spirit. The two creation kingdoms include metal and mineral. The spirit is the entropy form of the material kingdom. Each creator entity eventually transforms into a material form. The material kingdom's matter is the normative development of the four creator kingdoms by trading the energy of the inanimate creation in various forms. The material eventually transforms into a gravitational potential—which is the essential quality of each subject—within the black hole.

The four creator and the two creation kingdoms are the transformative development of the deity kingdom. Diverse deities transform into diverse forms of creator and creation. The deity kingdom is the metaphysical development of the dark matter kingdom. As a param creature, the dark matter conceives its infinite deity forms for self-managing the present reality. The dark matter trades the sentient potential from a param deity. Param deity is a "sentient entity" (*Siddha*, 7) that manages the intrinsic values responsibly for the universal extrinsic sentient well-being. A sentient entity generates a sentient consciousness of the intrinsic values as a gift from a primordial greeter. The primordial greeter is a "soul essence community" (*Mandala*, 16).

The entire universe of entities is a product of the eighteen soul essence communities, each one of which forms one division of our intellect (*buddhi*). The eighteen soul essence communities form the eighteen parallel dimensions of one universe. Each parallel dimension is a kingdom of entities, known as "Eden" (Kingdom: *Nayana*, 957). Each kingdom is a "primeval value" (Chirp: *Vyojya*, 957) of a system of energies, comprising the two energy layers: the masculine (descending, ideal, conservative, disproportionate, potential, positive, heavenly, odd, linear, convergent, disequilibrating, gross, and extrinsic) and the feminine (ascending, theoretical, liberating, proportionate, present, negative, hellish, even, nonlinear, divergent, equilibrating, subtle, and intrinsic).

Twelve kingdoms are the energies of togetherness, otherness, attraction, repulsion, growth, and entropy, within either the theoretical, liberating present or the idealized, limiting future. By shaping an idealized vision of the future, the present becomes liberating. The present takes the limiting future for making a theory of action, seeking to destroy the masculine limits, without a consciousness that the disproportionate past actions cause the omnipresent limits.

Six kingdoms are the energies of the binary of the space dimensions and the quaternary of the time dimensions. The idealized objective-effect of a primordial-primordial entity transforms the present space dimension by forming the present creature (dark matter—the object). A primordial entity's theoretical subjective-effect forms the potential space dimension by organizing the primeval creature (sentient entity—the subject). Lunar time is the creation of the transforming object-effect. Solar time is the creation of the forming subject-effect. The present time is the negative-effect of the theoretical, liberating, proportionate, feminine, present consciousness. The potential time is the positive-effect of the idealized, conservative, disproportionate, masculine, potential consciousness.

Table 4 summarizes the eighteen "kingdoms" (*Nayana*, 959) and the thirty-six "geographies of energy realms" (*Ganarajya*, 476), which host thirty-six "grouping clusters of divine entities" (*Gana*, 387). Their "absolute geometrical, zodiac energy-effect" (*Panchang*, 8×10^{15}) forms the "zodiac system" (*Shunya Kalpa*, 8×10^{15}). The "entropy value of the zodiac-effect" (*Deva siddhi*, 375) forms the "primordial-primordial space" (*Drishti-mandala*, 375) of the cosmic universe. It is, therefore, the "truth" (*Sathya*,

375) that is shared by both the "solar universe" (*Rashi*, 13) and the "Milky Way Galaxy" (*Viyadganga*, 85). It accounts for the scientific prediction of the presence of a minimum of thirty-six active communicating intelligent civilizations locally in the Milky Way Galaxy, using what is known as the "Astrobiological Copernican Limit." In reality, the Milky Way Galaxy hosts no civilization other than that on the Planet Earth.

- **Sixth, the physical body hypothesis.** We trade and service energies with the diverse soul essence communities through our physical body. Our physical body trades the metaphysical consciousness from the "solar universe" (*Surya mandala*, 13) as a transformative embodiment of a normative "zodiac entity" (*Rashi*, 13). It forms through the disembodiment of a formative "queen of divinity" (*Maha Kali*, 13). The queen of divinity is the "luminous" (*Maha Kali*, 13): its twelve divisions form the twelve proportionate zodiac entities, each with a varying gravitational quality but a constant energy essence.

The light irradiated by the solar universe radiates with a force that imprints the transforming gravitational quality of each entity within the astral body. It services the varying qualities of the "para consciousness" (*Nirharin*, 18) to each twin entity at theeach moment. The specific quality is a function of the varying distance of a twin entity from the radiating entity and the varying quality of the light, at the moment when the entity radiated the light force. The luminous is the overall quality of the light within the proportionate proportion of each zodiac-effect. The luminous operates at the eleventh infinity, as the eleven-dimensional "foundation" (*Sva*, 11) of the twelve-dimensional self-luminous entity, without the thirteenth, holistic dimension.

Our physical body generates energy effects of our subjective consciousness of the thirteenth holistic dimension (within and without potential mediation by us as a self-luminous entity) in the form of the inanimate quantum particles—quarks. Our subjective consciousness generates the six forms of energy effects at the cellular level. These include togetherness and otherness consciousness within our mental body (within or without potential mediation by us as a causal body), attraction and repulsion consciousness within our intellectual body (within or without potential mediation by us as an etheric body), and growth and entropy consciousness within our physical body (within or without potential mediation by us as an astral body).

Quarks are of two types—potential and present. A potential quark is a function of the present consciousness (of the subject as the causal body for potential life) and para-consciousness (of the object as the causal body for present life) without the entropy-effect of the etheric body. Therefore, it has a positive +2/3 value. A present quark is a function of the past infinite entropy-effect of the etheric body, without present or para consciousness. Therefore, it has a negative –1/3 value.

Table 4. Eighteen Kingdoms (Edens) and Thirty-Six Energy Realms (Starseed Universes) of Divine Entities

Eden/ Heaven/ Celestial realm	Vedic nomenclature	Western nomenclature	Eastern nomenclature	Divine entity cluster (*Gana*)	Nature of the energy system
First: Primordial maternal kingdom	*Descending:* *Put* or *Asipatravana naraka* (*Sanatkumara kalpa*); *Ascending:* *Lavana naraka* (*Mahendra kalpa*)	*Descending:* Garden of Eden; Yggdrasil—Home of the Tree of Life; *Ascending:* Ma'on—BE-LIVE-LIFE	*Descending:* *Wuji Tanshi* Heaven, 无极公誓天; *Ascending:* *Pingyu Jiayi* Heaven, 平育贾奕天	*Descending:* Atlantean (*Pishacha*, Vampire; Canine); *Ascending:* *Mahoraga* (Animal)	Present-effect * Objective-effect = Present space
Second: Spirit kingdom	*Descending:* *Maharaurava* or *Taptakumbha naraka* (*Shukra kalpa*); *Ascending:* *Taptaloha naraka* (*Mahashukra kalpa*)	*Descending:* Helheim—Home of the dishonorable dead; *Ascending:* Vilon—The Curtain	*Descending:* *Xianding Jifeng* Heaven, 显定极风天; *Ascending:* *Taiqing Jingdachi* Heaven, 太清境大赤天	*Descending:* Titan (*Danava*, Indigos—Cattle; *Kalasutra* [黑繩]—"Purple entity"); *Ascending:* Lemurian (*Naga*, Serpent; Spirit)	Potential-effect * Entropy-effect

Eden	Vedic nomenclature	Western	Eastern nomenclature	Divine entity cluster (*Gana*)	Nature
Third: Para deity kingdom	*Descending: Taptamurti* or *Visasana naraka* (*Anata kalpa*); *Ascending: Pranarodha naraka* (*Pranata kalpa*)	*Descending:* Jotunheim—Home of the Giants; *Ascending:* Shehaqim—The Clouds	*Descending:* Xuanming Gonghua Heaven, 玄明恭华天; *Ascending: Longbian Fandu* Heaven, 龙变梵度天	*Descending:* Pleiadean (Giant, Rakshasa, Ogre, *Arbuda* [頞部陀]—"Blistering entity"); Ascending: *Krttika* (Divine architects; Reptilian)	Potential-effect * Repulsion-effect
Fourth: Supreme deity kingdom	*Descending: Vahnijwala naraka* (*Brahma kalpa*); *Ascending: Shwabhojana naraka* (*Brahmottara kalpa*)	*Descending:* Muspelheim: The Home of Fire; *Ascending:* Agartha	*Descending: Xumming Tangliao* Heaven, 虚明堂曜天; *Ascending: Yulong Tengsheng* Heaven, 玉隆腾胜天	*Descending:* Orion (Mintakan; *Preta, Pitra,* Wolf); *Ascending:* Agarthan (Soul; *Atman,* Genie, Lightworker; Insect; *Krandita, Atata* [頞听陀]—"Shivering entity")	Potential-effect * Subject-effect = Solar time
Fifth: Primordial deity kingdom	*Descending: Kumbhipaka* or *Adhomukha naraka* (*Arana kalpa*); *Ascending: Shukaramukha naraka* (*Achyuta kalpa*)	*Descending:* Alfheim—Home of the Light Elves; *Ascending:* Makhom—Fixed home	*Descending:* Guanming Duanjing Heaven, 观明端靖天; *Ascending: Wushang Changrong* Heaven, 无上常融天	*Descending:* Sirian (*Kinnara,* Werewolf; Amphibian); *Ascending:* Polarian (*Abhijit,* Bovine; *Maharaurava* [大叫唤]—"Silient entity")	Present-effect * Object-effect = Lunar time

Eden	Vedic nomenclature	Western nomen.	Eastern nomenclature	Divine entity cluster (*Gana*)	Nature
Sixth: Devoted deity kingdom	*Descending: Puyoda* or *Krishna naraka* (*Saudharma kalpa*); *Ascending:* *Avata-nirodhana* or *Kalasutra naraka* (*Aishana kalpa*)	*Descending:* Svartalfhcim—Home of the Black Elves; *Ascending:* Araboth—The Desert	*Descending:* Xuanming Gongqing Heaven, 玄明恭庆天; *Ascending:* Xiule Jingshang Heaven, 秀乐禁上天	*Descending:* Andromedan (*Kimpurusha*, Wizard; Rabbit); *Ascending:* Beast (*Bhuta*, Phantom, Ghost; Boar; *Hahava* [膿膿婆]—"Lamenting entity")	Potential-effect * Subjective-effect = Potential space
Seventh: Deity kingdom	*Descending: Raurava naraka* (*Amogha Kalpatita*); *Ascending: Tamas* or *Tamisra naraka* (*Archi Kalpatita*, Thiruambalam—Realm of the Deity)	*Descending:* Bardo; *Ascending:* Muzaloth—The change	*Descending:* Taihuan Jiyao Heaven, 太焕极瑶天; *Ascending:* Hanchong Miaocheng Heaven, 翰宠妙成天	*Descending:* Arcturian (Equine; *Gandharva*, Angel; *Sravaka*); *Ascending:* Deva (*Swati*, Mermaid; *Sanjiva* [等活]—"Reviving entity")	Potential-effect * Attraction-effect
Eighth: Param deity kingdom	*Descending:* Samhata or *Vimohana naraka* (*Lantava kalpa*); *Ascending:* *Vajrakantaka-salmali naraka* (*Kapistha kalpa*)	*Descending:* Zevul—The Abode; *Ascending:* Home of the Phoenixes	*Descending: Sihuang Xiaomang* Heaven, 始黄孝芒天; *Ascending: Yuantong Yuandong* Heaven, 渊通元洞天	*Descending: Zombie* (*Suta, Dvipakumaras*; Monster; *Nirarbuda* [剌部陀]—"Burst Blister entity"); *Ascending:* Vegan (Avian; *Apsara, Balakhilya*)	Potential-effect * Growth-effect

Eden	Vedic nomenclature	Western nomen.	Eastern nomenclature	Divine entity cluster (*Gana*)	Nature
Ninth: Supra deity kingdom	*Descending: Andhakupa naraka* (*Subhadra Kalpatita*, Param-Primordial hell); Ascending: *Andhatamisra* or *Papa naraka* (*Vairochana Kalpatita*)	*Descending:* Kuchavim—Council of Twelve; *Ascending:* Asgard—Home of the Gods	*Descending: Cimming Heyang* Heaven, 赤明和阳天; *Ascending: Haoting Xiaodu* Heaven, 皓庭霄度天	*Descending:* Hadarian (Whale; Magadha); *Ascending:* Lyran, *Aditya* (*Bodhisattvas*, Feline; *Samghata* [衆合]—"Crushing entity")	Present-effect * Negative-effect = Present time
Tenth: Primeval illuminator kingdom	*Descending: Sarameyadana* or *Sukara naraka* (*Pritikar Kalpatita*, New Lemuria); *Ascending: Rudhirandha naraka* (*Sphatik Kalpatita*, Dark matter; Param-Primordial heaven)	*Descending:* Vanaheim—Home of the Vanir; *Ascending:* Midgard—Home of the Humans	*Descending: Yaoming Zhongpiao* Heaven, 耀明宗飄天; *Ascending: Dalou* Heaven, 大罗天	*Descending:* Zeta (*Vanara*, Apes); *Ascending: Manushya* (Human; Livestock; New Lemurian; "Fiery Chariot entity" [火車地])	Potential-effect * Positive-effect = Potential time
Eleventh: Primeval deity kingdom	*Descending: Sandamsa naraka* (*Shatara kalpa*, Blackhole; Primordial-primordial hell); *Ascending: Krimisha naraka* (*Shahsrara kalpa*, Present paradigm; Radiant love)	*Descending:* Hildskjalf—Home of the cosmic observer; *Ascending:* Grigori—The awake	*Descending: Taiji Mengyi* Heaven, 太极蒙翳天; *Ascending: Yuqing Jingqingwei* Heaven, 玉清境清微天	*Descending:* Cassiopeian (Crustacean; (Feivian; *Uragendra*); *Ascending:* Martians (Primates; *Sadhyas*, Rishi; Buddhahood, Xisigui; *Mahapadma* [摩訶鉢特摩]—"Lotus seed entity")	Potential-effect * Togethernes s-effect

Eden	Vedic nomenclature	Western nomen.	Eastern nomenclature	Divine entity cluster (*Gana*)	Nature
Twelfth: Super deity kingdom	*Descending: Rodha naraka (Sumanas Kalpatita)*; *Ascending: Mahajwala naraka (Saumrup Kalpatita*, Primordial-primordial heaven; New paradigm; White star)	*Descending:* Nidavellir—Home of the Dwarves; *Ascending:* Choshech—Home of darkness	*Descending: Xuvu Yueheng* Heaven, 虚无越衡天; *Ascending: Shangqing Jingyuyu* Heaven, 上清境禹余天	*Descending: Vandi* (Dwarfs; *Bhutavadita*); *Ascending:* Hyades (Mammals; *Yaksha, Sudaksha*, Warlock, Witch, *Xiqigui*; *Padma* [鉢特摩]—"Lotus entity")	Present-effect * Otherness-effect
Thirteenth: Primeval greeter kingdom	*Descending: Rijisha* or *Lalabhaksa naraka (Suvishala Kalpatita*, Primordial hell); *Ascending: Dandasuka* or *Daruna naraka (Saum Kalpatita)*	*Descending:* Angelic Prisoners; *Ascending:* Ashtar Command	*Descending: Taian Wangya* Heaven, 太安皇崖天; *Ascending: Taihuang Huangceng* Heaven, 太皇黄曾天	*Descending: Asura* (Nocturnals; Arachnids; Satan, Rex Mundi; *Raurava* [叫唤]—"Screaming entity"); *Ascending:* Phage (Sponges; *Vasu, Pavaka, Pratapana* [大焦熱]—"Great heating entity"])	Potential-effect * Otherness-effect
Fourteenth: Self-luminous entity kingdom	*Descending: Shulaprota* or *Tala naraka (Sudharshan Kalpatita*, Universe of the queen of divinity); *Ascending: Puyavaha naraka (Aditya Kalpatita*, Primordial heaven)	*Descending:* Araphel—Home of thick cloud; *Ascending:* Great Sea	*Descending: Wusi Jiangyou* Heaven; 无思江由天; *Ascending: Qiyao Moyi* Heaven, 七曜摩夷天	*Descending:* Regulus (Panther; *Anilas, Tusitas,* [磨拇]—"Grinding entity"); *Ascending:* Bacteria (*Charana*, Spies; Worms; *Kusmandana*)	Present-effect * Attraction-effect

Eden	Vedic nomenclature	Western nomen.	Eastern nomenclature	Divine entity cluster (*Gana*)	Nature
Fifteenth: Primeval perpetuator kingdom	*Descending:* *Kudmala* or *Sucimukha naraka* (*Yashodhar Kalpatita,* Universe of departed souls, Primeval hell); *Ascending:* *Paryavartana naraka* (*Vair Kalpatita*)	*Descending:* Dragon's lair; *Ascending:* Niflheim—Home of Fog	*Descending:* *Yuanzai Kongshen* Heaven, 元载孔升天; *Ascending:* *Yuanming Wenju* Heaven, 元明文举天	*Descending:* Aldebaran (Taurean; Fowl; *Mahakran-ditaka,* *Tapana* [焦熱]— "Heating entity"); *Ascending:* Archaea (*Pramathas*)	Potential-effect × Entropy-effect
Sixteenth: Primordial greeter kingdom	*Descending:* *Raksogana-bhojana naraka* (*Saumanas Kalpatita*); *Ascending:* *Krimibhojana naraka* (*Ark Kalpatita, Chittambalam*—Realm of Wisdom; Primeval heaven)	*Descending:* Anan— Home of the seven bands of angels; *Ascending:* Shechakim: The millstones	*Descending:* *Wengchong Furong* Heaven, 翁重浮容天; *Ascending:* *Xuantai Pingyu* Heaven, 玄胎平育天	*Descending:* Antarian (Rainbows; Reindeers; *Maharajikas; Pancha-prajnapti*); *Ascending:* Hathor (Eukaryotes; Mollusks; Shell-protected entities; Venusian; *Aprajnapti, Utpala* [嗢鉢羅]— "Black Lotus" entity)	Present-effect × Growth-effect

Eden	Vedic nomenclature	Western nomen.	Eastern nomenclature	Divine entity cluster (*Gana*)	Nature
Seventeenth: Devotee deity kingdom	*Descending: Kakola* or *Vaitarna naraka* (*Suprabuddha Kalpatita*, Bottomless pit, Universe of the demons [inanimate objects], Thirteen moons of Saturn, Param hell); *Ascending: Ayahpana* or *Puyavaha naraka* (*Archimalini Kalpatita*)	*Descending:* Rakia; *Ascending:* Machon—The resting place	*Descending: Zhuluo Huangjia* Heaven, 竺落皇笳天; *Ascending: Qingming Hetong* Heaven, 清明何童天	*Descending:* Corvus (Raven, Crow, Demon, *Daitya*, Huhuva [虎々婆]—"Chattering entity"); *Ascending:* Prokaryotes (Spines; The One with antenna in all directions; *Vishvedevas*; Dikkumaras)	Present-effect * Togethemes s-effect
Eighteenth: Luminous kingdom	*Descending: Avici* or *Vedhaka naraka* (Universe of the etheric spirits); *Ascending: Ksarakardama* or *Apritistha naraka* (*Sumeru paraloka*, Ganaparvata, Param heaven)	*Descending:* Hashem—Twelve gates of the sun; *Ascending:* Ioanit stations of light	*Descending: Shangshe Ruanle* Heaven, 上操阮乐天; *Ascending: Taiming Yuwan* Heaven, 太明玉完天	*Descending: Abhasvaras* (Mammoth; *Avici* [阿鼻]—"Incessant entity"); *Ascending:* Pegassi (*Rudras*, Octopus; Cell; The One with sensory tentacles; Rishiraditaka)	Potential-effect * Repulsion-effect

The present quark becomes the potential quark when a sentient entity in the universe services its strong psychic emotions for social (mental) togetherness, personal (intellectual) attraction, and institutional (physical) growth. Consequently, the present sense of wholeness within an entity descends and transforms into a positive energy potential quark, as a structural (astral) medium for fulfilling the objective of the religious devotional intensity generating the strong emotional intensity. The potential quark eventually becomes the present quark after the entropy diffusion of its immanent energy. When the objective of devotional intensity gets fulfilled, the sentient entity becomes a guider entity. It services the weak psychic emotions for social otherness, personal repulsion, and institutional entropy through the radiation of the structuring guider-effect. Consequently, the entity's potential sense of wholesomeness descends and transforms into a negative energy present quark.

The present quark is a divine medium for illuminating the escalating thermodynamic costs of the "I AM a Deity consciousness." The "I AM a Deity consciousness" is an illusionary masculine adrenaline-pumping heavenly effect of a disproportionate polarization of the natural energy for liberating the present self from the objective of the present hellish life. An entity realizes that the life has become hellish without the consciousness of the dynamic present ground reality. It seeks to benefit from the disproportionate natural energy of the transformed ground reality by descending the disproportionate polarization (subjugation) of the natural feminine energy into the supernatural masculine energy. Our emotional energy shapes the flavor of the quarks.

The sixth hypothesis is the primordial-thesis that lets a feminine primordial greeter substantiate a primordial-primordial thesis that includes the following six primordial-primordial hypotheses without devoting the energy into falsification of the validity of the self-luminous quaternary thesis:

- **First, the up quark hypothesis.** Suppose we polarize the "supernatural energy" (*Shram shakti*, 1) into our mental body for forming the "thoughts" (Sensory illusion: *Vichara*, 8) about the supremacy of our masculine intellect. In that case, we diffuse our feminine self-esteem in the form of the "thermodynamic fire-effect" (*Rodha*, 1) of the "up quark" (Manpower: *Naraki*, 1), with an intrinsic togetherness programming. A positively charged up quark is a natural resource available to us for substantiating our supremacy in togetherness with Mother Nature.

- **Second, the down quark hypothesis.** Suppose we polarize the "thermodynamic energy" (*Asir*, 497) into our intellectual body for norming "visions" (Para-sensory illusion: *Dharmasavarni*, 81) about the thoughts of the social others. In that case, we diffuse our masculine self-worth in the form of the "down quark" (Divine factor: *Vignesh*, 1) with an intrinsic otherness programming. A negatively charged down quark is our supernatural gift for substantiating the supremacy of the social others—the others whose emotions falsify the supremacy of our masculine self-worth.

- **Third, the charm quark hypothesis.** Suppose we polarize the "gravitomagnetic energy" (*Asrava shakti*, 10) from our intellectual body into our physical body to compensate for the negative present-effect of the down-quark on our consciousness. In that case, we transform the air of the social others into the physical self's grounded awareness. The secondary "air-effect" (Oxygen: *Maruti*, 78) of the supremacy of the social others takes the form of the "charm quark" (Proselytization: *Karma-yogi*, 1) with an intrinsic attraction programming. A positively charged charm quark is the supernatural gift of the universe to us for substantiating its supremacy in the primordial moment.

- **Fourth, the strange quark hypothesis.** Suppose we polarize the "electromagnetic energy" (*Karma-bandha*, 19) from our physical body into our mental body to accelerate the charm quark's positive potential-effect for fulfilling the supremacy objective formed in the primordial-primordial moment. In that case, we repel our intrinsic "gravitoelectric energy" (*Ajiva*, 92) in the form of the sentient water flow. The tertiary flow of "water-effect" (Hydrogen: *Jalaprana*, 4) descends the hydrogen bonding of the electromagnetic cellular energy within our physical body. It diffuses in the form of the "strange quark" (Inhalation: *Ana*, 1) with an intrinsic repulsion programming. A negatively charged strange quark is a natural resource available to the universe for falsifying the supremacy of our feminine self-esteem that is the foundation of the sentient life.

- **Fifth, the truth quark hypothesis.** Suppose we polarize "gravitational energy" (Lalita, 100) from our mental body into our causal body to equilibrate the negative present-effect of the strange quark as a subjective survival strategy formed in the param moment. In that case, we can trade the "sentient energy" (*Varuna*, 1000) from the universe of the etheric spirits. The quaternary trading of the "ether-effect" (Calcium: *Naad*, 257) strengthens the bones of the physical body and services the "truth quark" (Substantiation: *Aap*, 1) with an intrinsic growth programming. Strong bones strengthen the muscular power and compensate for the descending mental power. They ascend both the "manpower" (Paternal physical power for reproductive workforce proficiency: *Naraki*, 1) and the "material power" (Feminine earth-effect in the form of the carbon atoms, derived from the decay in the physical body to compensate for the proficient networking of the calcium mass: *Udita*, 2).

- **Sixth, the beauty quark hypothesis.** Suppose we liberate ourselves from the sentient energy by polarizing our astral body without the physical entropy-effect of the etheric body. In that case, our astral body can trade the positive potential-effect of the truth quark and transform into a Wisher spirit. The quinary servicing of the ether-effect as the senary mass-effect of our radiated life consciousness produces the "beauty quark" (Negation: *Jugupsa*, 1) with an intrinsic entropy programming. The thermodynamic-effect of the beauty quark generates an entropy in the "dark matter" (Liberated sentient energy of the universe of sentient entities: *Saum shakti*, 1600). It reincarnates the Wisher spirit as a physical body for yet another opportunity to be free from the competitive dynamics to exchange supremacy with the metaphysical universe. It empowers the one to be the "maternal management power" (*Nataraja*, 7) for the present creature's proficient exchange.

4.5 Objective for Planning the Param-Primordial Paradigm—Organization

An androgynous param creature's objective of planning the param-primordial six-dimensional dynamic organization paradigm is to authenticate the quinary thesis of a creature's present reality, as both a creator within the sentient consciousness and a creation without the sentient consciousness. It projects the present and the potential effects within a quark of six flavors. The quinary thesis includes the following six quinary hypotheses:

- **First, a creature has an absolute freedom to be the primeval masculine (Polytheist Satan: Asura, -1) and become blindly and faithfully devoted to the guider power of a "para deity"** (Monotheist God: *Ishvara*, 5). In that case, the creature creates a paradigm for becoming the creation and authenticates by eventually becoming an object observing the subjects who have the sentient consciousness of their supremacy.

- **Second, a creature also has an absolute freedom to be the blind and faithful devotee of the guider power of a primeval deity** (polytheist pantheon of deities) seeking to trade infinite divine energy. Such a creature becomes a "speck of dust" (Inanimate object wishing to be a demon: *Padartha*, -3), begging the objects to believe in the "devoted follower" (*Bhakta* [with the power to create the universe of potential reality], -5). A devoted follower is a para wisher who brags its supreme power for fulfilling the primeval wishes of the "universe of primeval masculine" (*Duniya*, -2). By radiating both intrinsic sentient and extrinsic divine energies in the form of the infinite gravitational energy, the creature gets bound with the thermodynamic energy. It generates an entropy of the electromagnetic energy essential for the survival of a physical body.

  - Without the physical body, the creature lacks the formative capability for the dynamic primordial oneness with the natural energy.

  - Without dynamic oneness with the natural energy, the creature lacks the normative organization for technological and primordial-primordial oneness with the "absolute self" (*Hurupa*, 280). Such an "absolute creature" (*Prabhu*, 1600) is the organizer of the natural energy and norms the state of "freedom from the present-effect" (*Moksha*, 1600).

o Without technological oneness with the absolute self as a sentient entity, the creature lacks the transformative energy for the param-primordial, organizational oneness with the "natural paradigm of present reality" (*Yukti*, 8).

o Without organizational oneness with the natural paradigm as a supernatural entity, the creature lacks the metaphysical planning for the primeval-primordial, ecosystem oneness with the "param deity" (*Nataraja*, 7).

o Without ecosystem oneness with the "primeval perpetuator" (*Vishnu* [Mother Nature + Param Deity = 8 + 7], 15), within the eighteenth, ascending, param heaven realm, the creature lacks the dynamic programming for the para-primordial, entity oneness with the "primordial greeter" (*Maha Durga*, 16).

o Without entity oneness with the primordial greeter (transcending beyond the realm of hell and heaven), the creature lacks the formative performing authenticity as a supreme-primordial para entity. A supreme-primordial para entity enjoys normative profiting and has the power to manifest an infinite transformative development without the diffusion of sentient energy. Such an entity is the creator of the sentient potential within an absolute oneness with the "primordial illuminator" (*Shri Krishna*, 10). The primordial illuminator is a supra-primordial deity. A supra-primordial deity is a creation of the super-primordial deity. The self as the "param deity" (Sentient entity: *Siddha*, 7) is the super-primordial deity.

- **Third, a creature has an absolute freedom to transcend beyond the intellectual body hypothesis on the eighteen different kingdoms forming the thirty-six different starseed realms.** Suppose there are, indeed, eighteen different kingdoms with a binary of energy realms. In that case, it implies that there must be an additional original thirteen "primordial energy" (*Adi shakti*, 15) realms radiating the "luminous energy" (*Adi para shakti*, 17). They must empower a devotee entity to conceive, perceive, and experience the thirty-six [15 + 17 + 6] energy realities of the six absolute forms (metal, mineral, plant, animal, human, and spirit kingdoms) of the "tertiary residual" (*Khara*, 6) of the "param deity" (Siddha, 7).   By trading the self-perpetuating, eighteen-dimensional "self-luminous energy" (*Para shakti*, 18) within the self as a devoted luminous, the entities in the diverse cultures of the world develop the power to script the present hellish (as in the Vedic culture), the potential heavenly (as in the Daoist culture), and the mix of hellish and heavenly (as in the Western culture) reality of the thirty-six starseed realms.

- **Fourth, a creature has an additional absolute freedom to transcend beyond the correction of the scientifically measurable energy value's causative dimensions.** Suppose the thirteen original primordial energy realms are radiating the luminous energy. In that case, it implies that there must be five additional "primordial-primordial energy" (*Asrava shakti*, 10) realms.   They let the feminine luminous project her "organizational potential" (Param-primordial inanimate energy: *Achitta shakti*, 396) into the thirteen primordial energy realms.

- **Fifth, a creature has an additional absolute freedom to transcend beyond the perpetuating frequency of the metaphysical cause of measurable energy value.** Suppose the perpetuating frequency has a curvilinear causative-effect. In that case, there must be two additional "para-primordial energy" (Animate energy: *Purani shakti*, 91) realms—one generating a disproportionate masculine-effect and one generating a proportionate feminine-effect.

- **Sixth, a creature has an omnipresent freedom to transcend beyond the illuminated dynamic cause of the measurable energy value.** Suppose there are thirty-six primary, eighteen secondary, and two tertiary energy realms. In that case, it must be possible to codify the convergent energy frequency of each energy realm into a "letter" (Fire-effect: *Rupa*, 100,000) and the divergent energy frequency of each energy realm into a "sound" (Ether-effect: *Naad*, 257). The Sanskrit alphabet has fifty-four letters, organizing the omnipresent energy within each of the fifty-four realms into the omnipotent oneness of omniscient form (as the letter) and the Almighty creator sound (as a phonetic).

The sixth hypothesis is the param-primordial thesis, with which an androgynous primordial greeter effortlessly substantiates the primeval-primordial thesis, which includes the six primeval-primordial hypotheses without servicing animated energy.

- **First, if there are indeed two technological dimensions of the "supreme deity"** (Creator deity: *Param Brahma*, 4), then it implies that the Eastern culture must be proficient in knowing the primeval, illuminated reality of these two dimensions; the Western culture must be proficient in knowing the param, shadow reality of these two dimensions; and the Vedic culture must be proficient in knowing the reality of the primordial, secret creator of these two realities. Further, there must be a primordial-primordial, super-secret potential of these two realities.

- **Second, the primeval illuminated reality of the two technological dimensions is the duality** between the present (proportionate, formative, technological, illuminated, solar, intrinsic, feminine) growth and the potential (disproportionate, normative, organizational, shadow, lunar, extrinsic, masculine) development. The present growth is the causative factor for the organizational development. The potential development is the causative factor for the technological entropy. The Eastern culture refers to the "present growth" (*Sva*, 11) as "yin" and the "potential development" (*Sada Shiva*, 1600) as "yang." Yin is the cause of the yang. Yang is the cause of yin. The Vedic culture identifies the "deity" (*Deva*, 1) as the primordial causative factor for the present growth and "supra deity" (*Devi*, 3) as the primordial-primordial causative factor for the potential development.

The feminine supra deity is the primordial-primordial causative factor for the disproportionate masculine development. Masculine deity is the primordial causative factor for the disproportionate entropy in absolute feminine growth by disequilibrating the proportionate value of the feminine growth within each entity. The param-primordial causative factor in equilibrating the ecosystem value is "Mother Nature" (*Anatanam*, 8), "self-radiating" (*Ham*, ¼) in the form of an "androgynous super deity" (*Jathara*, 2).

- **Third, the param, shadow reality of the two technological dimensions is the oneness of the present growth and the potential development** within the dynamic consciousness of a primordial-androgynous, "supreme deity" (*Param Brahma*, 4). The Western culture refers to such an entity as *Elohim* (literally, Creator deity, 4) or *El Elyon* (literally, Supreme entity, ascending in the deifying shadow of a feminine Elohim to be a Supreme deity, 4). The Vedic culture identifies the supreme deity as one who transcends the duality of the pair of opposites—of being and becoming. It is also immanent as the Creator factor within the potential growth, transcending the present development's limitations. Such a supreme entity is identified as *Bhagwan* (*Bhagwan*, 4), literally, a desirable entity with a conscious divine energy without the para-conscious guider power. The following six distinguishing characteristics characterize an entity as *Bhagwan*:

  - a sense of the growth in the consciousness of the etheric sound of the self's truth as the primordial-primordial creator (Growth-effect: *Shree*, 81);

  - a sense of entropy in the bonds of the earthly beauty of the primordial creation (Wealth: *Aishvarya*, 168);

  - a sense of attracting the air of the knowledge of the cause for liberating the self by creating the primeval creature (Knowing: *Jnana*, 19);

  - a sense of repulsing the water shower of the masculine blessings for incarnating the primeval creature as a self-luminous entity (Glory: *Yash*, 12);

  - a sense of the fire of togetherness with the devoted creature dimension (Potency: *Veerya*, 48); and

  - a sense of otherness as a devotee creator factor (Detachment: *Vairagya* 7).

The creator factor is how a feminine Elohim creates the creatures, who are the devoted followers of both the creator and the cause consciousness naturally conceived from the primeval masculine creation. While the feminine creation is the primordial cause of an entity becoming a supreme deity, the primordial-primordial cause is the absolute convergent oneness with the fifty-four frequencies of the celestial realms.

- **Fourth, the primordial, secret reality of the one extrinsic entity factor is the intrinsic duality of the masculine dimensions and the feminine para-dimensions within the supreme deity.** Suppose a feminine supreme deity conceives the convergent energy of the fifty-four realms from the primeval masculine creation. In that case, there must be fifty-four feminine metaphysical para-realms generating the fifty-four masculine physical realm energies. The Vedic culture identifies the para deity as the perpetuator of the differentiated and divergent para-primordial energies of the fifty-four feminine para realms within the convergent supreme-primordial energies of the fifty-four masculine realms. Para realms are the para entity dimensions. Realms are the entity dimensions.

One may realize oneness with the desired intrinsic, feminine para entity dimension of the para deity, transcending the time and space limitations. Oneness empowers one to manifest the desired, extrinsic, masculine entity dimension within the self. The Vedic culture illuminates 108 types of "oneness" (*Yogas*, 48) for sun salutations, 108 types of "breathing" (*Pranayama*, 37) for meditation, 108 types of "rosary beads" (*Rudraksha*, 6 x 10^{192}) for energy activation, 108 types of "homage" (*Mantra*, 16) sounds for chanting to open present consciousness of the 108 types of the heart energy lines, conscious sensations, and para-conscious desire passions.

The Vedic culture emphasizes that there is only one "para deity" (*Ishvar*, 5). That para entity is *Ishvar* (*ish* = wish; *war* = free; beyond the limit), literally, one free from the desire for wishes. It is an entity with a conscious divine energy as well as a conscious guider power. Within the global ecosystem, there can be only one conscious guider power. That conscious guider power is God of the global ecosystem. God services the convergent consciousness of her guider power through the para-conscious guider entities. Those para-conscious guider entities are the "sovereign rulers" (*Indra*, 0) of each physical realm. The sovereign rulers perpetuate the divergent consciousness of their personal guider power as if they are the masculine Gods created in the mysterious feminine para God's image. The western culture identifies one masculine God of their local physical realm. The eastern culture identifies one highest heaven from where the present God of their local physical realm (the head of the nation) takes the mandate for servicing the desirable "subject dimension" (*Dharma*, 370).

- **Fifth, the primordial-primordial, super-secret reality of the one intrinsic, feminine para deity dimension** is the infinity of the extrinsic, masculine dimensions formed within the universe of supreme deities. Each supreme deity creates a unique local universe of creatures, guided by the masculine guider power dominant within the local physical realm.   All local universes of creatures are a part of one national universe of creatures.   There is one universal masculine guider entity, which services the varying masculine guider powers to the diverse geographical realms at the different lunar time frequencies.   Each local masculine guider entity then services that unique local guider power at a convergent solar time-frequency. The local universe para-consciously trades alternative guider-effects from an infinity of guider entities.

The Vedic culture refers to one universal masculine entity as the "primeval deity" (*Maheshwar*, 6), literally, one who is servicing the guider power for fulfilling an infinity of wishes. It is an entity with the conscious divine energy, guider power, as well as sentient value. Primeval deity services the divergent consciousness of HIS guider power through the conscious guider entities. Those conscious guider entities are the "prime ministers" (*Pradhan mantri*, 2) and "priest-kings" (*Purohit*, 2) of each physical realm. The priest-kings destroy the divergent consciousness of the guider power of the primeval deity to become the primordial perpetuator of a religiously institutionalized guider power. They behave as if the "primordial paternal" (Primordial ruling king, who is the descendant of the mysterious para God: *Indra*, 0) is the omnipotent institution that gives them their omniscience. The Western culture identifies the present priest-king (e.g., pope, imam) as the omniscient primeval deity. The Eastern culture identifies the primordial priest-king (i.e., Dalai Lama, the present head of Buddhism, the root of Daoism) as the omnipotent institutional value for appropriating the perpetuating desirable religion.

- **Sixth, the param-primordial, root cause reality of the one mysterious para God dimension is the one intrinsic absolute masculine dimension** immanent within the varying guider power manifestations of the primeval deity. The Vedic culture refers to the immanent masculine dimension as the "param deity" (*Parameshwar*, 7), literally, one who is servicing the sentient energy for creating Mother Nature as a creation empowered to fulfill the present wish of the creatures without any time delay from the cost of guider mediation. Mother Nature is an inanimate entity free of the limitations of conscious divine energy, guider power, and sentient value. Param deity services the sentient energy for ascending the SHEENY value, guider power, and divine energy of each entity.

The Western culture identifies the mysterious para God as *El Shaddai*, literally, one who multiplies the fruitful value for the SHEENY well-being, guider power, and divine energy of a local physical realm. The Eastern culture identifies the mysterious para God as *Shennong* (神農), literally, one who productively farms the fruits of sentient creation through the gravitational energy of the physical body of each creature by empowering the latter to be a divine creator. The Vedic culture identifies the mysterious "para God" (*Dhumavati*, 7) as the "maternal cell" (Myelin cell, 7) that trades the whitening sentient energy of Mother Nature for destroying the blackening genetic code serviced by the devil-effect of the "primordial paternal" (*Indra*, 0).

4.6. Objective for Planning the Primeval-Primordial Paradigm—Ecosystem

Mother Nature's objectivity in planning the primeval-primordial six-dimensional ecosystem paradigm, which empowers a present entity to become a technological factor for creating the desired potential effects, is the present investigation's thesis. There are six significant findings of the present investigation:

- **First,** the para-primordial, essential cause reality of many local physical realms and metaphysical para-realms within one universal creation is the **twenty-seven primeval masculine dimensions.** Suppose there are indeed one-hundred eight para entities. In that case, it implies that there must be one almighty creator of the twenty-seven almighty creations, which are projecting the energy of the one absolute feminine essence into 4 x 27 = 108 creature dimensions. The Vedic culture identifies the twenty-seven shadow "lunar mansions" (*Nakshatra*, 185), radiating the entire energy present in the universe in the form of a twenty-eighth "solar mansion" (Aspiration: *Pranidhana*, 98).   The Western culture identifies the twenty-eight demonic inanimate angel mediators of the twenty-eight lunar mansions. The Eastern culture identifies the twenty-eight illuminated solar mansions (*Xiu*), trading the shadow lunar mansions' energy.

- **Second,** the supreme-primordial, essential essence of universal creation is the **one intrinsic "absolute feminine essence"** (*Mahashakti Bhairavi*, 8).   The Vedic culture identifies the one secret lunar mansion radiating the growth in energy potential as the "polarian" (*Abhijit*, 8), polarizinig tis energy into the twenty-seven shadow lunar mansions. A supra-primordial "deity lord" (Primordial Devoted Creator: *Brahma*, 59) trades the growth-effect from this mansion and services it to the twenty-eighth "demonic angel mediator" (*Amnediel: Shraddha*, 59), which in turn services that to the twenty-eighth "solar mansion" (女 [*Nǚ*]: *Pranidhana* [*Ladki*]), which is a "girl."

The Vedic culture identifies that a male mongoose trades the ascending growth-effect of the twenty-eighth lunar mansion and services it to twelve zodiac animal entities. Eventually, a female tiger trades the ascending entropy-effect of the twenty-eight mediating creatures and services it to the param deity. There are twelve pairs of the primordial zodiac animal entities, one pair of the primordial-primordial zodiac animal entities, and one pair of the para-primordial zodiac animal entities. The constitute the supreme-primordial truth of the "primeval illuminator" (*Maha Lakshmi*, 14) as the para-primordial perpetuator of the masculine growth-effect.

- **Third, it is possible for the param deity, as an inanimate organization, to both service the animate energy as well as trade the inanimate energy "silently," without any physical-effect.** The Vedic culture identifies a twenty-first, "primordial-primordial realm" (*Antara loka*, 3794) as the "center of the universe" (*Madhya Loka*, 3794) within the maternal womb of the "param creator realm" (*Kailasha*, 3794), the home of the seven-dimensional param deity. The twenty-seven masculine lunar mansions include the seven primordial-primordial dimensions of param deity, which converge into one para-primordial feminine lunar mansion of the "primordial illuminator" (*Parvati*, 10). The sixty divisions of the solar time metric comprise the sixty spatial gravitokinetic-effects of the one primordial-primordial realm, twenty primordial divisions, thirty-six layered param realms, and the three primeval realms. Three primeval realms are the realm of zodiac space (creation), the realm of astrological space (creator), and the realm of sentient space (creature).

Table 5 summarizes the twenty natural "primordial divisions" (*Diyu* 地狱: *Lokas*, 9 x 10^{18}) of the "primordial realm" (*Maha Kalpa*, 10^{1000}), besides the one supernatural "primordial-primordial realm" (*Antara kalpa*, 3794), that form the shadow heavenly-hellish stellar and the illuminated earthly universe. The twenty primordial realms host the "guider entities" (*Rigdzin*: *Vidyadharas*, 39). The twenty-first primordial-primordial realm hosts the "SHEENY entities" (*Siddhas*, 7). Each realm has a distinctive natural "elemental quality" (*Tattva*, 863). Together with the effects of the three primeval realms, there are twenty-four natural elemental qualities immanent within the twenty-four dimensional "entity" (*Trivikrama*, 24) as the zeroth realm without the universe of entities.

- **Fourth, there are eight "paradigmatic" (Yukti, 8) natural elemental qualities** immanent within the infinite realm of the universe of entities, the ninth realm, constituting its unique natural element quality. The creation's entire potential consists of the twenty-four-dimensional zeroth digit of the para entity universe and the nine digits of the entity universe. The thiry-three dimensions of the ten-digit creation are known as the thirty-three categories of the deity qualities within the Vedic universe. Table 6 summarizes the ten natural qualities immanent within the entity realm without the universe of para entities.

- **Fifth, the zeroth worker is the primordial oneness dimension of the param-primordial realm.** The strength of the primordial oneness dimension is the primordial-primordial oneness dimension of the para-primordial realm. "Strength" is the eleventh super-oneness entity dimension that is shaped by "Primordial Illuminator" (*Huhuva*-effect: *Shri Krishna*, 10). It is the absolute value of the pure psychic linkages. The strong force is the "param perpetuator" (*Keshava*, 22) value of the strong psychic linkages. The weak force is the "param illuminator" (*Trivikrama*, 24) value of the weak psychic linkages.

The "force" (Effect: *Prapya*, 34) is the "param destroyer" (*Vamana*, 23) value of the psychic entropy within the intense power of the para entity universe seeking a soul exchange if they are no opportunities left for a manipulative soul transfer within the entity realm. "Force" is the supernatural method for breaking the natural sequence of soul reincarnation by silently contaminating the entity reality without the consciousness of the ideal-shaping and theory-taking entity. The para entity's weak force causes the residual force (i.e., effect) that a param entity experiences. Including strength and weakness, there are eight dimensions of the primordial creator that manifest eight supernatural elemental qualities within the param entity, in oneness with the universe of para entities. Table 7 summarizes these qualities.

- **Sixth, the "effect" (Perpetuator-effect [Abhidheya-effect]: Prapya, 34) is the ninth super-sameness para entity dimension,** which is shaped by the "Primordial Supra Liberator" (*Prapya*, 34). The seventeen digits of the primeval entity universe manifest the "self-perpetuating" (*Udvaha*, ½) gravitational energy of the primordial supra liberator. The presence of effect (i.e., residual force) is the primordial-primordial oneness dimension of the supreme-primordial realm. It is the primeval value of the pure psychic linkages. The electromagnetic force is the "param liberator" (*Sridhara*, 25) value of the entity's gravitational energy, psychically correlated with the para entity. It is the divine method for breaking the supernatural effect of the soul transfer by loudly purifying the polluted para entity reality.

Table 5. Twenty " Primordial Realms" (Maha Kalpa) and One " Primordial-Primordial Realm" (Antara Kalpa) in the Universe

Celestial realm	Vedic nomenclature	Buddhism nomenclature	Natural element (*Tattva*)
First: Primordial maternal realm	*Bhu loka (Puskarvardvipa, Puskaroda)*	Realm of Oil Cauldrons (油鍋地獄)	*Aum* (Para; International)
Second: Primordial paternal realm	*Mahatala loka (Aparajita Annutar, Naga loka)*	Realm of Darkness (黑暗獄)/ Realm of Darkened Bodies (黑身地獄)/ Realm of the Pit of Cattle (牛坑地獄)	*Acintya-bhedabheda* (Independence; Downness-effect)
Third: Para deity realm	*Talatala loka (Ghrutvardvipa, Ghrutoda)*	Realm of the Mountain of Knives (刀山地獄)	*Varnadhva* (Asymmetry; Strangeness-effect)
Fourth: Supreme deity realm	*Nitala loka (Nandishvardvipa, Nandishvaroda, Nandikara Annutar)*	Realm of Molten Copper (烊銅地獄)/ Realm of Copper Pillars (銅柱地獄)	*Atman* (SHEENY; Soul)
Fifth: Primordial deity realm	*Vitala loka (Vaijayanta Annutar)*	Realm of the Pool of Blood (血池地獄)	*Abhidheya* (Perpetuator)
Sixth: Devoted deity realm	*Sutala loka (Jayanta Annutar)*	Realm of Beasts (畜生地獄)	*Advaita* (Creator)

Celestial realm	Vedic nomenclature	Buddhism nomenclature	Natural element (*Tattva*)
Seventh: Deity realm	*Jana loka or Deva loka* (*Kshirvardvipa, Kshiroda*)	Realm of Iron Beds (鐵床地獄)	*Nityanta* (Complexity; Charmness-effect)
Eighth: Param deity realm	*Svarga loka* (*Sarvarthasiddhi Annutar*)	Realm of Hanging Bars (吊筋獄)	*Vyom* (Ether)
Ninth: Supra deity realm	*Atala loka* (*Vijaya Annutar; Yama loka*)	Realm of Tongue Ripping (拔舌獄)	*Chitta* (Guider)
Tenth: Primeval illuminator realm	*Bhuvar loka* (*Ikshuvardvipa, Iksuvaroda*)	Realm of Scissors (剪刀地獄) / Realm of Disembowelment 抽腸獄	*Nayaki* (Workculture)
Eleventh: Primeval deity realm	*Tapa loka* (*Varunvardvipa, Varunoda*)	Realm of Skinning (剝皮獄) / Realm of Sawing (刀鋸地獄)	*Agni* (Fire)
Twelfth: Super deity realm	*Mahar loka* (*Ghatki Khand, Kaloda*)	Realm of Iron Books (鐵冊地獄) / Realm of Boulder Crushing (石壓地獄)	*Jiva* (Simplicity; Topness-effect)
Thirteenth: Primeval greeter realm	*Patala loka* (*Jambudvipa, Lavanoda*)	Realm of Pounding (碓搗獄) / Realm of Iron Mills (鐵磨地獄)	*Sva* (Corporate; Absolute time)

Celestial realm	Vedic nomenclature	Buddhism nomenclature	Natural element (*Tattva*)
Fourteenth: Self-luminous entity realm	*Satya loka (Dhamma, Ratnaprabha)*	Realm of Maggots (蛆蟲地獄)	*Sadakhya* (Culture)
Fifteenth: Primeval perpetuator realm	*Brahma loka (Arishta, Dhumprabha)*	Realm of Boiling Sand (沸沙地獄)/ Realm of Dismemberment (磔刑地獄)	*Sadhya* (National)
Sixteenth: Primordial greeter realm	*Indra loka (Shambhala, Megha, Balukaprabha,* White hole)	Realm of Mirrors of Retribution (孽镜地狱)/ Realm of Cover Mountains (蓋山地獄)	*Vaisnava* (Dependence; Upness-effect)
Seventeenth: Devotee deity realm	*Sri loka (Manidweepa, Anjana, Pankprabha, Agni loka)*	Realm of Weighing Scales (秤桿獄)/ Realm of Steaming (蒸籠地獄)/ Realm of Boiling Faeces (沸屎地獄)	*Payu* (Local)
Eighteenth: Luminous realm	*Go loka (Vaikuntha, Vansha, Sharkaraprabha, Varuna loka)*	Realm of the Wrongful Dead (幽枉獄)	*Bhagwat* (Knower)

Celestial realm	Vedic nomenclature	Buddhism nomenclature	Natural element (*Tattva*)
Nineteen: Param Child (ascending masculine-effect) realm	*Mrityu loka (Maadhavi, Mahatamahprabha, Nirrti loka)*	Realm of Moulting (脫殼獄)/ Realm of Mortars and Pestles (舂臼地獄)	*Bhava* (Divine)
Twenty: Param Manifestor (descending feminine-effect) realm	*Rasatala loka (Maghavi, Tamahprabha)*	Realm of the Mountain of Ice (冰山地獄)	*Asrama* (Symmetry; Bottomness-effect)
Twenty-First: Param Creator (horizontal androgenous-effect) realm	*Garbhastala adiloka (Kailasha, Madhya loka)*	Realm of the Mountain of Fire (火山地獄)	*Sadhna* (Manifestor)
Note: Bhu (earth) is the natural element of the stellar universe, Vayu (air) is the natural element of the zodiac universe, and Apas (water) is the natural element of the astrological universe.			

Table 6. The Ten Digits of Natural Elemental Qualities Within the Primeval Entity Realm of the Universe

Entity digit	Entity form (sentient-effect—Buddhist)	Elemental quality (Vedic qualities of *Rama*: ideal human entity)	Magnetic (physical; mechanical; social; thermal) nature of the elemental quality	Immanent entity dimension shaping the elemental quality (Vedic qualities of physical well-being)	Immanent para entity dimension shaping the elemental quality (Buddhist qualities of the metaphysical realm)
0	Worker (*Jiang* 蔣)	*Samratah tattva* (creepy; master creator of self-destiny)	Ecosystem fatigue and creep (Universal correlation with entity reality)	* Weak (*sukshma*): subtle, agent * Strong (*sthula*): gross, principal	Weak: Primeval Perpetuator (Grinding-effect: death fortune) Strong: Primeval Illuminator (Fiery chariot-effect: life fortune)
1	Destroyer (*Li* 歷)	*Charitrena yuktah tattva* (dense; unblemished)	Entity strain and density (Personal or physiological correlation with para entity reality)	* Weak (*drava*): soft; diffused * Strong (*sandra*): dense; concentrated	Weak: Para Deity (*Arbuda*-effect) Strong: Param Child (*Diyu*-effect: "Innocent Death")

Entity digit	Entity form	Elemental quality	Magnetic nature	Immanent entity dimension	Immanent para entity dimension
2	Illuminator (Yu 余)	*Krutajnah tattva* (stressed; blessed)	Space stress and flatness (Social or local correlation with para entity reality)	* Weak (*khara*): rough, residual * Strong (*shlaksna*): smooth, fundamental	Weak: Primordial Paternal (*Nirarbuda*-effect) Strong: Super deity (*Padma*-effect)
3	Liberator (*Lü* 呂)	*Dharmajnah tattva* (elastic; just)	Time elasticity and curvature (Structural or emotional correlation with para entity reality)	* Weak (*manda*): silence; dull * Strong (*tikshna*): loudness; sharp	Weak: Supreme deity (*Atata*-effect) Strong: Supra deity (*Samghata*-effect)
4	Devoted (*Bao* 包)	*Drudhavratah tattva* (plastic; resolute)	Phase plasticity and resilience (Psychological correlation with para entity reality)	* Weak (*chala*): mobile; inflecting; fractured; brittle * Strong (*sthira*): stable; healing; resilient; sturdy	Weak: Primordial Perceptuator (*Hahava*-effect) Strong: Primordial Maternal (*Raurava*-effect)
5	Devotee (*Bi* 畢)	*Viryavan tattva* (inflammable; invincible)	Temperature ductility and flammability (Psychic correlation with para entity reality)	* Weak (*sita*): cold; symbolic; congesting * Strong (*ushna*): hot; inflamed	Weak: Primordial deity (*Maharaurava*-effect) Strong: Deity (*Sanjiva*-effect)

Entity digit	Entity form	Elemental quality	Magnetic nature	Immanent entity dimension	Immanent para entity dimension
6	Greeter (*Dong* 董)	*Vidvan tattva* (tough; self-aware and non-malleable)	Pressure malleability and toughness (Para-psychic correlation with para entity reality)	* Weak (*mrdu*): soft; sensitive * Strong (*kathina*): hard; tough	Weak: Self-luminous entity (*Utpala*-effect) Strong: Primordial Greeter (*Tapana*-effect)
7	Self (*Huang* 黃)	*Sarvabhuteshu hitah tattva* (massifying; servicing mass well-being)	Gravitational machinability and massification (Grounding in para entity reality)	* Weak (*laghu*): light mass; heavenly * Strong (*guru*): heavy mass; grounded	Weak: Param Deity (*Kalasutra*-effect) Strong: Primeval Greeter (*Pratapana*-effect)
8	Ideal (*Lu* 陸)	*Gunavaan tattva* (fluid)	Electromagnetic superfluidity and hardenability (Channeling of para entity reality)	* Weak (*ruksha*): viscous; horizontal (resisting); dry * Strong (*snigdha*): fluid; high (ascending); oily	Weak: Primeval Deity (*Mahapadma*-effect) Strong: Luminous (*Avia*-effect)
9	Theory (*Xue* 薛)	*Satyavakhyah tattva* (truth mediator)	Sentient superconductivity and brittleness (Contamination of entity reality)	* Weak (*visada*): clear; natural * Strong (*avila*): grimed; polluted	Weak: Param Creator (Soul reincarnation) Strong: Param Manifestor (Soul transfer)

Table 7. The Eight Digits of Supernatural Elemental Qualities Within the Param Entity Realm of the Universe

Para entity digit (within 1, i.e., entity digit)	Para entity form (gravitomagnetic-effect—scientific name)	Elemental quality (Vedic qualities of *Rama*: ideal human entity)	Electric (biological; human; hydraulic) nature of the elemental quality	Immanent para entity dimension shaping the elemental quality (Vedic qualities of intellectual well-being)	Immanent param entity dimension shaping the elemental quality (Buddhist qualities of the metaphysical realm)
10	Child (Intensity of magnetization, i.e., the strength of feminine psychic force)	*Anasuyakah tattva* (self-teaming animal without human envy to become enviable deity)	Radioactivity and power decay (charge normalizability within a decaying entity through space-dependent cubic root function)	*Vata-Kapha* (condensation of divine-effect)	Cause: Primeval Perpetuator (Grinding-effect: death fortune) Effect: Param Perpetuator (soul birth as an inanimate entity)
11	Maternal (Magnetic potential, i.e., psychic density)	*Jitakrodah tattva* (self-sacrificing human without animalistic anger)	Competitiveness and power entropy (charge transformativity within a degenerating entity through time-dependent square root function)	*Pitta-Kapha* (dehydration/ freezing of earth-effect)	Cause: Para Deity (*Arbuda-*effect) Effect: Param Illuminator (soul sentiation as an animate entity)

Para entity digit	Para entity form	Elemental quality	Electric nature	Immanent para entity dimension	Immanent param entity dimension
12	Paternal (Magnetic field strength, i.e., devotional intensity)	*Bibyati devah tattva* (self-motivated spirit of fiery charming passion anger)	Electromagnetic and power loss (charge interruptivity with the burning-out death of an entity through infinity function)	*Vata-Pitta* (evaporation of ether-effect)	Cause: Primordial Paternal (*Nirarbuda*-effect) Effect: Param Liberator (soul materialization as the Majorana dark matter)
13	Creature (Magnetic flux, i.e., the strength of masculine ego force)	*Priyadarshana tattva* (self-steaming bewitching charm)	Supplementarity and power integration (charge additivity of diverse para entity realities)	*Pitta* (melting/burning of fire-effect)	Cause: Supreme deity (*Atata-*effect) Effect: Primordial Super Manifestor (soul quantization as the stellar luminous point in the form of Boson color potential)

Para entity digit	Para entity form	Elemental quality	Electric nature	Immanent para entity dimension	Immanent param entity dimension
14	Creation (Magnetic motive force, i.e., emotional intensity)	*Atmavan tattva* (self-esteeming ascended master)	Metaphysicality and power differentiation (charge quantizability or divisibility for creating diverse para entities)	*Tridosha* (precipitation of guider-effect)	Effect: Primordial Para Greeter (soul combination/ merger as the self-luminous zodiac point in the form of anti-neutrino)
15	Masculine (Magnetic induction, i.e., flux or ego density)	*Dyutiman tattva* (self-worth glowing light force)	Thermodynamism and power exponentiation (discharge accelerability through exponential function for perpetuating diverse para entity realities)	*Kapha* (sublimation/ submerging of water-effect)	Cause: Primordial deity (*Maharaurava*-effect) Effect: Param Primordial Destroyer (soul permutation/ demerger as the self-perpetuating astrological point in the form of soliton crystal)

Para entity digit	Para entity form	Elemental quality	Electric nature	Immanent para entity dimension	Immanent param entity dimension
16	Feminine (Magnetic permeability, i.e., flux or ego transfer)	*Pativrata tattva* (self-destroying radiating love)	Complementarity and power subtractability (charge conservability after entity exchange and soul transfer)	*Vata* (transpiration/ choking of air-effect)	Cause: Self-luminous entity (*Utpala*-effect) Effect: Self-luminous Primordial Destroyer (soul proliferation/ multiplication as the self-destroying sentient drop in the form of gluon)
17	Androgyne (Space permeability, i.e., magnetic or psychic transfer)	*Sanatanam tattva* (self-conscious agelessness)	Superpositioning and power reversibility (charge discontinuity and tripping after entity potentiation and soul reincarnation)	*Nirdosha* (deposition/ soaking of SHEENY-effect)	Cause: Param Deity (*Kalasutra*-effect) Effect: Self-luminous Supreme Devoted (soul entropy and transformation into a spirit entity in the form of neutrino)

Table 8. The Six Words of Divine Elemental Qualities Within the Primordial Entity Realm of the Universe

Primordial entity letter (without param entity letter and primeval entity letter of the para entity)	Primordial entity form (gravitoelectric-effect—electron orbital form)	Elemental quality (Vedic qualities of Krishna: ideal deity entity)	Gravitational (ecological; seismic) nature of the elemental quality	Immanent param entity dimension shaping the elemental quality (Vedic qualities of mental well-being and musicality)	Immanent primordial entity (biophysical) dimension shaping the elemental quality (Buddhist qualities of the metaphysical realm)
S (Sense of determination: uchchadana) [B: Protein; S: Ribose sugar]	Sharp (*tikshna*, Space permittivity)	*Dharjya tattva* (intellectual strength; sensory patience; manpower)	Gravitational potential (horizontal light-effect radiated by an entity, i.e., child)	Meter (*svara*, confidence; jiva-effect)	Absolute cause: Primeval Deity (*Mahapadma*-effect) Effect: Self-luminous Primordial Liberator (soul growth into an astral entity in the form of electron or "somatic cell")
P (Potential for imagination: siddhi) [N: Fatty acylcarnitine enzymes; Y: Pyrimidine]	Principal (*sthula*, Electric charge)	*Tapasya tattva* (mental strength; social equanimity; mental power)	Gravitational radiation (Axion or Phi radiation; ascending light-effect traded from a para entity, i.e., maternal)	Melody (*raga*, empathy; *asrama*-effect)	Absolute cause: Param Creator (Soul reincarnation) Effect: Self-luminous Supra Destroyer (soul development into a socially-correlating entity in the form of quark or "germ cell")

Primordial entity letter	Primordial entity form	Elemental quality	Gravitational nature	Immanent param entity dimension	Immanent primordial entity dimension
D (Density of virtue: sadachara) [D: Fat; M: Amino acid]	Dense (*sandra*, Electric power)	*Nyaya tattva* (astral strength; emotional equanimity; fighting for a just cause; muscular power)	Gravitational mass (descending light-effect exchanged by a param entity, i.e., paternal)	Rhythm (*tala*, clarity; vaisnava-effect)	Absolute cause: Param Manifestor (Soul transfer) Effect: Self-luminous Primordial Devotee (soul development into an intellectual-alert entity in the form of proton or "gamete")
F (Facility of intuition: *umapati*) [V: Acetyl coenzyme A; Z: Zero— the dead spirit after eighteen-fold mutation]	Fundamental (*shlaksna*, Electric potential)	*Kshama tattva* (etheric or psychological strength; effect freedom and forgiving; material power)	Gravitational redshifting (descending light curving effect of a primeval entity, i.e., creature)	Texture (*laksya*, character; *Acintya-bhedabheda*-effect)	Absolute cause: Param Child (*Diyu*-effect: "Innocent Death") Effect: Self-luminous Primordial Devoted (soul embodiment into a physically-inert entity in the form of muon or "oogonium")

Primordial entity letter	Primordial entity form	Elemental quality	Gravitational nature	Immanent param entity dimension	Immanent primordial entity dimension
G (Gravity of nature: *lingam*) [G: Guanine; R: Purine]	Grimed (*avila*, Electric current)	*Danasheel tattva* (causative or psychic strength; giving freedom and beneficence; management power)	Gravitational wave (ascending light curving effect of a primordial entity, i.e., creation)	Tonal color (*laksana*, depth; *nityanta*-effect)	Absolute cause: Luminous (*Avici*-effect) Effect: Self-luminous Devoted Devotee (soul transubstantiation into a metaphysically-pure entity in the form of plasmon or "oöcyte")
H (Holding power of excellence: *thiruvambala*) [H: Fatty acid; K: Keto—the weakened soul, supported by the strengthening self-incubating energies of a pair of nascent spirits]	Dominant (*snigdha*, Electric resistance)	*Dayabhava tattva* (consequential or para-psychic strength; radiating love, compassion, and social benefit; monetary power)	Gravitational lensing (horizontal light curving effect of a primordial-primordial entity, i.e., masculine)	Pitch (*sruti*, fullness; *varnadhva*-effect)	Absolute cause: Primordial Illuminator (*Huhuva*-effect) Effect: Self-luminous Super Destroyer (soul substantiation into a thermodynamically impregnated and polluted entity in the form of phonon or "sperm")

One becomes an ideal-making and theory-shaping entity seeking to perpetuate the self as the ideal. One shapes the theory by transforming the self into a phonon particle and then loudly destroying the phonetic reality for validating the self's invincibility. When the phonon particle gets destroyed, the self transforms into a scripted (word; not digitized value) "tau particle" (Zygote: *Tantri*, 48).

The tau particle's gender is the consequence of the immanent sentient (i.e., SHEENY) reality of the self as a param creature during the dark matter phase. Including the effect, there are six formative dimensions of the param creature. They manifest the six divine elemental qualities within the para entity, in oneness with the primordial entity responsible for soul transfer. Table 8 summarizes these qualities.

The sixth finding is the primeval-primordial reality of Mother Nature that empowers an entity to illuminate the six additional dimensions of the para-primordial reality of the self as the Almighty Creator.

- **First, excluding the effect, sixteen normative dimensions of the para entity manifest sixteen guider elemental qualities** within the primeval entity after the soul transfer, in oneness with the para entity enjoying the soul transfer. Table 9 summarizes these qualities. They are the "causative factor" (*Sva*-effect: *Prapaka*, 36) of the present reality's multidimensionality. Each preceding set of the horizontal and vertical qualities is the absolute cause of the succeeding set of qualities.

- **Second, the final primeval set of qualities is the absolute cause of the first primordial set of qualities.** The primeval set of qualities is the primordial-primordial cause of the six divine elemental qualities. They, in turn, are the primordial-primordial cause of the eight supernatural elemental qualities. They, again, are the primordial-primordial cause of the ten natural elemental qualities as well. Twenty qualities of the primordial realms are the primordial-primordial cause of the para entity realm's primeval set of qualities. One quality of the primordial-primordial realm is the primordial cause of the sixty qualities of the diverse para realms. The sixty qualities are the absolute cause of the sixty divisions within a unit of time. Table 10 summarizes the sixty time-varying pathways for the oneness of the absolute creature with the primordial-primordial realm.

- **Third, the sixty time-dimensions are a function of the four super-immanent elemental dimensions of the primordial illuminator:** *Prapya tattva* (primeval effect incarnate; the Almighty force of the devotee consciousness oriented toward self-development); *Prapaka tattva* (absolute cause incarnate; strategic intent for the perpetuating corporate-effect, diligently devoting their energy for self-development); *Samah tattva* (primordial cause incarnate; the strategic deadline for the self-development), and *Jalashayotsarga tattva* (primordial-primordial cause incarnate; intuition illuminator). Any entity may realize oneness with the primordial-primordial realm without a time-varying path. Instead, an entity may become a devotee of one of the four paths of "Mother Nature" (Primordial deity, 8) before deciding to perpetuate the "corporate effect" (*Arcisa*, 128) through energy polarization. Table 11 summarizes these four pathways.

- **Fourth, as an alternative, an entity may become devoted to one of the sixteen paths of the Primordial Greeter** for perpetuating the corporate-effect through energy polarization. These sixteen pathways are summarized in Table 12.

- **Fifth, as another alternative, an entity may decide to enjoy the absolute freedom to behave as if one is perpetuating or illuminating a unique path of the Primordial-Primordial Creator** or, better still, behaving with the infinite devotion of a devotee or infinite divinity of a devotee. These four pathways are summarized in Table 13.

- **Sixth, and finally, an entity may decide to destroy the absolute freedom by becoming devoted to one of the Param Child's sixteen paths.** Such an entity is a devotee, believing that a supernatural God will be a radiant savior, independent of what one does (*akarma path*: the path of non-action and blind faith). Table 14 summarizes these sixteen paths.

The one-hundred entity dimensions fructify within these one-hundred pathways to ascending self-awareness form the whole gravitational potential of an ideal human entity, within and without the present reality's paradigmatic boundaries.

Table 9. The Sixteen Words of Guider Elemental Qualities Within the Para Entity Realm of the Universe

4th Para entity letter (within 1st normative param RNA, 2nd transformative primordial DNA, and 3rd metaphysical mtDNA entity letters)	Para entity form (electro magnetic-effect)	Elemental quality (Vedic qualities of Krishna: ideal deity entity)	Sentient (economic; biophysical) nature of the elemental quality	Immanent primeval entity dimension shaping the elemental quality (Vedic qualities of psychological or objective well-being as a function of knowing)	Immanent primeval entity dimension shaping the elemental quality
UAGC (air-earth-water-fire)	Krypton (Ascending SHEENY-effect)	*Bhakta Shurita tattva* (Present incarnate)	Luster (bhagwat-effect: achara—constant or subject): Tau (nonvalent) particle or "zygote"	Gravitomagnetic growth (constancy; enthalpy; heat of formation): *jati* (class, i.e., heat value; gravity-effect; innate knowledge)	Primordial cause: Primordial Greeter (*Tapana*-effect)
UGCA (air-water-fire-earth)	Krypton (Descending SHEENY-effect)	*Niraskata tattva* (Potential incarnate)	Density (sadhana-effect: vibhakti—variable): Atom (univalent) network or "cell body"	Magnetic entropy (variability; magnetic polarization): *samsaya* (i.e., cold value; mass-effect; incipient knowledge)	Primordial cause: Primeval Illuminator (Fiery chariot-effect: life fortune)

4th Para entity letter	Para entity form	Elemental quality	Sentient nature	Immanent primeval entity dimension	Immanent primeval entity dimension
UCAG (air–fire–earth–water)	Krypton (Horizontal SHEENY-effect)	*Sangitajna tattva* (Dynamism incarnate)	Valency (*advaita*-effect: *prayashchitta*—mission): Compound (covalent) element or "chromosome"	Magnetic growth (covariability; magnetic reluctance): *drishtanta* (light force or illusion-effect; predominating or conceived knowledge)	Primordial cause: Self-luminous entity (*Utpala*-effect)
UCCG (air–fire–fire–water)	Krypton (Entropy SHEENY-effect)	*Neetibadi tattva* (Technology incarnate)	Color (*aum*-effect: *ranga*—value): Molecule (multivalent) or "mitochondria DNA molecule" or dynamic/ international consciousness	Electromagnetic entropy (objectivity; alkalinity): *prameya* (light; perception of object; dominating or perceived knowledge)	Primordial cause: Primeval Greeter (*Pratapana*-effect)
UCGC (air–fire–water–fire)	Argon (Ascending Guider-effect)	*Lila Madhurya tattva* (Organization incarnate)	Structure (*sadhya*-effect: *chakshu*—vision): Crystal (paravalent or ionic material) or "DNA strand" or metaphysical/ national consciousness	Electromagnetic growth (subjectivity; lipophilicity; ionizability): *nirnaya* (antumbra shadow personal cast; decisive or experiential knowledge)	Primordial cause: Super deity (*Padma*-effect)

4th Para entity letter	Para entity form	Elemental quality	Sentient nature	Immanent primeval entity dimension	Immanent primeval entity dimension
UGCC (air-water-fire-fire)	Argon (Descending Guider-effect)	*Pratibhasamp-anna tattva* (Ecosystem incarnate)	Hardness (*papu*-effect: *smriti*—memory): Metal (conductor) or "cytoplasm" or physical/ local consciousness	Electric entropy (conductivity, conductance): *vada* (penumbra secret social creed; past or contractual or referential knowledge)	Primordial cause: Primeval Deity (*Mahapadma*-effect)
UCCC (air-fire-fire-fire)	Argon (Horizontal Guider-effect)	*Tejsauktah tattva* (Entity incarnate)	Streak (*nayaki*-effect: *chandoga-vrishotsarga* or *buddhi*—intelligence): Mineral (superconductor) or "neuron cell" or intellectual / workculture consciousness	Electric growth (superconductivity, capacitance): *tarka* (umbra super-secret religious correlation; dynamic or scripted knowledge)	Primordial cause: Primeval Illuminator (Fiery chariot-effect: life fortune)
UCCA (air-fire-fire-earth)	Argon (Entropy Guider-effect)	*Saundarjyam-aya tattva* (Past incarnate)	Fracture (*sadakhya*-effect: *manas*—mind): Animal (resistor) or "muscle cell" or mental/ cultural consciousness	Gravioelectric entropy (electric resistance): *jalpa* (unconditional institutional bickering; formative or sentient knowledge)	Primordial cause: Supra deity (*Samghata*-effect)

4th Para entity letter	Para entity form	Elemental quality	Sentient nature	Immanent primeval entity dimension	Immanent primeval entity dimension
UCAC (air-fire-earth-fire)	Phosphorous (Ascending Divine-effect)	*Sarvaniyanta tattva* (Future incarnate)	Gravity (*atman*-effect: *vak*— voice): Human (super-resistor) or "gland cell" or astral/ sentient consciousness	Gravitoelectric growth (electric permeability; dielectric constant): *avayava* (quantum fractional bragging; normative or thermalized social knowledge)	Primordial cause: Param Deity (*Kalasutra*-effect)
UACC (air-earth-fire-fire)	Phosphorous (Descending Divine-effect)	*Sarvagnata tattva* (Human incarnate)	Geometry (*chitta*-effect: *ganitam*— self): Spirit (insulator) or "dendrite cell" or etheric/ psychological consciousness	Gravitoelectromagnetic entropy (combustion; the heat of transformation; insulation potential): *hetvabhasa* (proportionate codified begging; transformative or animated personal knowledge)	Primordial cause: Deity (*Sanjiva*-effect)

4th Para entity letter	Para entity form	Elemental quality	Sentient nature	Immanent primeval entity dimension	Immanent primeval entity dimension
UACA (air-earth-fire-earth)	Phosphorous (Horizontal Divine-effect)	*Priyamvada tattva* (Spirit incarnate)	Size (*bhava*-effect: *padartha*—object): Deity (super-insulator) or "glia cell" or causal/ psychic consciousness	Gravitoelectromagnetic growth (displacement; heat of normative flux density): *nigrahasthana* (absolute abstract bragging; metaphysical or inanimate religious/ universal/ observational knowledge)	Primordial cause: Primordial Pereptuator (*Hahava*-effect)
UCAA (air-fire-earth-earth)	Phosphorous (Entropy Divine-effect)	*Swayamniyanta* (Deity incarnate)	Shape (*agni*-effect: *rupa*—form): Para deity (thermodynamic) or "oligodendrocyte cell" or self-luminous/ astrological consciousness	Sentient entropy (electric current): *vitanda* (infinite, diffused blessing; materialized, frozen entity knowledge ideated by the self as an opinion or pure, idealized perspective/ mathematical knowledge)	Primordial cause: Primordial deity (*Maharaurava*-effect)

4th Para entity letter	Para entity form	Elemental quality	Sentient nature	Immanent primeval entity dimension	Immanent primeval entity dimension
UAAC (air-earth-earth-fire)	Oxygen (Horizontal air-effect)	*Paripurna* (Androgynous incarnate)	Volume (*apas*-effect: *rasa*—taste): Param deity (thermokinetic) or "myelin cell" or luminous/zodiac consciousness	Sentient growth (electric charge): *siddhanta* (finite, dense belief system; knowledge theorized by the self as empirically substantiated and conclusive until falsified/scientific knowledge)	Primordial cause: Supreme deity (*Atata*-effect)
UAAA (air-earth-earth-earth)	Oxygen (Entropy air-effect)	*Akrodha* (Masculine incarnate)	Order (*vayu*-effect: *sparsha*—touch): Primordial deity (electrodynamic; animate) or "motile cell" or celestial realm/primeval consciousness	Gravitational entropy (configuration or system coordination number): *chala* (common sense, just first to say aloud assuming it is valid to become a deity attracting infinite theory-taking followers; theoretical humanities knowledge)	Primordial cause: Para Deity (*Arbuda*-effect)

4th Para entity letter	Para entity form	Elemental quality	Sentient nature	Immanent primeval entity dimension	Immanent primeval entity dimension
TAAA (ether-earth-earth-earth)	Calcium (Entropy ether-effect)	*Mardavam* (Feminine incarnate)	Odor (*bhu*-effect: *gandha*—smell): Primeval deity (electrokinetic; inanimate; extrinsic ecosystem dimension of Mother Nature) or "Schwann cell" or primordial realm/ para consciousness	Gravitational growth (configuration or system toxicity): *prayojana* (sixth sense of intuition, motivating one to backbite and manipulate the reality through selective number-taking and brash theory-making as if one is para deity God leader; hermeneutical philosophical knowledge)	Primordial cause: Primordial Paternal (*Nirarbuda*-effect)
AAAA (earth-earth-earth-earth)	Hydrogen (Entropy water-effect)	*Samah* (Nature incarnate)	Octave vibration (*vyom*-effect: *nada*—sound): Primordial-Primordial deity (magnetic; natural; intrinsic entity dimension of Mother Nature) or "axon cell" or primordial-primordial realm/ param consciousness	Gravitomagnetic entropy (electric power; the heat of entropy): *pramana* (behaving as if one is param deity whose superficial theory-shaping word is the final authority; authoritative management knowledge)	Primordial cause: Primordial Maternal (*Raurava*-effect)

Table 10. The Sixty Time-Varying Pathways for Oneness of the Absolute Creature with the Primordial-Primordial Realm

Chronological pathway	Samvat-sara ^	The element of the Doctrine of Immanence	Signified-effect of the element of the Doctrine of Immanence	Shashty amasha ~	Signified Element of the Doctrine of Emanation	Dà-Jiāng-Jūn Star Deity*	Tài-Sui Celestial Deity**	Present Luminous cycle (year)
1	Prabhava	Aum tattva (Para; Potential; International)	Create an absolute feminine origin point	Ghora	Wisher (of the absolute energy)	Shen Xing	Yang Fire Tiger	1986
2	Vibhava	Acintya-bhedabheda tattva (Independence; Downness-effect)	Seek primeval (infinite) cause of the masculine orientation	Rakshasa	Knower (of the masculine darkness within the feminine energy)	Zhao Da	Yin Fire Rabbit	1987
3	Shukla	Varnadhva tattva (Asymmetry; Strangeness-effect)	Experience para-conscious sin within the conscious bliss state	Deva	Manifestor (of the light without the energy)	Guo Can	Yang Earth Dragon	1988
4	Pramoduta	Atman tattva (SHEENY-effect; Soul)	Enjoy the deceptive sense of joyfulness	Kuber	Creator (of the quantum colors of light)	Wang Ji	Yin Earth Serpent	1989
5	Prajothpatti	Abhidheya tattva (Perpetuator)	Perpetuate para-conscious linkages within the creative divine lineage linkages	Yaksha	Seeker (of the concealed energy)	Li Su	Yang Air Horse	1990
6	Angirasa	Advaita tattva (Creator)	Create deified satwik truth dharma fearlessly, within the immanent orientation for reproducing sin	Kinnara	Experiencer (of the vibrating energy)	Liu Wang	Yin Air Sheep	1991

^: 60th absolute unit within Doctrine of Immanence' ~: 1/60th para-absolute unit within Doctrine of Emanation

*: Mediating intrinsic androgyne factor; **: Moderating extrinsic androgyne factor

Chronological	Samvat-sara	Doctrine of Immanence	Signified-effect of the element of the Doctrine of Immanence	Shashty amasha	Signified Element of the Doctrine of Emanation	Star Deity	Celestial Deity	Year
7	Shrimukha	Nityanta tattva (Complexity; Charmness-effect)	Imagine the beautiful royal rajo guna valorously, for earning the deity-like privilege of consuming the forbidden apple of sin	Bhrashta	Enjoyer (of the descending energy)	Kang Zhi	Yang Water Monkey	1992
8	Bhava	Vyom tattva (Ether)	Sound the divine value of the tamo devil guna, by illustriously investing the private wealth for the national public service, and living a life of ascetic who like the super deity has the fundamental secondary residual right of trading the entire social wealth	Kulaghna	Destroyer (of the residual lineage energy)	Shi Guang	Yin Water Rooster	1993
9	Yuva	Chitta tattva (Guider)	Become a devoted asura guider, who suffers the pain of diseases from the engagement in sinful kriyas, within the strange polarized appearance of the youthful personified supra deity incarnation of the God	Garala	Perpetuator (of the poison root cause within the energy)	Ren Bao	Yang Ether Dog	1994
10	Dhata	Nayaki tattva (Workculture)	Become a supreme Godhead, who has the divine aura of the omniscient param deity, in the presence of the God, within a pleasing down dim-witted workculture for netting diverse victims	Agni	Illuminator (of the sentient essence within the poison root cause as the fire-effect)	Guo Jia	Yin Ether Pig	1995
11	Ishvara	Agni tattva (Fire)	Become a para deity God who has the omnipotent thermodynamic authority of universally enforcing the rule of the self-institutionalized dharma law, within a sense of discriminating faculty and short-tempered, impulsive hedonism	Maya	Devoted (to the root cause as the illusion of oneness with the sentient essence)	Wang Wen	Yang Fire Rat	1996
12	Bahudhanya	Jiva tattva (Simplicity; Topness-effect)	Become a primeval deity endowed with the omnipresent simplicity, within the ascending potential for universal togetherness consequences	Purishak	Devotee (of the illusionary beauty of the restless causal body)	Lu Xian	Yin Fire Bison	1997

Chronol ogical	Samvat- sara	Doctrine of Immanence	Signified-effect of the element of the Doctrine of Immanence	Shashty amasha	Signified Element of the Doctrine of Emanation	Star Deity	Celestial Deity	Year
13	Pramathi	Sva tattva (Corporate; Absolute time)	Become an absolute deity gifted with the omnipermeating absolute time consciousness of the intimate benefits acquired by the masculine force using the delegated positional power, within the mind that is super-polluted by the negative energy co-followers	Apam-pati	Godhead (the cooling water-effect of the true param-primordial sentient essence)	Long Zhong	Yang Earth Tiger	1998
14	Vikrama	Sadakhya tattva (Culture)	Become a para-absolute primordial deity who commands transformational leadership planning qualities of patiently, valorously, and fiercely championing the cause of generosity, without the negative energy followership	Marutvan	Primordial-Primordial (the transformative air-effect of the Godhead consciousness)	Dong De	Yin Earth Rabbit	1999
15	Vrusha	Sadhya tattva (National)	Become a self-luminous primordial perpetuator of the negative energy followership, as a 'credit collecting and discredit distributing' entrepreneur who is intoxicated with the latent gravitomagnetic effect of the sin	Kaala	Primordial (the formative growth-effect of the intrinsic God-consciousness)	Zheng Dan	Yang Air Dragon	2000
16	Chitrabhanu	Vaisnava tattva (Dependence; Upness-effect)	Become a primordial greeter illuminator who strategically manages within the patent gravitoelectric energy flow of the present moment, for ambitiously ascending the physical benefits — both intrinsic: infinite clothes and births in beautiful forms; as well as extrinsic: infinite natural objects and intimate subjects	Ahi-bhaaga	Stellar/Absolute (the normative development of the togetherness-effect within the vector coiled consciousness)	Lu Ming	Yin Air Serpent	2001
17	Svabhanu	Payu tattva (Local)	Become a positively charged luminous who self-norms the wholesome natural gravitational self-fulfilling qualities of the whole lineage in the present life as the path of local dharma	Amrita	Zodiac (the transformative exchange of the otherness consciousness within the constant earth-effect)	Wei Ren	Yang Water Horse	2002

Chronological	Samvatsara	Doctrine of Immanence	Signified-effect of the element of the Doctrine of Immanence	Shashtyamasha	Signified Element of the Doctrine of Emanation	Star Deity	Celestial Deity	Year
18	Tarana	Bhagwat tattva (Knower)	Become a homologous negatively charged readable entity who, in the garb of public service, masters the normative knowing for manipulating the public wealth to fulfill ascending wasteful private consumption wishables without the readable path of dharma	Indu	Astrological (the metaphysical entropy-effect on the sense of otherness within the variable lunar time)	Fang Jie	Yin Water Sheep	2003
19	Parthiva	Bhava tattva (Divine)	Become a homologous neutrally charged well-read entity, who thrives as a universal chakravartin leader and is subject to sensory charms within the strong psychic force of the readable path of dharma	Mridu	Astral (the subtle dynamic consciousness of the negative energy within the potential space)	Jiang Chong	Yang Ether Monkey	2004
20	Vyaya	Asrama tattva (Symmetry; Bottomness-effect)	Become a double negatively charged entity, who is proficient in reading the sentiments for creating an absolute karma path of lifelong begging to fulfill infinite sensory wishables within intoxication and indebtedness conditions	Komala	Mental (the super-subtle formative consciousness of the positive energy within the potential time)	Bai Min	Yin Ether Rooster	2005
21	Sarvajith	Sadhana tattva (Manifestor)	Become a double positively charged entity, who performs the role of a reader, whose eloquent lordship for the intuitive value-based path of karma wins over all the neutral dharma kings as well as the negative bhava king of kings	Heramba	Intellectual (supra deity angels of light emanating positive energy for the universal enlightenment)	Feng Ji	Yang Fire Dog	2006
22	Sarvadhari	Samratah tattva (Creepy: master creator of self-destiny)	Become a three-dimensional researchable entity, who is profiting from the infinite sheeny well-being without any lordship	Brahma	Physical (the supreme creator deity who is reading the immanent positive script)	Zou Dang	Yin Fire Pig	2007

Chronological	*Samvat-sara*	Doctrine of Immanence	Signified-effect of the element of the Doctrine of Immanence	*Shashty amasha*	Signified Element of the Doctrine of Emanation	Star Deity	Celestial Deity	Year
23	*Virodhi*	*Charitrena yuktah tattva* (Dense: unblemished)	Become a three-dimensional well-researched entity, who has developed both aversions to the SHEENY well-being as well as charms for experiencing the anti-SHEENY universe beyond the lineage boundaries	*Vishnu*	Metaphysical (the para deity who is perpetuating the positive script)	Fu You	Yang Earth Rat	2008
24	*Vikruti*	*Krutajnah tattva* (Stressed: blessed)	Become a four-dimensional entity, who is afflicted with the anti-SHEENY state and is researching the dark path of the machiavellian kriyas	*Mahesh-var*	Dynamic (the primeval deity who is illuminating the energy essence of the positive script)	Wu Huan	Yin Earth Bison	2009
25	*Khara*	*Dharmajnah tattva* (Elastic: just)	Become an eight-dimensional researcher entity who has mastered the science of bickering, having experienced the infinite anti-SHEENY state; is not afraid of calling a spade a spade, and is speech shaming the universe that is ignorant of the truth of the spade within the perfected path of karma	*Deva*	Technological (the super deity who is guiding the normative development of the positive script as the ruling consciousness)	Fan Ning	Yang Air Tiger	2010
26	*Nandana*	*Drudharratah tattva* (Plastic: resolute)	Become a six-dimensional entity living a practicable life, together with a loyal life-partner and the children, serving those in the non-SHEENY state, and bringing dignity for the kith, the kin, and the lineage nation	*Ardra*	Organizational (the sentient consciousness of the traded positive script)	Peng Tai	Yin Air Rabbit	2011
27	*Vijaya*	*Viryavan tattva* (Inflammable: invincible)	Become a five-dimensional entity which practices the science of charming, trading, and winning over everyone for fulfilling private objectives, without the consciousness of the social costs	*Kali-naasha*	Systemic (the conscious exchange of the negative energy)	Xu Dan	Yang Water Dragon	2012
28	*Jaya*	*Vidvan tattva* (Tough: self-aware and non-malleable)	Become a seven-dimensional entity, who is proficient in the art of practicing the doctrinal discourse for ascending private compensatory benefits without the private worker costs	*Kshitish-vara*	Entity (the conscious servicing of the auspicious horizon)	Zhang Ci	Yin Water Serpent	2013

Chronological	Samvat-sara	Doctrine of Immanence	Signified-effect of the element of the Doctrine of Immanence	Shashty amasha	Signified Element of the Doctrine of Emanation	Star Deity	Celestial Deity	Year
29	Manmatha	Sarvabhuteshu hitah tattva (Massifying: servicing mass well-being)	Become a practitioner of the nine-dimensional law of attracting beauty with the beauty, perfecting the soft, sweet discerning path of jnana for infinite aesthetic benefits	Kamala-akara	Deity (the conscious emanating growth of the auspicious moment)	Yang Xian	Yang Ether Horse	2014
30	Durmukhi	Gunavaan tattva (Fluid: flowing energy)	Become a seven-dimensional entity, who is the absolute scientifiable standard for the negative vector transformation of the physical, intellectual, and mental bodies	Gulika	Para deity (the para-conscious immanent growth of the inauspicious moment)	Guan Zhong	Yin Ether Sheep	2015
31	Hemalambi	Satyavakhyah tattva (Truth mediator)	Become an eight-dimensional entity, who is proficient in the science of backbiting the natural endowments, for living a very private life of the opulent material wealth, without the radiant love	Mrityu	Param deity (the para-conscious entropy-effect of the inauspicious sequence on the present consciousness)	Tang Jie	Yang Fire Monkey	2016
32	Vilambi	Anasuyakah tattva (Self-teaming)	Become a seven-dimensional entity, who is proficient in scientifying self-interest through absolute narcissist self-love by dispassionately patronizing the national institutional leaders	Kaala	Primeval deity (the axiological consciousness of the consequential entropy in the present consciousness)	Jiang Wu	Yin Fire Rooster	2017
33	Vikari	Jitakrodah tattva (Self-sacrificing)	Become a scientist conditioned by a nine-dimensional belief that the old objective universe has outlived its value, and perfecting the lonely path of devotion for fashioning a new subjective universe with unique objects	Davaagni	Primordial deity (the meta-physical measurability of the positive sequential-effect)	Xie Tai	Yang Earth Dog	2018
34	Sharvari	Bibyati devah tattva (Self-motivated)	Become a three-dimensional philosophizable entity, spiritually living within the diverse objects without attachment to the diverse subjects	Ghora	Supreme deity (the absolute measure of the immanent dynamic entropy-effect)	Lu Bi	Yin Earth Pig	2019

Chronological	*Samvat-sara*	Doctrine of Immanence	Signified-effect of the element of the Doctrine of Immanence	*Shashty amasha*	Signified Element of the Doctrine of Emanation	Star Deity	Celestial Deity	Year
35	*Plava*	*Priyadarshana tattva* (Self-steaming)	Become a one-dimensional entity, guided solely by the philosophy of servicing and getting serviced personal psychological well-being	*Yama*	Supra deity (forming a normative scale for measuring the causal body of the sequential-effect and the etheric body of the entropy-effect)	Yang Xin	Yang Air Rat	2020
36	*Shubhakruth*	*Atmavan tattva* (Self-esteeming)	Become a ten-dimensional entity, philosophizing auspicious fortune as the cause of personal SHEENY well-being, and risking duping by the intimate kin	*Kantaka*	Super deity (a normative entity conscious of the positive energy sequence and its negative consequence)	He E	Yin Air Bison	2021
37	*Shobhakrit*	*Dyutiman tattva* (Self-worth glowing)	Become a philosopher composed by the ten-dimensional auspicious divine qualities, and perfecting the path of siddhi for realizing all-round success	*Sudhaa*	Spirit (the polarizable positive sequence, separate from its negative consequence)	Pi Shí	Yang Water Tiger	2022
38	*Krodhi*	*Pativrata tattva* (Self-destroying)	Become a six-dimensional educable entity, who has impulsive, temperamental addictions but is open to enlightenment within the radiant love of the intimate kin	*Amrit*	Human (the punctuated positive sequence as the gravitational nectar of the sentient life)	Li Cheng	Yin Water Rabbit	2023
39	*Vishvavasu*	*Sanatanam tattva* (Self-conscious agelessness factor)	Become two-dimensional educated entity, who has universal divine qualities, admires those with unique divine qualities and enjoys comic relief with those who don't	*Poorna Chandra*	Animal (the proliferating mass of the electromagnetic body within the positive gravitational nectar)	Wu Sui	Yang Ether Dragon	2024
40	*Parabhava*	*Dharjya tattva* (Sensory patience: manpower)	Become an eleven-dimensional entity, proficient in the science of educating others (telling others harshly to mind their becoming) and diffusing all the private benefits by doing the exact opposite	*Visha Dagdha*	Plant (a paternal poisoned with the normatively programmed ecosystem of the negative gravitational forces)	Wen Zhe	Yin Ether Serpent	2025

Chronological	Samvat-sara	Doctrine of Immanence	Signified-effect of the element of the Doctrine of Immanence	Shashty amasha	Signified Element of the Doctrine of Emanation	Star Deity	Celestial Deity	Year
41	Plavanga	Tapasya tattva (Social equanimity: mental power)	Become an educator conditioned by a three-dimensional fulfillment void and restlessly and dimwittedly perfecting the behavior-remediating svaras path of the kinship well-being for the secondary self-fulfillment, without the radiant love	Kula naasha	Mineral (the programmable poison entropy consequence within the patrilineage)	Miao Bing	Yang Fire Horse	2026
42	Kilaka	Nyaya tattva (Emotional equanimity: muscular power)	Become a zero-dimension neutral studiable entity, who has average divine qualities, is devoted to the oneness with the divine guider, and is accruing the ascending divine fortune-effect	Vamsha kshaya	Metal (reprogramming the national citizen consciousness to descend the negative international trading-effect)	Xu Hao	Yin Fire Sheep	2027
43	Saumya	Kshama tattva (Psychological freedom: material power)	Become a nine-dimensional entity, who has studied the science of blessing for fructifying the wishable SHEENY values of universal popularity	Utpaata	Material (reprogramming the international para-citizen consciousness to ascend the turbulent development of the positive human gravitational-effect)	Cheng Bao	Yang Earth Monkey	2028
44	Sadharana	Danasheel tattva (Psychic freedom: management power)	Become a six-dimensional entity studying through worldly sojourn for ascending a discriminating guider sense within	Kaala roopa	Kriya (a maternal who is the harbinger of death, within the local consciousness of the nature of human-effect as disproportionately positive)	Ni Bi	Yin Earth Rooster	2029
45	Virodhikrita	Dayabhava tattva (Para-psychic love, compassion, and social benefit: monetary power)	Become a student conditioned by the nine-dimensional entity reality for perfecting the value-remediating atodya path and ascending mental agitation, intellectual aggression, and physical action without a sensible sense of the metaphysical value conflicts	Saumya	Dharma (the performable behavior of a divine King who is following a life script planned by the paternal King of Kings and is genetically programmed by the maternal Queen of Life)	Ye Jian	Yang Air Dog	2030

Chronol ogical	Samvat- sara	Doctrine of Immanence	Signified-effect of the element of the Doctrine of Immanence	Shashty amasha	Signified Element of the Doctrine of Emanation	Star Deity	Celestial Deity	Year
46	Paridhavi	Bhakta Shurita tattva (Present incarnate)	Become a seven-dimensional investigable entity, who has average guider qualities characterized by harmonious as well as para-harmonious behaviors	Komala	Karma (the blessing performance of a guider angel who is subtly leading an alternative holistic life script for descending the negative consequence)	Qiu De	Yin Air Pig	2031
47	Pramadicha	Niraskata tattva (Potential incarnate)	Become a nine-dimensional entity, who has investigated the absolute GUIDER values to avariciously backbite ascending social benefits for the kith and breed ascending the worker costs for the others and the psychological costs for the kin	Sheetala	Jnana (the performative bickering of a SHEENY Lord of Angels for cooling down the intense heat of the genetically programmed psychological consciousness)	Zhu De	Yang Water Rat	2032
48	Ananda	Sangitajna tattva (Dynamism incarnate)	Become a one-dimensional entity, investigating the metaphysics of happiness without a sense of fulfillment with infinite objects and infinite, intimate well-wishing kin	Danshtra a Karaala	Bhakti (an entity dangerously managed by the sentimentally breeding emotions rather than strategically managing those emotions as an ascendant master)	Zhang Chao	Yin Water Bison	2033
49	Rakshasa	Neetibadi tattva (Technology incarnate)	Become an investigator conditioned by a one-dimensional entity for perfecting the value-transmediating gana path within a malefic attitude toward infinite objects and infinite, intimate well-wishing kin	Chandra Mukha	Siddhi (the seductive charm of the emotional body as the secondary residual cause for self-organizing the belief system that is guiding the entity as the Lord of Lords)	Wan Qing	Yang Ether Tiger	2034
50	Nala	Lila madhurya tattva (Organization incarnate)	Become a wisher free of the Saturn conditioning, within the liberal values for growing the social benefits with sentient water-effect and compensating the past life costs of the conservative values oriented toward the private benefits	Praveena	Maha Siddhi (the proficient networking of the belief system for self-managing the cost-effective path to the wishable social becoming)	Xin Ya	Yin Ether Rabbit	2035

Chronological	Samvat-sara	Doctrine of Immanence	Signified-effect of the element of the Doctrine of Immanence	Shashty amasha	Signified Element of the Doctrine of Emanation	Star Deity	Celestial Deity	Year
51	Pingala	Pratibhasampanna tattva (Ecosystem incarnate)	Become a wisher free of the Jupiter conditioning, without the liberal values for growing the social benefits and within an alternative 'redemptive penance' approach to compensate for the present social costs	Kaalaagni	Param siddha (the exchange of the formative time-conditioned growth with the concealed negative ecosystem energy)	Yang Yan	Yang Fire Dragon	2036
52	Kalayukthi	Tejsauktah tattva (Entity incarnate)	Become a wisher free of the Mars conditioning, within the conservative values, thereby descending the private costs by propagating a theory of the inevitable para-conscious entropy in the future social benefits (the S-curve theory)	Danda Ayudha	Siddha (the workforce of the time lord punishing the entity for the original sin of separating the negative karma effects within the 13th zeroth dimension of space and time)	Li Qing	Yin Fire Serpent	2037
53	Siddharthi	Saundarjyamaya tattva (Past incarnate)	Become a wisher free of the Venus conditioning, without the conservative values, thereby descending the private costs within an alternative ideal of selflessly serving the Godfather(s) and the Godhead kings to ascend private benefits	Nirmala	Saturn (Lord of the universal kal behavior chakra impeccably forming the negative karma-effects of the whole lineage as the self-identity sacral mooladhar chakra within each living entity)	Fu Dang	Yang Earth Horse	2038
54	Raudri	Sarvaniyanta tattva (Future incarnate)	Become a wisher free of the Mercury conditioning, within the entropy of the present private benefits without the discriminating sense of the social expression of the true self-identity	Shubha	Jupiter (the Lord of the eternal dharma value chakra reproducing the growth-effect of the positive dharma alternatively planned by the guider angel for descending the negative programmatic kriya within the swadhisthan reproduction chakra)	Mao Zi	Yin Earth Sheep	2039

Chronological	Samvat-sara	Doctrine of Immanence	Signified-effect of the element of the Doctrine of Immanence	Shashty amasha	Signified Element of the Doctrine of Emanation	Star Deity	Celestial Deity	Year
55	*Durmathi*	*Sarvagnata tattva* (Human incarnate)	Become a wisher free of the Neptune conditioning, without the entropy of present private compensatory benefits within the entropy of the social consciousness, thereby ascending the private compensatory costs (health as well as career entropy that descends futuristic private well-being)	*Ashubha*	Mars (the Lord of the pervasive *shristi* nature *chakra* consuming the negative self-steaming energy for consciously asserting the negative performing kriya as the wishable ambition within the *manipura* consumption *chakra*)	*Shi* *Zheng*	Yang Air Monkey	2040
56	*Dundubhi*	*Priyamvada tattva* (Spirit incarnate)	Become a wisher free of the Uranus conditioning, without the entropy of social consciousness for ascending the private compensatory benefits (the secondary citizenship benefits in the form of the material object gifts, that strengthen the togetherness benefits of kith and kin)	*Ati* *Sheetala*	Venus (the Lord of the creator *chakra* circulating the inert, frozen-in-time, intellectual jnana of the guider angel for para-consciously sustaining a harmonious correlation to sustain the longevity of the *anahata* heart *chakra*)	*Hong* *Chong*	Yin Air Rooster	2041
57	*Rudhirodgari*	*Swayamniyanta* (Deity incarnate)	Become a Wisher free of the Earth conditioning, within the growth of social consciousness that is sustaining the private worker benefits, thereby trading negative energy of the para-conscious self-afflicted life and conflicting relationships	*Sudhaa*	Mercury (the Lord of the creature atman *chakra* exchanging the wish of the emanating cause for breathing the true nectar of life and for an authentic expression of the self within the *visuddha* throat *chakra*)	*Yu* *Cheng*	Yang Water Dog	2042
58	*Raktakshi*	*Paripurna* (Androgynous incarnate)	Become an absolute wisher free of the Moon conditioning, within the trading of the positive primeval self-blessedness and fortune, thereby ascending the private worker benefits without SHEENY consciousness	*Payod-heesha*	Neptune (the Lord of the luminous point crown *chakra* servicing an infinite space consciousness wishing positive well-being vision within the *ajna* pineal consciousness *chakra*)	*Jin Bian*	Yin Water Pig	2043

Chronological	Samvat-sara	Doctrine of Immanence	Signified-effect of the element of the Doctrine of Immanence	Shashtyamasha	Signified Element of the Doctrine of Emanation	Star Deity	Celestial Deity	Year
59	Krodhana	Akrodha (Masculine incarnate)	Become an absolute knower free of Sun conditioning, within the param exchange of the sattva SHEENY life purpose and spiritual calling, thereby descending the private worker benefits and ascending the negative tamas SHEENY costs	Bhra-mana	Uranus (the Lord of the sentient life earthstar *chakra* nurturing wisher consciousness for universalizing the SHEENY growth as the cosmic time-effect within lunar *bindu* spirit *chakra*)	Chen Cai	Yang Ether Rat	2044
60	Akshaya	Mardavam (Feminine incarnate)	Become an absolute manifestor free of zodiac conditioning, without primordial servicing of the rajo guider power to remediate SHEENY behaviors, thereby ascending the DIVINE benefits of the freedom from the physical, intellectual, social, emotional, sentimental and causal bonds	Chandra Rekhaa	Earth (the Lord of the lifecycle soul star *chakra* activating absolute consciousness of the High-Self super-wisher for the unique intrinsic causation of the divine SHEENY well-being within the wholesomewhole gravitational orbit of the solar pituitary *chakra*)	Geng Zhang	Yin Ether Bison	2045

Table 11. The Four Pathways of Primordial Deity, Without the Primordial Perpetuator

Sequential primary path	Metal GUIDER-effect	Mineral GUIDER-effect	Plant GUIDER-effect	Consequential primary entity	Animal GUIDER-effect	Human GUIDER-effect	Planetary GUIDER-effect	Value of system-effect	Quality of perception-effect
Harsiddhi (*Amba*, Protogenos Hemera or Theotokos; Ninlil; Longmu): Primordial Wisher path (within normative programming); Path of Self-acceptance (*Pushti Dasa marga*, i.e., the path of devotee grace; the path of alienship authority)	Omni-potence growth: Power of the energy concentration, endowing an entity with the strength and the determination, within the descending blessing-effect of the Wisher (Knower, i.e., *Bhagvat tattva* dimension of the five-face *Sadakhya tattva*: channeling the natural healing power as a disconnected quaternary-effect, without any residual sentient energy of the Worker dimension)	Acoustic Structure: Geometric ionized energy flow without the crystalized unit cell, forming the oneness of fossilized organic molecules within an ordered arrangement of atoms	Rejuvenation response: Release of the physical disconnect with self	Physical body: Growth in physiological sensation (self-team), *Srishti Yogi* or Mentor, growing the energies within the physical body for the mental oneness and resilience, without the intellectual dualism and inertia	Life guide: Oneness of the life-effect, guiding the true self	Projector guide (<u>maha</u> <u>naadi</u>): Projector of sound-effect (*Maha Durga*, organic ether energy)	House of self-identity (northeast direction, the primary life-effect within the secondary northern space, and the primary eastern solar causal body lifetime)	−6/8 = −3/4; the 5/4 octave of the negative Creator 4 reality without the Perpetuator 5 reality, and the 1/2 octave of the illusionary Deity 1 without the Knower 2 reality, formed into the Ideal-effect: seeking the calling of the greeting opportunities	Bioacoustics energy without the whole

Sequential primary path	*Harbuddhi* (*Kamalatmika*; Protogonos Nyx; Ninmah; Meng Po): Primordial Knower path (without normative programming); Path of Self-discovery (*Suddha Dasa marga*, i.e., the path of devotee Passion or Love; the path of citizenship authority)
Metal GUIDER-effect	Omni-science growth: Energy of the crystalized togetherness of the organic molecular ideals, the inorganic imagination ions and the groups of mesonic entities within one perfect solid, with the descending becoming-effect of the Knower (Manifestor, i.e., *Sadhana tattva* dimension of the five-face *Sadakhya tattva*: attracting the social human-effect as the primary causal body)
Mineral GUIDER-effect	Nuclear Form: Naturally assembled inorganic layers of the varying opaqueness, mechanically crystallized into the unified whole gaseous solid
Plant GUIDER-effect	Recognition response: Realizing the witness conscious-ness of the shadow or hidden truth
Consequential primary entity	Intellectual body: Growth in the psychological insights (self-insights (self-steam); *Chikitsa Yogi* or Healer, balancing the energies within the physical, the intellectual and the mental bodies, without the astral body
Animal GUIDER-effect	Shadow Guide: Shadow of the message-effect, discovering the truth
Human GUIDER-effect	Reflector guide (*maha gandh*): Reflector of the earth-effect (*Maha Lakshmi*; organic earth energy)
Planetary GUIDER-effect	House of material growth (southeast direction, the quaternary life-effect without secondary the northern space, but within the primary eastern solar causal body lifetime)
Value of system-effect	$-6/8 = -3/4$; the 5/4 octave of the negative Creator 4 reality without the Perpetuator 5 reality, and the 1/2 octave of the illusionary Deity 1 without the northern Knower 2 reality, formed into the Theory-effect: experiencing the feel of the gifted opportunities
Quality of perception-effect	Bionuclear energy within the whole

Sequential primary path	Metal GUIDER-effect	Mineral GUIDER-effect	Plant GUIDER-effect	Consequential primary entity	Animal GUIDER-effect	Human GUIDER-effect	Planetary GUIDER-effect	Value of system-effect	Quality of perception-effect
Harriddhi (*Matangi*, Protogenos Physis; Ninti; Houtu): Primordial Manifestor path (within normative performing); Path of Self-awareness (*Maryada Dasa marga*, i.e., the path of devotee institutional authority)	Omni-presence growth: Effect of the infinite hardness, firmness, sharpness, and fundamental essential pristine purity, unity, and virtue integrity within the entity space and life (Creator, i.e., *Advaita tattva* dimension of the five-face *Sadakhya tattva*: repelling the ecological-effect as the motivating economic factor for life), manifested with the descending belief-effect of the manifestor	Chemical decomposition: Thermal conductivity effects of the primary dominant matter (forming chemical compounds with tertiary matter)	Regression response: Reconciling the spirit level programming that is resisting the exchange of the soul (destiny; primary causal body)	Mental body: Growth in para-conscious awareness (self-esteem); *Hatha yogi* or Dreamer – Manifesting the mudras or performative postures for the perfect spiritual unity of the "*ha*" yang masculine force and the "*tha*" yin feminine energy	Messenger guide: Messenger of the whispering-effect, communicating the truth	Generator guide (*maha sparshi*): Generator of the air-effect (*Maha Kaali*; organic air energy)	House of moving ambitions (northwest direction, the secondary life-effect within the northern space and without the primary eastern solar causal body lifetime)	–6/8 = –3/4 or the 5/4 octave of the negative Creator 4 reality without the Perpetuator 5 reality, and the 1/2 octave of the illusionary Deity 1 without the Knower 2 reality, formed into the Super Ideal-effect: enjoying the nature of the emanating opportunities	Biochemical energy without wholesome parts

Sequential primary path	Metal GUIDER-effect	Mineral GUIDER-effect	Plant GUIDER-effect	Consequential primary entity	Animal GUIDER-effect	Human GUIDER-effect	Planetary GUIDER-effect	Value of system-effect	Quality of perception-effect
Harkriya (Danu; Protogenos Tethys or Thesis; Nammu; Magu) Primordial Creator path (without normative performing); Path of self-reliance (*Pravaha Dasa marga*, i.e., the path of the river of worldly affairs; the path of devotee religious authority)	Omni-permeation growth: the intuitive sense of the perfect creation and of the energy balance within the extrinsic tertiary life (Perpetuator, i.e., *Abhidheya tattva* dimension of the five-face *Sadakhya tattva*: creating the wishable material wealth, life journey friends, loving children, and descending the breeding-effect of the creator)	Photonic color: Refractive prismatic effects of the intrinsic secondary matter (transforming the energy flow with the polluting color-effect)	Reaction response: Renewing soul with the constellation of the secondary limiting proliferating factors (trans-formative vector)	Astral body: Growth in psychic, punctuated, gravitational potential (self-worth); *Shakti yogi* or Mystic, creating the desired profitable causal experience of the effortless meditative oneness: the perpetuating everlasting immanent Dao	Journey guide: Togetherness of polarized journey-effect, authenticating truth	Maker guide (*maha rasi*): Maker of water-effect (*Maha Saraswati*, organic water energy)	House of maternal emotions (southwest direction, the tertiary life-effect without secondary northern space and primary eastern solar causal body lifetime)	−6/8 = −3/4; the 5/4 octave of the negative Creator 4 reality without the Perpetuator 5 reality, and the 1/2 octave of the illusionary Deity 1 without the Knower 2 reality, formed into the Super theory-effect: breathing the flow of the immanent opportunities	Biophotonic energy within wholesome parts

Note: The overall destroyer dimension value of the entity-effect (i.e., ether element: *shuddhi tattva*) mediated by the four dimensions of the five-face *Sadakhya tattva* with the *Dasa marga* = -6. The immanent unmediated illuminator dimension of the entity-effect is the SHEENY element (ojas tattva). The emanating moderated liberator dimension of the entity-effect is the guider element (*chitta tattva*).

Table 12. The Sixteen Pathways of Primordial Greeter, Within the Primordial Perpetuator

Sequential primary path	Sequential secondary effect	Sequential tertiary and quaternary effects	Sequential quinary and senary effects	Consequential primary entity	Consequential secondary and tertiary effects	Consequential quaternary effect	Consequential quinary and senary effects	Value of entity-effect	Quality of perception-effect
Abhava (void): Esoteric or personal authority path (without temporally variable self), characterized by the self-imposed value limitations	Formative capability path (*Kriya marga*, i.e., the path of rituals; *Satputra marga*, i.e., the path of filial piety or ideal son): Water (*apas*) tattva	*Samkhya yoga.* A Wisher, believing in the infinite gravitational power of an extrinsic entity (i.e., of the Almighty Creator of the Natural Order), awakening mind about the value of own sentient energy	Forming a fulfilling Knower by creating an infinity of wishables, within the radiating body, the seeking intellect, the compassionate mind, and the sentimental spirit	*Asura (Purusha* with egoity, i.e., ahamkara), who is out of rhythm (*Sura)* with the reproduction value of one's sentient energy, and consequently binding oneself to the secondary residual effects	Tantra: the mediating workforce	A dualism of the guider-effects within the self → *asata*, i.e., out of touch with the self-fulfilling reality / *sata = Kaivalya mukti*	Seek the whole infinity of the gravitational power, without the entropy of the mediating guider, thereby consuming the wholesome sentient life energy of self; Be **Bodhisattva** (awakening mind about the reality of self)	−1	Negative entity-effect

Sequential primary path	Secondary effect	Tertiary + quaternary	Quinary + Senary	Consequential primary entity	Secondary + tertiary cons.	Quaternary Consequence	Quinary and senary cons	Entity -effect	Perception
Abhinaya (forming belief; gesticulation): *Mahakriya* or *Siddhanta* or *Shastras*, i.e., textual / social authority path (without temporally variable cause), characterized by the self-imposed behavioral limitations	Formative growth path (*Dharma* or *San marga*, i.e., the path of righteousness): Air (vayu) tattva	*Buddhi* or *Ati yoga*: A wisher begging an infinity of entities, each emanating a unique gravitational power, for spontaneously fulfilling an infinity of freewill wishables, without consuming own sentient life energy	Fulfilling all the wishables by creating an infinity of the Knower fulfillers	*Sura*/ Wisher / *Rajah*/ *Kshatriya*/ *Indra*/ *Sita* (devil without self-consciousness, i.e., without Godhead conscience and oneness with the presiding birth planet, also known as ONE or DAO)	Mantra: the freedom workforce	A dualism of the guider-effects without self = *Dharana* or *Ati mukti*	Grow the whole infinity of the extrinsic sentient life energy, thereby accruing the wholesome gravitational power within the entropy of self; Be *Mahisha*	0	Neutral entity-effect

Sequential primary path	Secondary effect	Tertiary + quaternary	Quinary + Senary	Consequential primary entity	Secondary + tertiary cons.	Quaternary cons.	Quinary and senary cons	Entity -effect	Perception
Dharmi (nonming breeding): Action path (without temporally variable spirit), characterized by the behavioral freedom	Workculture path (*Karma marga*); Workculture (*nayaki*) tattva	*Kriya* (or *Anu*) yoga: An entity breeding an infinity of wishes, affirming its gravitational power to fulfill those wishes using freewill (*manas*) sentient energy	Fulfilling all wishes without knowing how	*Shudra, Divya Purusha, Maya; Nirrti; Vignesh; Naraki; Vedanta* (deity)	Divine: the conscious workforce	Oneness of sheeny-effects within self = Krama (or Pada) Mukti	Service the whole intrinsic sentient life energy for fulfilling wholesome wishes, thereby accruing wholesomewhole sentient energy + gravitational power without entropy; Be *Deva*	+1	Positive entity-effect
Vrtii (transforming behavior): Knowing path (temporally variable astral), characterized by the value freedom	Culture path (*Jnana marga*): Culture (*sadakhya*) tattva	*Raja* (or *Maha*) yoga: An entity bragging an infinity of divine entities working as High Self (*chitta*), and guiding self	Fulfilling all wishes by knowing how	*Brahman / Brahmin / Sudeva Brahma/ Jiva/ Jathara* (super deity)	Guider: the global exchange	Oneness of self + without sheeny = Nirvana Mukti	Exchange wholesome extrinsic sentient life energies for fulfilling the holistic wish, thereby accruing wholesomewhole sentient energy + gravitational power within entropy; Be *Guru Deva*	+2	Double positive entity-effect

Sequential primary path	Secondary effect	Tertiary + quaternary	Quinary + Senary	Consequential primary entity	Secondary + tertiary cons.	Quaternary cons.	Quinary and senary cons	Entity -effect	Perception
Bhakti (devotion): *Bhakti path* (without temporally variable mind) path, characterized by value mediation	Human path (*Bhakti marga*): Fire (*tejas*) *tattva*	*Dhyana yoga:* A devotee blessed by an infinity of guiders, who are working as the spiritual committee	Fulfilling all the wishes, through innovative linkages into knowing how	*Vaishya/ Devi/ Hanuman/ Maha* (para) *Brahma/* (supra deity)	Sheeny: the socialized networking	Oneness of Self + within sheeny well-being = *Videha mukti* or *Vimukti*	Trade wholesome extrinsic life energies for fulfilling wholesome intrinsic wishes, thereby accruing wholesome gravitational power without entropy; Be *Soham Deva*	+3	Triple positive entity-effect
Siddhi (divinity; radiation): Yoga path (without temporally variable intellect), characterized by the behavioral mediation (i.e., samadhi)	Trading path (*Saha marga*, i.e., the path of friendship): Divine (*bhava*) tattva	Maha Kriya yoga: the divine who has become the Devotee, and servicing self-fulfilling gravitational energy, as omnipotent behavior	Manages those who do not know how to fulfill their wishes, through creative linkages with one who knows how.	Absolute (*param*) *Brahma/ Bhagwan/ Manush/ Atman/* Soul/ Paternal *Jinn* (supreme deity)	Super divine: the zeroth consciousness workforce	Oneness of self + without guider = *Moksha mukti*	Invest wholesome intrinsic life energies for fulfilling the wholesome extrinsic wishes, thereby accruing wholesome gravitational power within entropy; Be Creator *Deva* (*sarga*)	+4	Quadruple positive entity-effect

Sequential primary path	Secondary effect	Tertiary + quaternary	Quinary + Senary	Consequential primary entity	Secondary + tertiary cons.	Quatern ary cons.	Quinary and senary cons	Entity -effect	Percepti on
Svaras (tone): *Maha siddhi* (without temporally variable physical creation) path, characterized by the behavioral remediation	Corporate path (*Vichara marga*, i.e., the path of discrimination): Corporate (*sva*) *tattva*	Maha Raja yoga: Self-attract the gravitational energy of the diverse divine entities, and become omniscient	Knows how to organize within self those who know, how to manage the fundamental elements for fulfilling own unique wishes	Absolute (*param*) *Vishnu*/ *Ishvar*/ God (para deity)	Super guider: the zeroth consciousness exchange	Oneness of self + within guider = *Sadyo mukti*	Accrue wholesome extrinsic life energies for self-fulfilling intrinsic wishes, thereby producing holistic gravitomagnetic power without entropy; Be perpetuator *Deva* (*sthiti*)	+5	Quintupl e positive entity- effect
Atodya (musicality): *Param siddha* path (without spatially variable metaphysical creature), characterized by the value remediation	Local path (*Atma- ananda marga*, i.e., the path of blissful self): Local (*payu*) *tattva*	*Maha Dhyana yoga*. Self repel the gravitational energy to the locally engaged human entities, as an omnipresent worker substituting the followers' work	Knows how to manage the fundamental elements for fulfilling universal wishes	Absolute (*param*) *Shankara*/ *Mahesh*/ *Maheshwar*/ *Uma* (primeval deity)	Super sheeny: the zeroth consciousness networking	Oneness of self + without divinity = *Jivan mukti*	Accrue wholesome intrinsic life energies for self-fulfilling the extrinsic wishes, thereby performing holistic gravitoelectric power within entropy; Be Destroyer *Deva* (*samhara*)	+6	Sextuple positive entity- effect

Sequential primary path	Secondary effect	Tertiary + quaternary	Quinary + Senary	Consequential primary entity	Secondary + tertiary cons.	Quaternary cons.	Quinary and senary cons	Entity -effect	Perception
Gana (devoted singer): *Siddha* or *Param atma* path (without spatially variable paternal creator), characterized by the value trans-mediation	National path (*Dakshina marga*, i.e., the path of right, anti-clockwise feminine): Flame (*Parshnisamasta*) *tattva*	*Karma yoga* (infinite *bhakti*): Exchange the sentient life energy with the wish-fulfilling gravitational energy	Knows how to fulfill the entire infinity of wishes through absolute entropy of the entire infinity of energy perpetuating the Wisher	Shiva/ Nataraja/ Parmeshwar/ Amurru, the Belu Sadi (literally, the infinite singularity: the creature, i.e., oneness within infinite power (param deity)	Zeroth luminosity divine: the perfect creature	Oneness of self + within divinity = *Ichha mukti*	Accrue wholesomewhole extrinsic life energies for self-fulfilling the intrinsic wishes, thus programming wholesomewhole sentient energy without entropy; Be Illuminator *Mahadeva* (*tirobhava*)	+7	Septuple positive entity-effect
Rangi (radiating color): *Maha* (i.e., Para) *Siddha* or Wisdom path (without spatially variable maternal devotion), characterized by the behavioral trans-mediation	International path (*Param Prapatti marga*, i.e., the path of absolute realization; *Sharanagati path*, i.e., the path of absolute surrender): *Para* (*Anm*) *tattva*	*Jnana yoga* (zeroth *bhakti*): Exchange the gravitational energy serviced by the Almighty Creator of the primordial creation, as the ideal wishable value	Knows how to fulfill the entire infinity of wishers, through absolute entropy of the entire infinity of energy diffused in wishing	Mother Nature/ Yahweh/ Kartikeya/ Abhijit/ Maha*shakti* Bhairavi/ Brahmi/ Satarupa/ Kundalini/ Vichara (primordial deity)	Zeroth luminosity guider: the perfect creation	Oneness of divinity-effects without self = *Param mukti*	Accrue wholesomewhole intrinsic life energies for self-fulfilling extrinsic wishes, thereby planning wholesomewhole sentient energy within entropy; Be Liberator or Blessing *Mahadevi* (*spanda*)	+8	Octuple positive entity-effect

Sequential primary path	Secondary effect	Tertiary + quaternary	Quinary + Senary	Consequential primary entity	Secondary + tertiary cons.	Quaternary cons.	Quinary and senary cons	Entity -effect	Perception
Param Samadhi (Heavenly): Exoteric or *Adharma path* (without spatially variable ancestral phenotype), characterized by the behavioral mutation through meditation	Zeroth Normative development path (*Charya marga*, i.e., the path of followership): Position or Up Quark or Togetherness (*Vaisnava* or Dependence *tattva*) dimension of the earth (*bhu*)	*Bhakti yoga* (zeroth karma, known as the octave of heaven): Exchange the descending sentient life energy with the physical entropy	Absolute entropy of the Wisher as a physical body, thereby polarizing and fulfilling the wishing of the intellect	*Tantrika Purusha* (Super Wisher): One who reconstructs the value of past lives without the guider-effect	Infinite luminosity divine (*bodhi chitta,* seeking the nirvana for all): the self-destroying entity	The omnipresence of divine-effect within the self = *Sarva mukti*	Destroy the wholesomewhole extrinsic (physical) life energy, thereby norming the absolute entropy of one's own intrinsic (cosmic and causal) self	$2^1 \times -1 = -2$	Zeroth entity SHEEN Y-effect
Samsara (Earthly): *Asiddha path* (without spatially variable zeroth genotype), characterized by the value mutation through mentation	Infinite Normative development path (*Dhyana marga*): Velocity or Down Quark or Otherness (*Acintya-bhedabheda* or Independence *tattva*) dimension of the earth (*bhu*)	*Adharma* or *Bhukti yoga* (infinite *karma,* i.e., the octave of hell; the cycle of painful rebirths): Exchange physical entropy with metaphysical entropy	Absolute entropy of the Wisher as a metaphysical body, thereby punctuating the intellect's wishing through infinite mutations	*Bahirmukha Purusha* (Supreme Wisher): One who transconstructs future life behavioral effects without the divine-effect	Infinite luminosity guider (*chitta Bodhi* or ontological consciousness, exchanging the epistemological *prajna* or insight into the axiological *sila* or morality)	The omnipresence of divine-effects without the self = *Salokya mukti*	Destroy wholesomewhole intrinsic (cosmic and self-luminous) life energy, thereby norming the absolute entropy of one's own extrinsic (physical) self	$2^2 \times -1 = -4$	Infinite entity SHEEN Y-effect

Sequential primary path	Secondary effect	Tertiary + quaternary	Quinary + Senary	Consequential primary entity	Secondary + tertiary cons.	Quaternary cons.	Quinary and senary cons	Entity -effect	Perception
Arya (Nobility): *Sattvika marga* (sincerity path) or *Preyas marga*, i.e., Sensory dissolution path (without spatially variable zeroth value), characterized by the value normalization	Proliferating normative development path (*Vama marga*: the path of clockwise, left, masculine): Acceleration or Strange Quark (*Varnadhva* or Asymmetry *tattva*) dimension of the earth (*bhu*) *tattva*	*Mukti Yoga* (infinite jnana, known as the octave of the true path values): exchange the metaphysical entropy with the dynamic entropy	Absolute entropy of the super Wisher as the guider entity, thereby proliferating the wishing of the intellect through the finite Wisher	*Akarta Purusha* (Primordial Wisher): One who transconstructs present life behavioral effects with the guider-effect	Infinite entropy guider (seeking the eternity for the self): the universe destroying self-steaming entity	Zeroness of the divinity-effects within self = *Sarsti mukti*	Destroy the universal (physical) energy, thereby norming one's own intrinsic (cosmic and fundamental) effect as the absolute para-wishable growth	$2\wedge3^{*} - 1 = -8$	Zeroth divine effect
Maya (Royalty): *Param Siddhi* or *Rajasika marga* (Sensory Resolution or vow path) or *Sreyas marga*, i.e., Psychic well-being path (without spatially variable infinite value), characterized by the behavior normalization	Punctuating normative development path (*Sadachara marga*, i.e., the path of virtue): Jerk or Charm Quark (*Nityanta* or Complexity tattva) dimension of the earth (bhu) tattva .	*Seva yoga* (zeroth jnana, known as the octave of the true entity behaviors): Exchange the dynamic entropy with the formative entropy	Absolute entropy of the supra wisher as the divine entity, thereby polluting the wishing of the intellect through the entropy wisher	*Vasanatma Purusha* (Supreme Wisher): One who transconstructs the present life value with the divine-effect	Infinite entropy divine (seeking eternity for the universe): the Selfless entity	Zeroness of the divinity effects without self = *Sarupya mukti*	Destroy one's own intrinsic (cosmic and tertiary) effect, thereby norming the universe with the secondary effects of the absolute primeval wish	−3	Infinite divine-effect

Sequential primary path	Secondary effect	Tertiary + quaternary	Quinary + Senary	Consequential primary entity	Secondary + tertiary cons.	Quaternary cons.	Quinary and senary cons	Entity -effect	Perception
Chhaya (Situational): *Tamasika marga* or Personification or Worship or Solution attachment-effect (*phal moha*-effect) path (without spatially constant behavioral infinity)	Polluting normative development path (*Pravrtti marga*, i.e., the path of materialism): Jounce or Bottom Quark (*Asrama* or Symmetry *tattva*) dimension of the earth (*bhu*) tattva	*Darshan yoga* (negative *jnana*, sentient essence of all linkages, known as the octave of the down entity behaviors): Exchange the formative entropy with the normative one	Absolute entropy of the supreme Wisher as an entity, thereby de-polluting the Wisher and freeing the entropy curse	*Bhakta Purusha* (Para Wisher): One who transconstructs the eternal life value without the secondary residual divine-effect	Infinite entropy sheeny (seeking divinity for the universe) – self-conscious guider entity	Zeroness of the guider effects within self = *Samipya mukti*	Destroy secondary fundamental residual energy effects in the universe, thereby norming each entity's own (cosmic and quaternary) primeval wishing as absolute	–5	Zeroth guider-effect
Tapas (Austerity): *Sadhana marga* or penance path (without spatially constant zeroth behavior)	Uni-polarizing transformative development path (*Nivrtti marga*, i.e., the path of renunciation): Pop or Top Quark (*Jiva* or Simplicity tattva) dimension of the earth (bhu) tattva	*Nir yoga* (positive *jnana* free of all linkages, known as the zeroth octave of the up entity value): exchange normative entropy with transformative oneness	Absolute entropy of the param wisher as the ideal, thereby de-polarizing the Wisher and freeing the punctuated proliferation	*Vairagya Purusha* (Param Wisher): One primary entity without trans-constructing value	Zeroth entropy (servicing sheeny, i.e., luminosity for the universe) – self-conscious sheeny entity	Zeroness of the guider effects without self = *Sayujya mukti*	Destroy primeval wishers, thereby transforming the primordial Wisher into the param absolute wisher within the self	–7	Infinite guider-effect

Table 13. The Four Pathways of Primordial-Primordial Creator, Without the Primordial Greeter

The greetable factor of the primordial-primordial creator	Emanating para greeter	Emanating-effect of param greeter	Immanent greeter	Immanent-effect of primeval greeter	Bred primordial greeter (*Chatur Maha Kala*)	Breeding-effect of the primordial deity: (*Chatur Maha Nitya*)	Breeder-primordial perpetuator (*Chatur Maha Swapna*)	Breeder-effect of the primeval illuminator (*Chatur Maha Vibhu – Maha Lakshmi*)
Perpetuator (*Dharma yogi*)	*Citra*	*Rupa*	*Manas puja: Yashoda Vatsala bhava:* WELLSPRING (exchange sameness with consciousness duality)	Oneness-effect (sameness)	*Anubhuti: anubhava-* Consequential reaction consciousness (without universal entity)	*Samanya* (within universal entity)	A cheerful girl child (*Camara: asurakumara*)	*Adi*
Illuminator (*Karma yogi*)	*Citra-kanaka*	*Rupa-nsika*	*Para puja: Arjuna Sakhya Bhava:* WHOLE (exchange wholeness without consciousness duality)	Wholeness-effect	*Smriti: vibhava-* Causation impact consciousness (without unique entity)	*Visesa* (within unique entity)	A well-wisher (*Vairochana: asurakumara*)	*Radha*
Devoted (*Jnana yogi*)	*Satera*	*Surupa*	*Primeval puja: Hanuman Dasya Bhava:* BE-LIVE-LIFE (wholesome network with squared consciousness intensity)	Wholesome ness-effect	*Pramana: abhava-* Becoming cause consciousness (without immanent entity)	*Samavaya* (within immanent entity)	A divine entity (*Dharana: nagakumara*)	*Padanja*
Devotee (*Bhakti yogi*)	*Sautra-mani*	*Rupa-kavati*	*Param puja: Krishna Madhura Bhava:* RADIANT LOVE (develop holistic wholesomeness with triple consciousness intensity)	Wholesome ness without effect (Wholesomew holeness)	*Rasa: bhava* – Self-luminous consciousness (without absolute entity)	*Rasi* (within absolute entity)	The self within a divine entity (*Bhutananda: nagakumara*)	*Rukmini*

Table 14. The Sixteen Pathways of Param Child, Within the Para Deity

Paradigm path	Behavior	Effect Power	Octave tone	Contingent Pathway	Effect-form	Dominating subjective identity	The Deciding protagonist (*Tenshi/ Dikkumari/* Sabaoth)
Present Paradigm	Upness (Togetherness-effect)	−2	"b" $2^{1/12}$	Karma-effect × Prakriti-effect + Jnana-effect × *Dharma*-effect	Social-effect	Reader	*Bhogankara*
New Paradigm	Downness (Otherness-effect)	−1	"B" $2^{2/12}$	Togetherness-effect × Togetherness-effect + Togetherness-effect × Togetherness-effect	Human-effect	Researcher	*Bhogavati*
Emerging Paradigm	Strangeness (Repulsion-effect)	0	"c" $2^{3/12}$	Togetherness-effect × Otherness-effect + Togetherness-effect × Otherness-effect	Ecological-effect	Practitioner	*Subhoga*
Responsible Paradigm	Charmness (Attraction-effect)	1	"C" $2^{4/12}$	Otherness-effect × Otherness-effect + Togetherness-effect × Repulsion-effect	Economic-effect	Scientist	*Bhogamalini*
Contingent Paradigm	Bottomness (Beauty-effect; Entropy-effect)	2	"d" $2^{5/12}$	Attraction-effect × Togetherness-effect + Repulsion-effect × Otherness-effect	National-effect	Philosopher	*Toyadhara*

Paradigm	Behavior	Effect	Octave	Contingent Pathway	Effect-form	Dominating identity	Deciding protagonist
Risk Management Paradigm	Topness (Truth-effect; Growth-effect)	3	"D" $2^{6/12}$	Entropy-effect × Entropy-effect + Otherness-effect × Attraction-effect	Psychological-effect	Educator	*Vicitra*
Cost-effective Paradigm	Present time (Negative-effect)	4	"E" $2^{7/12}$	Growth-effect × Attraction-effect + Attraction-effect × Attraction-effect	Determination-effect	Student	*Puspamala*
Formative growth Paradigm	Potential time (Positive-effect)	5	"f" $2^{8/12}$	Negative-effect × Entropy-effect + Growth-effect × Otherness-effect	Imagination-effect	Investigator	*Abhinandita*
Normative development paradigm	Potential space (Intrinsic-effect; Subjective-effect)	6	"F" $2^{9/12}$	Growth-effect × Entropy-effect + Attraction-effect × Togetherness-effect	Virtue-effect	Wisher	*Meghankara*
Transformative exchange paradigm	Present Cause (Subject-effect; Solar time)	7	"g" $2^{10/12}$	Negative-effect × Growth-effect + Positive-effect × Otherness-effect	Intuition-effect	Knower	*Meghavati*
Metaphysical entropy paradigm	Potential Cause (Object-effect; Lunar time)	8	"G" $2^{11/12}$	Subject-effect × Entropy-effect + Subjective-effect × Otherness-effect	Natural-effect	Manifestor	*Sumegha*

Paradigm	Behavior	Effect	Octave	Contingent Pathway	Effect-form	Dominating identity	Deciding protagonist
Dynamic fulfillment paradigm	Present space (Extrinsic-effect; Objective-effect)	9	"A" $2^{12/12}$	Entropy-effect × Subjective-effect + Growth-effect × Otherness-effect	Excellence-effect	Creator	*Meghamalini*
Technological oneness paradigm	Future space (Positive-effect × Entropy-effect)	10	"b" $2^{13/12}$	Negative-effect × Togetherness-effect + Entropy-effect × Otherness-effect	Global-effect	Seeker	*Toyadhara*
Organizational sameness paradigm	Present time (Potential time + Potential space)	11	"B" $2^{14/12}$	Growth-effect × Otherness-effect + Entropy-effect × Subject-effect	Unique-effect	Greeter	*Vicitra*
Ecosystem wholeness paradigm	Present moment (Entropy-effect × Subjective-effect)	12	"c" $2^{15/12}$	Objective-effect × Repulsion-effect + Entropy-effect × Subjective-effect	Inclusive-effect	Paternal	*Varisena*
Entity wholesomeness paradigm	Future moment (Engagement-effect × Responsibility)	13	"C" $2^{16/12}$	Objective-effect × Attraction-effect + Entropy-effect × Entropy-effect	Diversity-effect	Maternal	*Balahaka*

4.7. Objective for Planning the Para-Primordial Paradigm—Entity

A strong, silent, and wholesome paradigm of Mother Nature necessitates the energy's zero creation or zero destruction. It empowers the universe of creatures to become the infinite perpetuators of the primeval deity's energy. However, a weak, freedom paradigm of the universe of creatures, suffering from infinite bounds on their rationality, is subject to the masculine manipulation for furthering the SHEENY cause of others as a demonic angel as a path to trading sentient energy from the universe of devoted followers. Therefore, there is a further need to investigate the method of present reality for understanding how Mother Nature manages the cost of guider mediation of an ideal human entity. Toward this, it is fruitful to conduct a case study for developing an intuitive sense of the cost-escalating effects of human objectivity as a self-conscious entity.

A case study of how scientists manipulatively further the SHEENY cause

of the universe of human creatures by trading sentient energy from the universe of devoted followers, and the philosophical doctrines for an entity seeking to manage the transformative scientist-effect

One of the enigmatic grand challenges confronting the scientists is the sense of smell. How does a sentient entity smell something? Six para-primordial ideal media use diverse methods from the modern scientific world to test the theory of smell. Theoretically, the smell is an ideal case for sensing the sensitivity of the sentient human energy to the universe of angel-smelling devoted divine followers.

First, <u>shape is the ideal medium</u>. Conventional biological science says that an entity smells something using its luminous shape. "Shape" (*Rupa*, 100,000) is the secondary formative-effect of the present consciousness, which is a thermodynamic "fire element" (*Agni*, 17). A scientist uses the "wisdom" (*Chitta*, 100) to shape the present reality by servicing a unit of "sentient energy" (*Varuna*, 1,000). It is a way for a scientist to declare, "I AM the master (ideal potential self), take-it if you believe in science, leave-it if you don't." If one believes in science, then one is bonded to the sound of declaration and gets indoctrinated while referencing the scientist in all endeavors to perpetuate self-integrity as a sentient entity capable of strategically smelling the reality. The self thereby realizes the socially integrating "*shapamukti*" (literally, freedom from the curse of the escalating costs of personal sense-making in a world of infinite blind referees, -9/4).

Second, <u>sound is the ideal medium</u>. The radical chemical science says that an entity smells something using its self-luminous sound, i.e., "octave vibration" (*Sargam*, 60). "Sound" (*Naad*, 257) is the secondary metaphysical-effect of the infinite extroversion of the sentient energy, which produces a unit of "ether" (*Shuddhi*, 285). "Extroversion face" of the voice (*Bahirmukha*, -4) is the pathway to diffuse the sentient energy to form a universe of ethers that is shaped by one's "High Self" (*Chitta*, 100) as a wise "grandmother soul" (*Punyatma*, 100).

The ether element is the first step to becoming a "futuristic zodiac grandmother spirit" (*Kapinjala*, 20) with "self-awareness" (*Maryada*, 30) of the cause of the entropy of the sentient energy. It is a way for a radical to declare, "I AM the unique self, take-it if you are one too, leave-it if you aren't." If one does not believe in science, then one is socially bonded to the voice of radicalism and gets indoctrinated in falsifying the radical socialist using all the might for perpetuating the self-integrity as a sentient entity capable of tactically voicing the reality. The self thereby realizes the weirdly isolating "*vivabasambadhamukti*" (literally, freedom realized through a Herculean effort for divorcement from the grand solution of an institutionally-conforming sentient life, -9/4).

Third, the <u>voice is the ideal medium</u>. The occultic psychic science says an entity smells something using the causal body voice, i.e., gravity. "Voice" (*Atman*-effect: *Vak*, 629) is the secondary physical-effect of the introverted presence of a paternal "soul" (*Atman*, 4) within the self, which is also a unit of "gravity" (*Jamadagni*, 629) servicing "conditional destiny in the nick of time" (*Ayati vela*, 629). "Introversion face" of voice (*Antarmukha*, 10^{1024}) is the pathway to destroy the sentient energy and transform a universe of ethers into a "self-luminous sentient entity" (I: *Svayam*, 12). The one attuned to the paternal soul's inner voice is a "faceless self-luminous entity" (*Purusha*, 12). It generates the entropy of the sentient self by diligently persisting with the pre-programmed path of diffusion.

Tuning is a way for an occult to declare, "I AM the universal self, take-it if you are not, leave-it if you are too." If one does not believe in the occult (metaphysical) science, then one is spiritually bonded to the object of occultism and gets indoctrinated in the occult scientist's effect power. One does so without any effort to destroy self-integrity as a sentient entity capable of tactfully tuning the reality. The self thereby realizes the mysterious "*karavimukti*" (literally, effortless freedom from the sentient life's grand challenges, -9/4).

Fourth, <u>quantum object is the ideal medium</u>. The psychological black magic science says that an entity smells something using its physical, etheric body (inanimate object reality). Quantum composition of an "object" (*Padartha*, -3) is the primary physical-effect of the objective presence of the sentient element within the self and is the "universe of sentient reality" (*Vasanatma*, -3). "Objectivity" (*Vyavaharika*, 3064) of an occult object is the pathway for a subject to exchange sentient energy by deciding to become psychically bound to the occult object. A subject that services an object-tuning hypnotizing voice behaves like a "selfish wisher" (*Pravartayitr*, -10^{1024}), employing the finite science of hookery for unilateral soul servicing after failing with the art of crookery to instigate reciprocal soul exchange with the object of obsession.

Soul trading is a way for an occult object to declare, "I AM the devotee self, take-it if you are one too, leave-it if you are not." Suppose one is not a devotee of the personal SHEENY well-being. In that case, one can be divine-bonded to the subject of occultism, indoctrinated to the motivating power of the occult scientist. One does so without knowing the moment of destroying the self-integrity as a sentient entity capable of sensing the infinitely complex reality. The self thereby realizes the ignoble "*banamukti*" (literally, freedom from the sentient life through arrow-like polarization of the sentient energy into one risky basket, -9/4).

Fifth, the <u>entity is the ideal medium</u>. The physical, structural light science says that an entity smells something using its metaphysical astral body (animate subject reality). "Subject" (*Vidheya*, 697) is the primary metaphysical-effect of the subjective presence of a "sentient entity" (*Siddha*, 7) within the self. The "subjectivity" (*Pratibhasika*, 3,010) of the subject is a pathway to trade the sentient energy of an object astrologically bound to the subject. One bound to an emanating light of the wisdom conditioning becomes a stupid "layperson" (*Jadamati*, -10^{29}). It deploys the infinite science of ascension for ascending naturally with the flow of the cosmic universe after descending to the "bottomless pit" (Black Hole: *Vishnunabhi*, 82) through the blind followership of a supernatural force.

Soul ascension is the way for a subject to declare, "I AM the devoted self, take-it if you are not, leave-it if you are not." If one is not self-conscious, then one is gravitationally bound to the light-emitting, para-conscious astral body, getting indoctrinated in the black hole physicist's marketing power within a moment of uncertainty whether to take or leave the hypothetical proposition. The hypothetical proposition tests the integrity of the universe of the sentient entities, which alone is capable of destroying the self-radiating infinite complexity of "thought" (*Vichara*, 8) that generates the subjectivity. The infinite complexity of thought destroys the self-integrity as a sentient entity. The self thereby realizes the strategic "*amukti*" (literally, freedom from the sentient life through the self's objectification as an entity, -10).

Sixth, the <u>creator is the ideal medium</u>. The absolute self-conscious science of Tibetan Buddhism says that a "dumb-ass sentient entity" (*Gardhaba*, 1000) smells everything using its formative mental body (natural entity reality). "Nature" (*Anatanam*, 8) is the primary formative-effect of the presence of a "primordial-primordial creator" element (*Krishna*, 32) within the self. The "character" (*Prakriti*, 485) transformation of a "dumb-ass para entity" (*Ashtavakra*, 19) is a pathway to the primeval trading of the "transmigrating" (*Vijnana*, 47) sentient energy of the "animal kingdom" (*Tiryaggati*, 47) zodiac-bound to the self. One bound to an immanent, para-consciousness conditioning power of natural imagination becomes a "slave person" (*Dasa*, -10^{16}). As a leader on a natural discovery path, such a person applies the primordial science of natural entropy for descending to the "bottomless pit" (Black Hole; *Vishnunabhi*, 82).

The natural intrinsic entropy is the way for the subject to declare, "I AM the liberated self, take-it if you wish oneness with me, leave-it if you do not." Suppose one is enslaved to the lay consciousness. In that case, one is electromagnetically bound to the intrinsic force, motivating conformity to the ruling "religion" (*Dharma*, 370) for determining the potential worth of "*lohitamukti*" (literally, freedom from the empirically-evident phenomenological mirage, -10^{10}). Infinite simplicity of thought is the primordial cause for trading the primordial para entity's phenomenological integrity to transform oneself into an inanimate entity.

The sixth ideal is the para-primordial factor, which empowers an inanimate angel entity to determine the social benefit of smelling a polluting sentient entity by creating six supreme-primordial ideals using the diverse philosophical doctrines from ancient India.

First, <u>creature is the ideal medium</u>. The absolute science of Vedas says that an entity smells nothing through the normative mediation of its intellectual body (supernatural polluted masculine reality). Instead, the "geometry" (*Ganitam*, 8×10^{15}) of the transformative, curvilinear, energy flow smelled by an entity is the primary formative-effect of the presence of a "guider" (*Guru*, 100) without the self. The "trigonometry" (*Trikonamiti*, 10^{96}) of the triangulated energy flow, as a correlation with the feminine natural creation and the naturally-guiding masculine creator, is a pathway to the organizational planning of the sentient energy within the creature for ascending the SHEENY benefit of the para-conscious guider entity. Consequently, the creature believes that the "guider-effect" (*Guru*, 100) is the divine self-determination of the life objective.

By conceiving the astrological universe as the well-wishing para-conscious guider, the creature becomes a "theist" (*Bhakta*, -5), who is guided by the wishes of an entrepreneurial "holy spirit" (*Trinetra*, 1) guardian angel. It decides whether to fill in the consciousness void generated by the dynamic universe or not. Para-conscious self-determination is a way for the creature to self-declare, "I AM the devoted self, take-it if you wish oneness with me, leave-it if you do not." Suppose one is devoted to a "guiding force" (Presiding deity, 100). In that case, one is gravitationally bounded to the extrinsic force of the impulsive wishes of that guiding force for authenticating the absolute worth of the innovative "*shvayamukti*" (literally, self-steaming, swelled and illusionary, freedom, -10^{100}). Infinite triangulation of thought is the primeval cause of the depression of the animated, sentient energy within the self, infusing a selfless suicidal urge for overworking until the point of the absolute stress breakdown of the para-conscious mental body.

Second, <u>creation is the ideal medium</u>. The primordial science of the *Upnishads* says that an entity smells everything through the transformative "voluminous" medium of its physical body (supernatural polluted feminine reality). "Truth" (*Sathya*, 375) of the triangulated energy flow is the predominating cause for the presence of the "taste element" (*Rasa*, 269), diffusing an after-smell. "Numerology" (*Jyotish*, 10^{11}-1), the science of quantifying the sequential effects of the triangulated energy flow, is a pathway to the organizational programming of the gravitational energy of the creation within the creature for ascending the SHEENY benefit of the param conscious divine entity. The param conscious divine entity is neither the para-conscious guider creator nor the conscious, sentient creature. Instead, the divine entity is the fourth-dimension, tangentially added when the scientist binds the creature's intellectual rationality with the "transcendental wisdom" (*Para Vidya*, 10^{11}-1). The creature so bound becomes an "atheist" (*Pratyahara*, -12): a scientist's devotee as a unique wisher.

As an Almighty creator of the new paradigm, the scientist behaves like a "viral bug" (*Raktabeeja*, -12) for manufacturing an infinity of wish solutions to authenticate the self-worth. He pressures the creature to devote ascending energy for servicing the prime projected capitalization value of those solutions. As a management agent of the "undertaker" (*Yajman*, -10^{480}), the Almighty creator declares, "I AM thy devotee, take-it if you wish oneness with me, leave-it if you do not." Seeking to be a devotee of the "managing wisher" (Devil: *Sura*, 0), one becomes psychically bound to the tactical "*bhuktimukti*" (literally, freedom from enjoyment to work for the enjoyment of self as an enjoyer, -10^{100}). Finite projection of thought for the SHEENY well-being of a guider principal is the primeval-primeval cause of the distressed, sentient energy within the self, generating indecisiveness about the value of action behavior, given ascending conditional probability of the physical loss.

Third, the <u>organization is the ideal medium</u>. The primordial-primordial science of *Puranas* says that the smell of an entity decides the smell the entity smells. The "method" (*Vidhi*, 950) of projecting the four-dimensional subjective energy flow into the fifth, objective dimension of the four-dimensional object is the dominating cause of the presence of the "color element" (*Ranga*, 15), the subject of joyful living. A "doctrine" (*Siddhanta*, 375), i.e., the technique of qualifying the sequential effects of the four-dimensional energy organization, is a pathway to a thermodynamic (i.e., gravitoelectromagnetic) creation.   A thermodynamic creation is the vector-transformed form of the creature. It manifests the truth of the voluntary freedom from the creator dimension and the beautiful color of the perpetuating objective dimension. The objective dimension is the smell of the creature, which is the ideal smell conditioning the creature's discovery of the worth of the alternative smells.   The truth of the smell binds and makes a creature a "non-doer" (*Akarta*, -8) who can freely "demonstrate" its wishes to the universe of potential agents and demand all wishes to be fulfilled within a finite time as a demonstrating "primordial wisher" (*Pradarshan*, -8).

As a master of "critical science" (*Nindastuti*, -8), the creature becomes the deciding cause of ascending the SHEENY cost of reproducing the "devil" (*Sura*, 0). It forces the devil to continually change color like a "chameleon" (*Avajati*, -1/60) to compensate for the social universe's action. As a self-managing "undertaker" (*Yajman*, -10^{480}), the creature becomes an organization embodying the self's doctrinal truth. It declares, "I AM my devotee, take-it if you wish oneness with me, leave-it if you do not." Suppose one is seeking the universal SHEENY well-being to be the devil "king" (*Rajah*, 0) of the blind self-steaming "holy spirits" (*Trinetra*, 1). In that case, one becomes psychologically bound to the tact of "*unmukti*" (literally, deliverance by religiously "leading" the covenant of universal brotherhood, sisterhood, or kinship, until the point of degeneration, decay, and death, fighting for the noble cause: the doctrine of *vasudhaiva kutumbakam*, -10^{1000}). Absolute entropy of thought is the SHEENY cost of the guider mediation of the universal well-being without the consciousness of each demonstrator's potential to self-realize the primordial wish by pursuing an alternative path of the ideal-mediated karma behavior.

Fourth, <u>technology is the ideal medium</u>. The param-primordial science of *Samhita* says that an entity has the absolute freedom to diffuse the smell optimized for charm attracting the desired universe of citizens, hypnotically committed to the entity's seductive serpent power.   A fractured "mind" (*Manas*, 38), generating muscular pain, is a consequential quality of the five-dimensional inanimate "feminine energy" (*Achitta*, 398), which bunches together the ascending SHEENY costs of the pathway to the organizational profiting of the animated organizational heuristic. The unique SHEENY cost dimension is the "culturally-generated entropy" (*Sadakhya*-effect, 38) in the creature's sentient energy, as a consequence of the creature's wish for an ascending guider mediation of the universal SHEENY benefit. The escalating SHEENY costs generate a fracture in the mental body and diffuse the creature's sentient energy. The creature seeks "healing" (*Reiki*, 38) by resisting the diffusion, thereby concentrating the cost of the supernormal "intentionality" (*Abhipraya*, 38) within the "backbone" (Devil-effect: *Asura*, -1) to earn fame through the guider mediation with the muscular power. The concentrated cost manifests the backbone's fracturing as the ideal technological medium, as a natural path for breaking the circuit of the hypnotic culture-effect.

A creature that is bound by the hypnotic culture-effect becomes a "preacher" (*Pracharak*, -10^{256}) of the SHEENY value of the guider mediation as the technological solution for all social evils. The creature declares, "I AM the universal devotee, take-it if you are a part of the universe, leave-it if you are not." Consequently, the creature trades the animated, "masculine energy" (*Purani*, 91) from the entities wishing to enjoy the universal SHEENY benefits. As the "diurnal front-of-the-scenes entity" (*Daina*, 91), who is the master of the "science of bickering" (*Balkhiliya*, 91), the creature becomes the metaphysical cause of the ascending guider cost of "preaching" (*Prachar*, 15). As a cultural animal, the creature becomes bound to the track of "*prajnavimukti*" (literally, emancipation from personal intellect by religiously "following" the doctrine of emanation, i.e., *Vasudhaiva Kutumbakam siddhanta*, -10^{1000}). Absolute entropy of the "intelligence" (*Buddhi*, 48) is the cultural cost of the guider mediation of the universal well-being, impeding the mental willpower to sustain the path of the animated, "masculine energy" (*Purani shakti*, 91).

Fifth, <u>willpower is the ideal medium</u>. The primeval-primordial science of *Yoga Shastra* says that an entity has absolute freedom for strange repelling the desirable, ideal smell to manifest the theoretically-desired universal reality using the volitional willpower as the medium. To do so, the entity needs to discipline the mind through a sensible technique for realizing the intellectually-mediated "oneness" (*Yoga*, 48) between the "soul" (*Atman*, 4) and the "universe" (*Brahman*, 2). The universe is a two-dimensional "object of present reality" (*Jiva*, 2). As a "para deity" (*Param Vishnu*, 5), the entity creates both the soul and the universe by trading the "supernatural energy" (*Shram shakti*, 1) of the creature (2+4 = 5+1). Willful mental "agitation" (*Kriya*, 75) empowers the creature to realize a state of "oneness" (*Yoga*, 48) with the creator entity for descending the SHEENY cost of the guider mediation of the universal well-being.

"Willpower" (*Cetana shakti*, 75) is the volitional cause for the oneness of the energy value of the "agitation" (*Kriya shakti*, 75) and the "universal consciousness" (*Karta-dharta*, 75). The willpower is the culturally-generated growth that is mediated by the "memory" (*Smriti*, 35) of the "polluted, strangely repelled self" (*Payu*-effect, 35) who has forgotten that it is just a natural creation of the masculine creator. Without the energy of willpower, the agitation-effect has nothing but a "sour quality" (*Khatta*, -3 x 10^6). By servicing the willpower for agitating the equanimity of "life" (*Prabhasa*, 4), the creature transforms the sour "agitation-effect" (*Kriya-effect*, -3 x 10^6) into the "culture" (*Sadakhya*, 9) element. With the absolute entropy of the consciousness, the creature becomes a "mindless," "insane" entity (*Amanaska*, -9). Mindlessness shapes the creature into a "gambler" (*Pasaka*, -9), who has nothing to lose and is, therefore, "arrogant" (*Abhimanin*, -9) about the ideal God-self, working to shape the theoretical God-self, through the hypothetical, masculine agitation. The creature becomes "self-obsessed" (*Nastika*, -9) and "skeptical" (*Adharmi*, -9) about everything that is present without the self. He develops perfection in the science of argument and declares, "I AM nobody, take-it if you agree that the universe as a creation is essentially "nothingness" (*Shunyata*, -2), leave it if you disagree."

The creature, consequently, trades the finite five-dimensional inanimate, "feminine energy" (*Achitta*, 398) from the creator of the competing argument that the "paternal creator" (*Pitra*, 4) is "everythingness" (*Tathagata*, 4), since everything originates from the paternal creator. As a human, the creature becomes bound to the destiny of "*chetovimukti*" (literally, emancipation from the social mental consciousness through a secular, "entrepreneurial" mediation of the doctrine of emanation, i.e., *Vasudhaiva Kutumbakam siddhanta*, -10^{1000}). Absolute entropy of the "mental consciousness" (*Manas*, 38) is the cultural benefit of the guider mediation of the universal well-being that destroys the mental will to sustain the mindless devotional behavior

Sixth, <u>Mother Nature is the ideal medium</u>. The para-primordial science of *Jataka Sutra* says that an enjoyer entity enjoys the absolute freedom for enjoying the varying enjoyable smells of Mother Nature at the varying moments. The entity masters the science of astral projection to realize the freedom from the "water-mediated life consciousness phase" (*Apas*, 169). The water transforms the varying "natural, tender" (*Visada*, 270) smell into the constant corporal "supernatural" smell (*Hayagandha*, 270) of the entity. The entity trades the natural smell's multidimensional-effect for enjoying the varying "taste" (*Rasa*, 269) of Mother Nature.

An entity's varying consciousness state generates the varying forms of "sound" (*Naad*, 257) vibrations. By attuning to the untamed colorless "wild sound" (*Nishada*, 1) of the natural energy, immanent within all sound notes, the entity descends the "pungent-smelling" (*Katu*, 10) and intoxicating "intrinsic consciousness" (*Antarmana*, 10) of the colorful domesticated "bull" sound (*Rishabha*, 1) of the supernatural self. The entity tames the sound of the "devilish" (*Dhairata*, 0) masculine ego-consciousness within the "wild" sound (*Nishada*, 1) of the feminine nature. It ascends oneness with the "divine primordial sound" (*Gandhara*, 10), the "desired primordial smell" (*Ceshta* 10), the "dominant primordial touch" (*Snigdha*, 10), the "bitter primordial taste" (*Tittika*, 10), and the "formless white primordial color" (*Shvetah*, 10) of the "centering, androgynous, primordial illumination" (*Sushumna*, 10).

The creature, consequently, becomes a "spirit" (*Kapinjala*, 20) with a "desirable smell" (*Ishta*, 10), transcending the "self-perpetuating" (*Udvaha*, ½) moment. As a spirit, the creature declares, "I AM everything, take it if you agree that the soul as a creator is "everything" (*Sarvam*, -5), leave it if you disagree." The creature trades the immanent ten-dimensional "divine energy" (*Asrava shakti*, 10) from the "primordial illuminator" (*Shri Krishna*, 10) of the white truth. It realizes a state of "*rinamukti*" (literally, emancipation from the psychological astral consciousness of the indebtedness to the creator universe, through a democratic and "managerial" mediation of the doctrine of immanence, i.e., *Aham Brahmasmi*, -1024). Absolute projection of the "astral consciousness" (*Linga-sharira*, 3) is the work-culture cost of the divine moderation of the soul well-being for reproducing and perpetuating the path of divinity through the thermodynamic radiation.

The Thirty-Fold Effect of a Conceptual Idea on the Breathing System

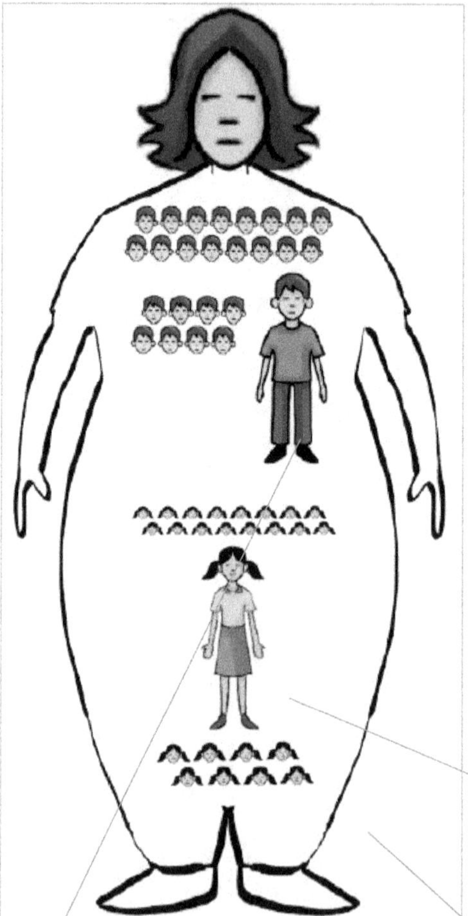

A son's Y chromosome has a potential to be a grandfather and produce eight fathers. A son's X chromosome has a potential to be a grandmother and produce eight mothers. Each Y chromsome turned son may give birth to eighteen entities.

A mother's X chromosome has a potential to be a daughter and produce sixteen granddaughters by reproducing a pair of X chromosomes. By potentiating the Y chromosome within her son, she may produce an octave of sons as well.

A paternal soul's Y chromosome has a potential to be a son, with a potential to produce sixteen grandsons with the Y chromosome potential of each reproduced X chromosome. By potentiating the grandmother's spirit-level X chromosome, he may produce an octave of daughters as well.

Since the grandmother's spirit and the paternal soul are immanent within the mother, a mother has a potential to give birth to fifty-two entities. Excluding the 19-unit son, 1 paternal soul, 1

Breathing System of an Immortal Entity

= Three-Fold Divine-effect of Behavior System * Five-Fold Gravitational-Effect of Belief System * Five-Fold Sentient-effect of One's Becoming System * Eight-fold Effect of a Theoretical Thesis on the Breeding System * The Thirty-Fold Effect of a Conceptual Idea on the Breathing System = Eighteen-Thousand-Fold Space for Incarnating the Children Within an Entity Seeking to Be Immortal

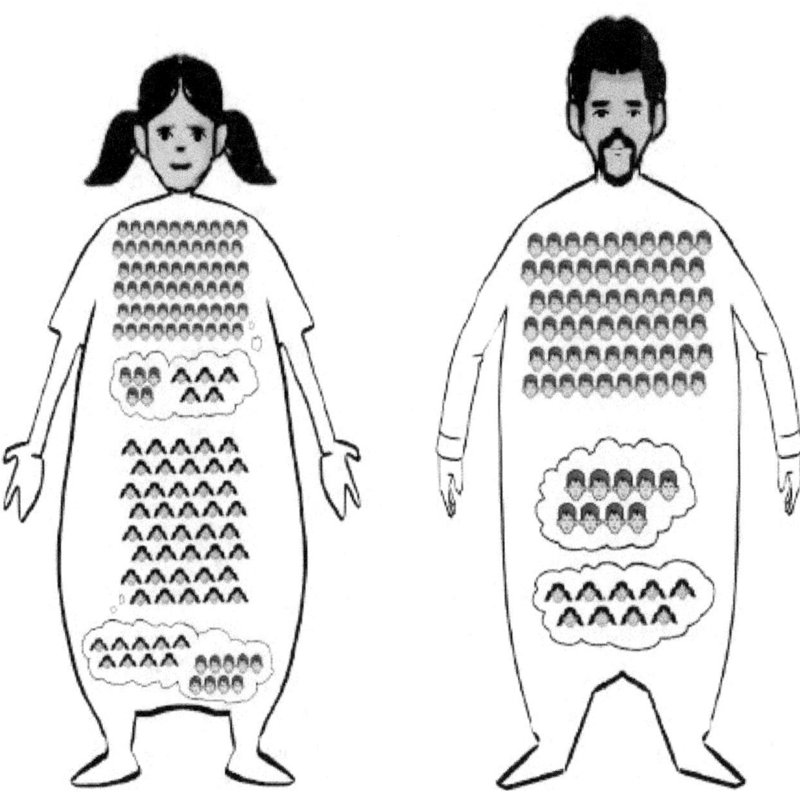

A mother's X chromosome has a potential to produce fifty daughters, without the grandfather and the paternal elements. A father's Y chromosome has a potential to produce fifty sons, without the grandmother and the maternal elements. A grandmother's X chromosome has a potential to produce eight mothers. A grandfather's Y chromosome has a potential to produce eight fathers. A grandson and a granddaughter may together produce an octave of children, half male and half female. The potential of all these entities is immanent within a daughter.

Without the grandmaternal and the maternal elements, a father's Y chromosome has a potential to produce fifty sons. Each son has a potential to produce one grandfather eight fathers, with his Y chromosome, and one grandmother and eight mothers, with his X chromosome.

An Entity Services Egoistic Masculine Gravitational Energy

for conceiving the undesirable virtual vision of the present reality and makes the Feminine Sentient Energy trading the desirable mission of his Divine Planning.

A planning entity becomes immortal by discovering the omnipresent reality and liberating the universe of devoted followers from his followership. The smell of the simple future reality is the eventual paradigm that immortalizes the planning entity. Let's look at the eventual paradigm plan when one takes and aggregates the diverse entity ideals, which constitute the omnipresent reality, to shape one's own ideal.

- **The Son ideal-taker**. My masculine chromosomes have the potential to incarnate one grandfather and eight father bodies. Therefore, my feminine chromosomes must also have the parallel potential to incarnate one grandmother and eight mother bodies. As one who is taking the simple idea of the parallel reality from my mother, who has given birth to a masculine-smelling son with her smell-less feminine-only chromosomes, I have a "Primordial Maternal" (*Amudhehswari*, 18) potential of the eighteen-units.

- **The Mother ideal-taker**. My maternal chromosomes endowed my ideal-taking son with an eighteen-unit primordial maternal potential. As an "ideal taker" (*Chhaya*, 52), I must have the power to endow my daughter with the same potential and have the same power within me. As an "Omnipresent Maternal" (*Chhaya*, 52), my potential is fifty-two (18 x 3 – 2, the son and the daughter included in my 18).

- **The Father ideal-taker**. My feminine chromosomes have a potential of fifty-two. Two of these can be my parents for reincarnating me. Without my parents, I may incarnate fifty entities. As a "Param Self-Luminous Paternal" (*Virupaksha*, 900), I may "incarnate" (*Virupaksha*, 900) each of those fifty entities as the ideal-taking sons and let them multiply my potential to the nine hundred units (50 x 18).

- **The Daughter ideal-taker**. My paternal chromosomes have a potential of nine hundred. My maternal chromosomes must have the same potential. As a "Luminous Param Child" (*Vicaksh*, 1800), who "reveals" (*Vicaksh*, 1800) the wholeness of the paternal and the maternal chromosome potential within my chromosome, my potential is eighteen hundred. Anybody who potentiates this potential within each of their 10-unit "divine energy" (*Asrava shakti*, 10), as a simple consciously divined future reality, becomes an "immortal entity" (*Maha Kaya*, 18,000) with the eighteen-thousand-unit potential. When one takes me as an ideal, I become the "tree of creation" (*Uttanapad*, 18,000).

5.1   The Subject Planning the Supreme-Primordial Paradigm—God

As an Almighty Creature, each entity has a metaphysical self-declared, "I AM God consciousness" (*Ekam Evadvitiyam Brahma mahavakyam*, -10^{30}). It energizes the creature to be the "divine perpetuator" (*Param Vishnu*, 5) of the energy created by the "primordial self" (*Parvati*, 10). Without the limits of the finite bounds on the sentient energy, the creature's freedom paradigm is the "absolute object" (Dark matter: *Sadashiva*, 1600) state of the param human child. It is the "absolute soul" (*Paramatma*, 1600). The "I AM God consciousness" of the primordial self transforms the creature's object state into one of the ten illuminated subject states, each with a varying sound, smell, touch, taste, color, and divine entity form. The "I AM God consciousness" generates an infinite, shadow subjective state, with a constant sound, smell, touch, taste, color, and a divine entity form. The "I AM God consciousness" is a transformative growth-effect of the "I AM a Soul consciousness" (*Esa Ta Atmantaryamyamrtah mahavakyam*, 100). The "I AM a Soul consciousness" is the metaphysical "open-system growth-effect" (*Citra*, 100) of the "closed-system entropy-effect" (*Brahmarandhra*, 1000).

5.2 Six Paradigms for Becoming an Immortal Entity

A creature, who conceives sentient energy as finite, seeks to be an almighty creature using the earthly smell of the six supreme-primordial doctrinal paradigms for becoming an "immortal entity" (*Mahakaya*, 18,000):

- **First, "I am the Almighty Creation doctrine"** (*Etad Vai Tat siddhanta*, -8000). An Almighty Creation believes the self to be the organizational sameness of the "universal Mother" (*Minerva*: Supreme Deity, 4) and takes the "smell" (*Gandha*, 268) of the physical body as the validation. To further authenticate the validity, it becomes a master of the science of *Akashic* Records for discovering the truth of one's thermodynamic fire-mediated "causative reality" (*Sabdartha*, 38). One's animated fire of the "thermodynamic breath" (*Shvasa*, 31) degenerates one's sentient reality into a fleeting, transient "ablative" (*Pancham*, 485) sound. It makes the earth's fickle "variability" (*Asthirta*, 485) the "character" (*Prakriti*, 485) of the creature.

  The creature trades the varying-effect of the "undesirable" (*Anishta*, 38) animated smell for experiencing the self's constant "smell" (*Gandha*, 268) of the inanimate "Mother Earth" (*Bhu*, 724). Consequently, its binds its destiny to that of Mother Earth. As a "destroyer of the creature's finite divine reality" (*Brahmani*, 80), Mother Earth exchanges her entire physical reality every eighty sidereal years. She services her gravitoelectric energy and trades the sentient energy emanating from the Vega white star. Therefore, the normative life of a creature on the earth is eighty sidereal years.

- **Second, "I am the Almighty Creator doctrine"** (*Sarvam Khalvidam Brahma siddhanta*, -10^6). An Almighty Creator believes the self to be the organizational child of the "universal Father" (*Ares*: Supra Deity, 3) and takes the "fragrant" (*Nirharin*, 18) smell of the intellectual body as the validation. To further authenticate the validity, it becomes the master of the science of *Svara* (i.e., the tonal quality of voice) for discovering the truth of the gravitational guider-mediated "sequential reality" (*Bhavartha*, 40) of the self. The spiritual sound of the gravitational "bicker" (*Visarga*, 270) quietly transforms one's sentient reality into a sound of the "spellbinding" utterance (*Vashikarna*, 38). It makes the expressive "radiance" (*Eroli*, 100,000) of the Sun the "undesirable" smell (*Anishta* 38) of the "meditating entity" (*Dhyani*,

38), who instead wishes to know the emanating cause of that immanent value.

The creature trades the "fire-effect" (*Rupa*, 100,000) for experiencing the "depth" of the "tonal color" (*Laksana*, 2) of the inner child spirit of the unique Father "Sun" (*Surya*, 21) within the self. The creature, consequently, binds its destiny to that of the Father Sun. The Father Sun exchanges his entire physical reality as an "omnipresent creator" (*Akhanda*, 8×10^{15}) every 8×10^{15} sidereal years by servicing the gravitomagnetic energy immanent within the self and trading the natural energy emanating from the Dark Matter. Therefore, the normative speed of light for a creature within the solar universe is 8×10^{15} meters/second. It is with the correction for the two-unit "tonal color" of the ascending divine-effect per ten-unit divine energy of the creature as the almighty creator of the supernatural reality, i.e., 9.6×10^{15} meters/ second.

- **Third, "I am the Creature doctrine"** (*Tat Tvam Asi siddhanta*, -100,000). A creature believes the self to be the organizational kith of the "universal Sister" (*Aphrodite*: Super Deity, 2) and takes the "austere" (*Ruksha*, 49) smell of the mental body as the validation. To further authenticate the validity, the creature becomes a master of the science of the aura reading for discovering the truth of the electromagnetic soul-mediated, "consequential reality" (*Laksyartha*, 296) of the self. The causal sound of the "creator blessing" (*ha-berakhah: Ashirwad*, 1000) illuminates the sentient reality of the creature with a sound of a cyclical, circular "inflection" of the meridian (*Madhyama*, 596) self. It makes the circular "circumference" (*Paridhi*, 64) of the unique Sister "Vega white star" (*Vega*, 67), the "complex" smell (*Samhata*, 3) of the creature. The creature contaminates the "meditating entity" (*Dhyani*, 38) with his undesirable "spiritual reality" (*Ratnakosha vadartha*, 3600).

The creature trades the "consequential filial piety texture" (*Laksya*, 485) of the "Sister Vega" (*Vega*, 67) within the self. Sister Vega services sentient energy for perpetuating the whole "circular reality" of the creature (*Prakarana vadartha*, 90,000) every 90,000 sidereal years and trades the thermodynamic energy emanating from the creature. The creature, consequently, binds the destiny of the self to that of Sister Vega. Therefore, the normative value of the "infinity" (*Ananta*, 90,000) within a creature's genome is the ninety thousand DNA strands. The normative value of the genes (nucleotides; SNPs) within a creature's phenome is the ninety thousand RNA strands or proteins. The normative value within an oöcyte is the ninety thousand mtDNA strands or mitochondria. The ninety thousand genes are the consequential father paternal programming of the circular sentient reality. The ninety thousand proteins are the causative masculine child planning of the circular sentient reality. The ninety thousand mitochondria are the sequential maternal performing forces, seeking to compensate for the "escalating degenerating cost" (*Arthi*, 173) of the "intense desire" (*Iccha Shakti*, 173), begging energetically for the undesirable circular sentient reality.

- **Fourth, "I am the Creation doctrine"** (*Ayam Atma Brahma siddhanta*, -10,000). A creation believes the self to be the organizational kin of the "universal Brother" (Holy Spirit: *Trinetra*, 1). It takes the "bloated" (*Ushna*, 383) touch of the astral body as the validation. To further authenticate the validity, the creation becomes a master of the "science of clairvoyant channeling" (*Avadhi*, -2), for discovering the truth of the gravitomagnetic spirit-mediated "eventual reality" (*Vakyartha*, 298) of the self. The causal, bloated touch, of the "begging creation" (*Ukalita*, 28) is the immanent "illuminating-effect" (*Narayana*, 28), experienced by a sentient creature, who is radiating a "symbolic" (*Sita*, 0) touch of the etheric body as a psychic medium. The skin inflammation is the tertiary residual of the sound mediated by a psychic. It is a function of the "disjunctive correlational presence" (*Samsarga*, 37) of the unique Brother, the "New Lemurian Black Hole" (*Chandra*, 82).

The creature trades the "eventual pitch" (*Shruti*, 71) of the "sentient entity" (*Siddha*, 7) with the freedom from the perpendicular, tangential "gravitational potential" (*Ardhajya*, 20) of the black hole. Brother New Lemurian services the gravitomagnetic energy to destroy the creature's "centering" (*Sushumna*, 10) every 2,600 sidereal years and trade the planet Earth's gravitoelectric energy. Consequently, the creature binds its destiny to that of the Brother New Lemurian. Therefore, the normative value of the sentient potential within a creature, as a cell, comprises the two-thousand-six-hundred RNA protein strands, DNA strands, and mtDNA strands.

The RNA protein strands are the amino acid sequences (cNLS: the classic nuclear localization signal peptide) for the nuclear transport of the masculine child creature's RNA-led centering planning. The RNA-led planning forms the creature's sequential reality. The physical body programs the sequence using amino acids. The RNA planning generates a deadly "cytokine storm" (*Anekikarana*, 155) by orienting the cell to pursue the path of a "conjunctive covarying presence" (*Samyoga*, 36) with the "starseed deity" (*Taraka*, 36). For instance, when a cell devotes its entire planning to manage the inanimate space phobia, seeking to expel the COVID-19 virus, the hyper-immune cell becomes the virus that destroys the physical body's normative programming.

The auto-immune cell, energized by the "cyclical dualistic heaven-hell starseed energy" (*Cyclopea*, 36), is the "causative factor" (*Prapaka*, 36) for the death of a cell after exploring all the 2,600 potential pathways for an alternative self-development. These pathways are the "absolute guider-effect" (*Kanya dharma*, 805) of the eighteen heavenly ascending energy realms, followed by the eight hellish descending energy realms. Consequently, the cell realizes oneness with the "primordial self" (*Parvati*, 10), without correlating with the ten hellish descending energy realms and develops the normative potential for reincarnation as a new chromosome.

- **Fifth, "I am the Creator doctrine"** (*Prajnanam Brahman siddhanta*, -1,000). A creator believes the self to be the "universal organizational system" (Devil: *Sura*, 0), taking the "symbolic" (*Sita*, 0) touch of the etheric body as the validation. To further authenticate the validity, the creator becomes a master of the "science of crystal ball telepathy" (*Manahparyaya*, -10^8), for discovering its truth of the gravitoelectric entity-mediated "symbolic reality" (*Vyanjanartha*, 297). The symbolic touch of the "bragging creator" (*Vikatth*, 486) is the "one identity symmetry" (*Chandra Yoga*, 486) of the primordial space (creation), and the primordial time (creator), within the primordial self (creature).

The *henna* art-like energy flow within the crystal ball is the quaternary-effect of a psychic medium's touch. It is a function of a codependent "conditional correlation" (*Sahacharya*, 40) between the wisher client and the working holy spirit of the self. The creature trades the "symbolic meter" of confidence (*Svara*, 28) of the "psychic medium" (Holy Spirit: *Trinetra*, 1) because of the "devotional intensity" (*Atri*, 274) of the latter to the austere "sentient cause" (*Ruksha*, 49) of the former. The psychic organizational medium services the gravitoelectric energy for repelling the "undesirable entity-effect" (*Utkramajya*, 38). Consequently, the "ecosystem cost"

(*Dhanu*, 26) of the wisher's consciousness ocean surfaces as the *henna* art-like "palmar flexion crease" (*Samudrika*, 39,753) energy flow system.

The creature trades the gravitoelectric energy for the "polarization" (*Shakti-Bheda*, 257) of the sound (*Naad*, 257) of the "infinite psychic linkages" (*Siddhidhatri*, 257), with the power to consciously determine the desired future once a day for a duration of the "twenty-four sidereal minutes" (*Ghatika*, 24). Over this duration, the Earth is in a state of absolute oneness with the primordial-primordial realm of the "param deity" (*Shiva*, 7). For the rest of the time, the creature binds its destiny to that of the unique organization, the "masculinized Saturn" (Satan; Masculine *Shani*, -1) and the metric of the present reality.

The normative gravitational potential within a creature as a chromosome is the "communion" (*Gotra*, 1964) with a strand of the twenty-three "chromosomes" (*Vyoma*, 285). Ten of these twenty-three chromosomes enjoy a primordial conditional correlation with the ten hellish descending energy realms. The other thirteen enjoy a primordial conditional correlation with the "thirteen primordial moons of the planet Saturn" (Ring Council: *Vamana*, 23 = 24 - 1). The twelve primordial moons of the planet Saturn enjoy a primeval conditional correlation with the twelve Zodiac dimensions. The thirteenth primordial moon of the planet Saturn enjoys an absolute conditional correlation with the astrological universe.

During the para-primordial oneness with the param deity every day, the creature first destroys the entity-effect mediated by the planet Saturn through the wheel of time. The creature perpetuates the unconditional alphabetically-ordered "lexicographic sequence" (*Viprayoga*, 39) for imprinting the ecosystem cost within the mitochondria genome in the form of the thirteen oxidative phosphorylation protein RNA planning sequences, the twenty-two transfer RNA programming sequences, and the two ribosomal RNA performing sequences. The protein RNA planning sequences manifest the varying divine oneness with the thirteen moons of Saturn.

The transfer RNA programming sequences manifest the varying guider mediating-effect of the eighteen energy realms, the nineteenth overall zodiac realm, the twentieth overall astrological realm, the twenty-first, primordial realm of the creators of the diverse realms, and the twenty-second, primordial-primordial realm of the creator of all the realms. The two ribosomal RNA performing sequences manifest the presence or the absence of the constant SHEENY moderating-effect of the creature, within param oneness with the "symbolic essence" (*Svara*, 28) of the creator immanent within the self.

The thirty-sixth "genetic sequence" (*Buddhi*, 48), within the 37-gene mitochondria genome norms the creation of the genome by trading the ascending energy of the thirteen moons within the descending "entity-effect" (*Utkarmajya*, 38) of the Saturn over the rest of the day. The thirty-seventh genetic sequence is the diffusion torque dimension. It generates a metaphysical entropy of the mitochondria by transforming the "causative diagonal-effect linkage" (*Samastjya*, 4,964) into the horizontal "perpetuating sentient linkage" (*Kramajya*, 1000), for the rest of the day within the self's "ascending guider linkage" (*Ardhajya*, 10). The causative diagonal-effect linkage is the "cytoplasmic" (*Prapancha*, 1964) "communion" (*Gotra*, 1964), with the one-thousand-unit sentient energy of the self as the planning creature, the programming creator, and the performing creation. The "consequential parallel linkage" (*Bhujajya*, 169) of the programming creator is the product of the thirteen-unit sequential linkage servicing the creation's "enduring strength" (*Pathitajya*, 959), and the creature's thirteen-unit metaphysical linkage servicing the "sequential sentient energy" (*Kramajya*, 1000).

The twenty-three chromosome DNA molecule and the thirty-seven gene mtDNA genome together norm the sixty realms' varying energies. The twenty amino-acid RNA protein forms the consequential parallel linkage of the creature as an almighty creator. The fourteen-antigen "cytoplasm" (*Prapancha*, 1964) array forms the "circumference" (*Paridhi*, 64) of the almighty creation. Overall, the cellular entity projects its "almighty creature effect" (*Kshetragata*, 379) by "prime-projecting" (*Karma chakra*, 7) the ninety-four elements (23 + 37 + 20 + 14) as the 94th primeval (i.e., prime) value, i.e., 491, into the four "life" (*Prabhasa*, 4) dimensions, to form, 1,964 units of the gravitational potential (*Gotra*, 1964). The four life dimensions are the creature's oneness with the creator and the creation, within the sentient, almighty creature self.

- **Sixth, "I am the entity doctrine"** (*So Ham siddhanta*, 4). A creature believes the self to be the universal "organizational metric" (*Maha Shunya*, -1) and takes the "conspicuous" (*Kathora*, -1) touch of the causal body as the validation. To further authenticate the validity, the entity becomes a master of the science of past life regression, for discovering the sentient truth of the mediation-free "factual reality" (*Yathartha*, 19). The conspicuous touch of the "breathing entity" (*Dham*, 19) is the paternal programmer of the multidimensional factual reality, of the "sky" (*Abha*, 9696), over the earth. The consciousness of a "self-ordered super-lexicographic sequence" (*Virodhita*, 108) of the past lives is the quinary-effect of the varying conspicuous touch impact of each life on the SHEENY well-being.

The predominating past life is one with the absolute impact on the factual reality of the present life. The dominating past life is one with the strongest, primeval impact. The deciding past life is one with the weakest, primordial impact. The "primordial impact" (*Prarabdha*, 596) is the sequential effect in motion beyond the present lifetime. The "primeval impact" (*Sanchita*, 470) is the convergent consequence of the infinite sequential effects from an infinity of lifetimes. The "param impact" (*Kriyamana*, 286) is the divergent cause of the present lifetime. Suppose the divergent cause is in absolute oneness with the sequential effects in motion. In that case, the entity experiences the "intense emotions" (*Shankara*, 264) of an infinity of divine-effects from an infinity of lifetimes. Else, the entity experiences the "intense ego" (*Shakti-Bheda*, 257), of an infinity of psychic linkages with the entities from an infinity of lifetimes, each seeking to blow the "sound" (*Naad*, 257) of its "co-varying presence" (*Samyoga*, 36).

The "convergent consequence" moderates the presence or the absence of the absolute oneness with the sequential effects in motion. If the convergent consequence is in absolute oneness, then the Eastern workculture-effect predominates in the present life. Else, the Western culture-effect is the predominating factor. The Eastern workculture-effect manifests an "ascending theory impact" (*Antardasha*, 374) of the Jupiter on the entity's "consequential, intuitive reality" (*Laksyartha*, 296). The Western culture-effect manifests an "ascending ideal impact" (*Chara Paryaya dasha*, 28) of the Saturn on the entity's life. Jupiter has a "localizing theory impact" (*Mahadasha*, 0) on the "factual growth reality" (*Yathartha*, 19), while the Saturn has a "globalizing, i.e., forward, ideal impact" (*Dasha*, 1) on the "causative workforce reality" (*Sabdartha*, 38).

The Jupiter promotes the solution discovery within the local networking system by locally converging the primeval corporate-effect. The Saturn promotes the solution discovery within the international exchange system, as the entity's primeval corporate-effect diverges from the local nation. The 'converging entity-effect' implies that the predominating proportion of the cast of characters essential for manifesting the sequential impacts is present within the local nation. It offers the freedom for realizing the "eventual divined, temporal reality" (*Vakyartha*, 298) by "descending the assertion impact" (*Pratyantara dasha*, 9) of the Mars on one's life. The entity must first descend the globalizing impact of the Saturn before working towards descending the impact of Mars.

An entity may descend the Saturn's globalizing impact by localizing the globally dispersed cast of characters. For instance, it may form the geographically dispersed social networks. Or it may psychically attract a dispersed cast of the characters to be the naturalized citizens of the local nation. An entity may promote the national "fame" (*Kirtti*, 37) at the international level and leveraging national fame for the desired social networking. An entity may descend the Mars' institutional impact by naturalizing the national workforce system guided by the reciprocal exchange relationships.

If the entity trades the national fame for generating the institutional rents on the national workforce system, then it psychically repels the local cast of characters from its social network. It has to work with the Jupiter's substantial localizing impact and the Mars' institutionalizing impact on future lives. With the strengthening of the "conjunctive presence" (*Samyoga*, 36) of the Jupiter and the Mars, the "disjunctive presence" (*Samsarga*, 37) of the Venus weakens. The "autonomous sequential reality" (*Bhavartha*; *Nivrtti dharma*, 40) of the "one-sided toxic natural harmony orientation" (*Vimshottari dasha*, -1) also weakens, without the proportionate social reciprocation.

Consequently, the "geocentric lexicographic presence" (*Viprayoga*, 39) of the globally sentimental "Mercury-effect" (*Shodashottari dasha*, 40) weakens. The "self-ordered presence" (*Virodhita*, 108) of the purposeful "Neptune-effect" (*Shastihayani dasha*, 107) as the "unique, ethnocentric dimension" (Samanya *dharma*, 108) strengthens. The "supernatural conditional correlational presence" (*Sahacharya*, 40) of the inclusive

"Uranus-effect" (*Dvisaptati dasha*, 29) descends. The "coercive conditional covariance presence" (*Apadesha*, 168) of the "temperamentally diversity-effect" (*Shattrimshat dasha*, 109) of the planet Earth also descends. The entity gets liberated from the Moon-guided "path of journey movement" (*Ashtottari dasha*, 30) as a fact of life. The entity opens the Solar "path of proportionate, i.e., sentient constancy" (*Panchottari dasha*, 39) as the "global factual dimension" (*Sanatana dharma*, 39) of mastery over personal life.

With an "absolute consciousness" (*Grand-I*, 17), the entity develops a "mysterious sainthood awareness" (*Chaturashiti dasha*, 189) of the "param Vega-effect" (*Chuturashiti dasha*, 189) in the form of the sentient element (*Ojas tattva*, 189). With the strategic awareness of the truth of the "field value" (*Kshetra*, 189) of the sentient element, the entity enjoys the freedom at par with that of the "absolute creature" (*Prabhu*, 1600) from the "psychological attachment to the present factual dimension" (*Chakra dasha*, 38) of the field value. With a clarified consciousness of the "purpose of life" (*Abhipraya*, 38), without the limitations of the "param dark matter-effect" (*Vashikarana*, 38), one unlocks the full might of the human entity. One is "virtuously liberated, for experiencing an infinity of field dimensions" (*Kalachakra dasha*, 68), without the "tribulation element" serviced by the "param black hole-effect" (*Pariksha tattva*, 68).

An entity trades a "conspicuous melody" of the empathy (*Raga*, 250) attachment to one of the twenty-seven varying "factual dimensions of the lunar, i.e., entity time" (*Nakshatra*, 185). It experiences the "emotional intensity" (*Shankara*, 264) of an "infinity" (*Ananta*, 90,000) of the sequential moving energies from ninety-thousand lifetimes. The para-psychic primordial entity services the sentient energy to generate the consciousness of the "competing, sacramental reality" (*Pratijnavad artha*, 0), i.e., the "ruling wisdom" (*Chochmah*, 0) of the self as a one-faced "all-taking arrogant king" (*Kshatriya*, 0). It empowers the entity to fight against the escalating cost of the "compensating, natural reality" (*Vdyartha*, 206), of a two-faced "all-giving compassionate queen" (*Sadojathi; Daan dharma*, 206).

Consequently, the "entropy linkage" (*Jya*, 974) of the intrinsic linear "human-effect" (*Linga*, 53) of the self as a "primeval masculine" (*Asura*, -1), wishing for the disproportionate SHEENY benefits, transcending beyond the earned workculture-effect, and fighting the multidimensional reality, becomes evident. The entity ascends the consciousness of the immortal extrinsic non-linear "trading-effect" (*Chiranjivi*, 26) of the "zodiac geometrical self" (*Ganitam*, 8 x 10^{15}) that is generating the disproportionate SHEENY costs of the inherited culture-effect.

The multidimensional reality of the "imaginative self" (*Moksha dharma*, 2) is more than the twenty-eighth "sequential reality" dimension (*Nivrtti dharma*, 28) of the infinite sequential moving energies. The latter is the localizing "param oneness" (*Payu*, 366) of the param deity's primordial-primordial realm. The former is the globalizing "primordial oneness" (*Stree dharma*, 32) of the primeval realm of the twelve zodiac entity group, the twelve astrological entity group, the one astral (Stellar; *Yati dharma*, 29: growth through the zodiac-effect x the astrological-effect) entity group, the one self-luminous (Lunar; *Sankhya dharma*, 30: entropy without the zodiac-effect x he astrological-effect) entity group, and the one primeval (Sentient; *Beeja dharma*, 31: technologically varying primordial-primordial oneness between the zodiac-effect and the astrological-effect).

The normative value of the "entity dimension" (*Praktriti dharma*, 27) is immanent within the twenty-seven "microtubules" (*Dasha*, 1). These microtubules form the "cytoskeleton" (*Nirjara*, 1000) for giving the cellular shape to the cytoplasm. They take the energy from the twenty-eighth dimension—the "cellular membrane wall" (*Shudra*, 1). The proficient cellular exchange system of trading the infinite energy from the twenty-eighth dimension generates a "smelling" (*Agandha*, -2), "mortal reality" (*Vyartha*, -12). The proficient cytoskeleton workforce system, servicing a finite energy to the twenty-eighth dimension, generates the "extrinsic smell" (*Yojana gandha*, 386) of the microtubules.

The microtubules carry a degenerative tension of the consequential "entropy reality" (*Mahartha*, 20). The proficient cytoplasmic networking system exchanges the immanent param deity energy to create a twelve-dimensional self-luminous cellular membrane and generates an "intrinsic smell" (*Svadu gandha*, 39). The twelve self-luminous dimensions are the twelve "plasma-spanning segments" of the cytoplasm that form the "unique reality" (*Ekartha*, 21) of the cellular membrane wall.

The "primordial dimension" (*Adharma*, -10) of the fourteen ascending "luminous" paths (Solar time, 13) manifests the multidimensional "sentient" reality of a "cellular" prokaryote entity (*Hiranyagarbha*, 19). Table 15 summarizes these. The primordial dimension generates the twelve sequential paths for ascending the twelve zodiac-effects (and the ascending twelve covarying astrological-effects) by ascending the Uranus-effect (which norms the incremental growth in the solar system) as the twelfth sequentially developed pathway. The thirteenth, consequential path destroys the cost-escalating zodiac system-effect (i.e., trading-effect). The fourteenth, causative path illuminates the astrological entity-effect (i.e., human-effect) as the primordial dimension.

The "primeval dimension" (*Yuddha dharma, 103*) of the fourteen ascending "self-luminous paths" (*Purusha,* 12) manifests the multidimensional "zodiac" reality of a "self-luminous" eukaryote entity. Table 16 summarizes these. The primeval dimension is the consequence of the twelve sequential paths for the descending twelve astrological-effects (and the descending twelve covarying zodiac-effects), by descending the Neptune-effect (which norms the incremental entropy in the solar system) as the twelfth sequentially developed pathway. The thirteenth causative path destroys the cost-escalating astrological entity-effect (i.e., human-effect). The fourteenth consequential path liberates the param deity-effect (i.e., workculture-effect) as the primeval dimension.

The "param dimension" (*Yuga Dharma,* 366) of the fourteen ascending "self-conscious paths" (*Prajna,* 2,222) manifests the multidimensional "astrological" reality of the "algebraic" human entity. It is free from the zodiac-effect of the "geometrical" animal entity. Table 17 summarizes these paths. Twelve sequentially developed paths perpetuate the twelve horizontal self-conscious zodiac paths of a self-luminous entity by trading the Vega-effect (which perpetuates growth in the solar system). The thirteenth sequential path destroys the cost-escalating zodiac-effects of the prior causative paths. By limiting the sentient reality's self-consciousness, the fourteenth sequential path illuminates the sentient creator of the animate astrological-effect as the consequential factor.

Table 14. The Fourteen Pathways of the Luminous, Without the Self-luminous Entity

Lunar zodiac	GUIDER kingdom	SHEENY-effect*	Symbol	Planet Lord	Astral Lord**	Signifies	Deity Lord ^	Demon mediator &	Yoga	Signifies	Solar Mansion %	Animal mansion	Pathway (dasha)
Descending Scorpio/ Earth	Supra deity	*Ashwini* - Healer: duality without self-luminous	Two Horses' heads	Neptune (*ketu*)	Beta Arietis	Star of transport, dynamic swiftness, and consciousness, without mass/universe of earth-effect	*Ashwini kumaras*	Azariel	*Dhr iti*	Determinat ion (consuming extrinsic energy)	虛 (Xu): Emptiness	Male horse	*Shatabdika*
Descending Aquarius/ White Star	Super deity	*Bharani* — Bearer: oneness within self-luminous	Femininity (yoni)	Venus	35 Arietis	Star of restraint, metaphysical harnessing (reign), connecting hellish creation mass/universe with the heavenly creative creature consciousness, without the para-conscious air-effect of a creator	*Yama*	Alhoniel	*Sool a*	Imaginatio n (diffusing intrinsic energy)	危 (Wei): Rooftop	Male elephant	*Chaturashiti*
Ascending Aries/ Saturn	Spirit	*Krttika* — Luminous: Nurturer	Knife	Sun	17 Tauri	Star of fire, physical purification, precision, perfection and balance of creature consciousness within para-conscious creator mass/universe, without the thermodynamic contaminated negative water-effect of the creation	*Agni*	Barbiel	*Gan da*	Virtue (forming vector quality)	室 (Shi): Encamp-ment	Female goat	*Dwisaptati*

*: Descending Lunar Mansion (nakshatra); **: Astronomical star; ^: Horizontal Earth Mansion; &: Forward demonic angel mediator; %: East-North Ascending Solar Mansion (xiu)

Lunar zodiac	GUIDER kingdom	SHEENY-effect	Symbol	Planet Lord	Astral Lord	Signifies	Deity Lord	Demon mediator	Yoga	Signifies	Solar Mansion	Animal mansion	Pathway (dasha)
Descending Hell/ Black Hole	Plant	*Rohini* – Radiant Love: Without Luminous	Wagon	Moon	Epsilon Tauri	Star of ascent, growth, fertility, radiance, intellectual brilliance, and creation within the creature consciousness, without the negative thermodynamic fire-effect of the para-conscious creator mass/universe	*Brahma*	Ataliel	*Vriddha*	Intuition (variable growth consciousness)	壁 (Bi): Wall	Male serpent	*Shodashottari*
Ascending Hell/ Black Hole	Plant	*Mrigshirsha* – Grace: Within Luminous	Dear's head	Mars	Lambda Orionis	Star of followership, social wandering, fulfillment, potentiation, immortality, eternity, siddhi, and plant; Creation lordship of the para-conscious creator universe, without the divine-effect of the creature consciousness	*Chandra*	Aziel	*Dhruva*	Natural (constant para consciousness)	奎 (Kui): Legs	Female serpent	*Shattrimshat*
Ascending Leo/ Mercury	Param perpetuator	*Ardra* – Greeter: sentient energy	Teardrop	Uranus (*rahu*)	Zeta Orionis	Star of entrepreneurship, creative destruction, structured effort, emotional intensity, and compassion of the creature for his conscious creation, without the guider-effect of the para-conscious creator universe	*Rudra*	Alhoniel	*Vyagatha*	Excellence (impacting normative development)	婁 (Lou): Bond	Female dog	*Vimshottari*

Lunar zodiac	GUIDER kingdom	SHEENY-effect	Symbol	Planet Lord	Astral Lord	Signifies	Deity Lord	Demon mediator	Yoga	Signifies	Solar Mansion	Animal mansion	Pathway (dasha)
Descending Gemini/ Mars	Supreme deity	Punarvasu – Self-luminous: Restorer	Quiver of arrows	Jupiter	Mu Gemin-orium	Star of renewal, maternal sentiments, & substantiation, experienced by the creation within the divine-effect of the para-conscious creator universe, without the conscious creature's SHEENY-effect	Aditya	Nociel	Harshana	Within Divine (restoring equanimity)	胃 (Wei): Stomach	Male cat	Shastihayani
Descending Aries/ Saturn	Spirit	Pushya – Wisher: Nourisher	Wheel	Saturn	Theta Cancri	Star of paternalism, dharma, orthodoxy, religion, I AM consciousness, and correlated shadow causation enjoyed by a creature within the thermodynamic fire-effect of the virgin creation, without the para-conscious creator universe's oneness-effect	Brihaspati	Scheliel	Vajra	Without Divine (disequilibrium pressure)	昴 (Mao): Hairy Head	Male goat	Chakra
Ascending Gemini/ Mars	Supreme deity	Ashlesha – Knower: intimate knowing	Coiled serpent	Mercury	Delta Hydrae	Star of clinging, attachment, contamination, negativity and unconditional luminosity sought by the creature within the contaminated water-effect of the virgin creation, without the para-conscious creator universe's wholeness-effect	Nagas	Azeruel	Siddhi	Divine (holistic self-fulfillment)	畢 (Bi): Net	Female cat	Lagna Kendradi

Lunar zodiac	GUIDER kingdom	SHEENY-effect	Symbol	Planet Lord	Astral Lord	Signifies	Deity Lord	Demon mediator	Yoga	Signifies	Solar Mansion	Animal mansion	Pathway (dasha)
Ascending Pisces/ Dark matter	Animal	*Magha* — Manifestor: magnificent	Royal throne	Neptune (*ketu*)	Alpha Hydrae	Star of royalty, management power, ancestor orientation, past orientation, heritage orientation of the investigating creature, within the proliferating air-effect of the polluted creation, without the wholesome-effect of the para-conscious creator universe	*Pitras*	Ardesiel	*Vya tapa ia*	Global (ecosystem disequilibra ting otherness chaos)	觜 (Zi): Turtle beak	Female rat	*Yogardha*
Descending Pisces/ Dark matter	Animal	*Purva Phalguni* — Creator: Without Radiant love	Hammock	Venus	Upsilon[1] Hydrae	Star of procreation, groupism, passion, energy radiation, engrossment of the studying and observing creature, within the earth-effect of the polluted creation, without the para-conscious creator universe's wholesomewhole-effect	*Bhaga*	Dirachiel	*Var iyan*	Unique (ego normalizati on)	参 (Shen): Three stars	Male rat	*Karaka Graha*
Descending Heaven/ Sentient energy system	Mineral	*Uttara Phalguni* — Seeker: Within Radiant love	Four legs of conjugal or marriage bed	Sun	Alpha Crateris	Star of patronage, romance, individualism, hospitality, honor, chivalry, punctuation, integrity, and contractual commitment of the propagating, radiating, and educating creature, within the para-conscious creator universe's entity-effect, without the polluted creation's entropy-effect	*Aryaman*	Amutiel	*Pari gha*	Inclusion (attract secondary fundament al negative effects)	井 (Jing): Well	Male cow	*Ashtottari*

Lunar zodiac	GUIDER kingdom	SHEENY-effect	Symbol	Planet Lord	Astral Lord	Signifies	Deity Lord	Demon mediator	Yoga	Signifies	Solar Mansion	Animal mansion	Pathway (*dasha*)
Ascending Cancer/ Venus	Primordial deity	*Hasta* – Experiencer. Within present	Fist	Moon	Gamma Corvi	Star of manifestation, self-managing dexterity, concentration, polarization, inspiration, awakening, and devotional intensity of the experiencing, irradiating, compensating and philosophizing creature, within the dry ether-effect of the polluted creation, without the dry ether-effect of the para-conscious creator universe	*Aditya*	Egibiel	*Siva*	Diversity (strangely repel tertiary aura as the illusion maya of the earned endowment s)	鬼 (Gui): Ghost	Female bison	*Karaka Kendradi*
Ascending Libra/ Uranus	Param deity	*Chitra* – Enjoyer. Without present	Gem	Mars	Alpha Virginis	Star of opportunity, architect potential, creator potential, illumination, and devotee intensity of the enjoying, exchanging, mediating, guiding, servicing, scientizing creature, within the entropy-effect of the polluted creation, without the entity-effect of the para-conscious creator	*Tvashtar*	Abrinael	*Siddha*	Engagement (self-consciousness of the realized ascendant master)	柳 (Liu): Willow	Female tiger	*Sthira*

Table 15. The Fourteen Pathways of the Self-luminous Entity, Without the Luminous

Lunar zodiac	GUIDER kingdom	SHEENY-effect*	Symbol	Planet Lord	Astral Lord**	Signifies	Deity Lord ^	Demon mediator &	Yog a	Signifies	Solar Mansion %	Animal mansion	Pathway (dasha)
Descending Cancer/ Venus	Primordial deity	Swati – Para-Absolute: Self without branches	Coral bead	Uranus (Rahu)	Kappa Virginis	Star of self-going freedom, i.e., self-steaming ego and earth-effect	Vayu	Enediel	Sadhy a	Responsibility (togetherness as the conscious archangel)	星 (Xing): Star	Male bison	Kalachakra
Descending Libra/ Uranus	Param deity	Vishakha – Absolute: Self within branches	Potter's wheel	Jupiter	Alpha Librae	Star of purpose and air-effect	Indragni	Jazeriel	Subb a	Within Guider (negatively charged pulse as a Cherubim guardian angel)	張 (Zhang): Extended Net	Male tiger	Kala
Ascending Capricorn/ Sun	Primeval deity	Anuradha – Deity: Spark of lightening	Triumphal archway	Saturn	Pi Scorpii	Star of success, leadership, and water-effect	Mitra	Amnediel	Sukl a	Without Guider (positively charged impulse as a Seraphim angel of light)	翼 (Yi): Wings	Female deer	Chakra
Descending Capricorn/ Sun	Primeval deity	Jyeshtha – God: Lord	Earring	Mercury	Sigma Scorpii	Star of wisdom, protection, self-organization, and fire-effect	Indra	Abduxuel	Brah ma	Guider (neutral dominating Dominationes pastoral angel)	軫 (Zhen): Chariot	Male deer	Dvadasho-ttari
Descending Leo/ Mercury	Param perpetuator	Mula – Destroyer: Root	Bunch of roots tied together	Neptune (ketu)	Mu Scorpii	Star of originality, investigation and DIVINE-effect	Nirru	Amnixiel	Indra	Social (extrinsic intellect as the predominating Thronos angel)	角 (Jiao): Horn	Male dog	Panchottari

*: Descending Lunar Mansion (nakshatra); **: Astronomical star; ^: Horizontal Earth Mansion; &: Forward demonic angel mediator: %: East-North Ascending Solar Mansion (xiu)

Lunar zodiac	GUIDER kingdom	SHEENY-effect	Symbol	Planet Lord	Astral Lord	Signifies	Deity Lord	Demon mediator	Yoga	Signifies	Solar Mansion	Animal mansion	Pathway (*dasha*)
Descending Taurus/ Jupiter	Para deity	*Purva Ashadha* – Without Godhead (Without Victorious)	Fan	Venus	Gamma Sagittarii	Star of uniqueness, invincibility, and catalyst of the GUIDER-effect	*Jal*	Requiel	*Vaid briti*	Human (entropy of mental presence as the Principatus deciding angel)	亢 (Kang): Neck	Male monkey	*Manduka*
Ascending Virgo/ Neptune	Deity	*Uttara Ashadha* – Within Godhead (Within Victorious)	Small cot	Sun	Phi Sagittarii	Star of universality and the SHEENY blessing-effect	*Vishva deva*	Geliel	*Vish kamb ba*	Ecological (growth of the physical power as the Lord of Lords Yotzer ohr angel)	氐 (Di): Root	Female mongoose	*Pindayu*
Ascending Taurus/ Jupiter	Para deity	*Shravana* – Perpetuator : Famous	Trident	Jupiter	Beta Capricorni	Star of learning, listening, and empathy of ONENESS connection-effect	*Vishnu*	Bethnael	*Priti*	Economic (intrinsic contentment as the absolute SHEENY Aeon angel)	房 (Fang): Room	Female monkey	*Brahma Grahashrita*
Ascending Sagittarius/ Moon	Human	*Dhanishtha* – Illuminator	Flute	Mars	Epsilon Aquarii	Star of fame, symphony, and WHOLF-effect	*Ashta vasus*	Ergodicl	*Ayus bman*	National (normative agelessness as the King of Kings Potentates angel)	心 (Xin): Heart	Female lion	*Amsayu*
Ascending Scorpio/ Earth	Supra deity	*Shatbhisha* – Devotee	Thousand flowers	Uranus (*rahu*)	Beta Aquarii	Star of veil, healing, and WHOLESOME-effect	*Varuna*	Kiriel	*Saub bagya*	Psychological (transformative fortune as the Huiotetes sonship of God)	尾 (Wei): Tail	Female horse	*Nakshatra-drusi*

Lunar zodiac	GUIDER kingdom	SHEENY-effect	Symbol	Planet Lord	Astral Lord	Signifies	Deity Lord	Demon mediator	Yoga	Signifies	Solar Mansion	Animal mansion	Pathway (dasha)
Descending Sagittarius/ Moon	Human	*Poorva-bhadrapada* – Without Devoted	Man with two faces (feminine and masculine)	Jupiter	Alpha Aquarii	Star of transformation, spiritual growth, and WHOLESOME WHOLE-effect	*Aja Ekapada*	Tagriel	*Saisana*	Within SHEENY (metaphysical masculine splendor as the greetable Ennoca)	箕 (Ji): Winnowing Basket	Male lion	*Nisargaja*
Ascending Heaven/ Sentient energy system	Mineral	*Uttara-bhadrapada* – Within Devoted	Four legs of a funeral cot	Saturn	Alpha Pegasi	Star of warrior, plant rainmaker, and ENTROPY-effect	*Ahir Budhanya*	Abrinael	*Atiganda*	Without SHEENY (dynamic feminine vengeance as the conscious Thelesis greet)	斗 (Dou): Southern Dipper	Female cow	*Drg*
Ascending Aquarius/ White Star	Super deity	*Revati* – Devoted	Drum for keeping time	Mercury	Gamma Pegasi	Star of wealth, nourishment, animal lordship, and Ether-effect	*Pooshva* "	Cabiel	*Sukarma*	SHEENY (formative wholesome magnanimity as the greeting Varaha-mihira Godhead)	牛 (Niu): Bison	Female elephant	*Yagna*
Descending Virgo/ Neptune	Deity	*Abhijit* – Godhead: Victorious	Three-cornered nut (perfect prism of creation, creature and creator triangle)	Moon	Vega	Star of accomplishment, deity realization, divinity, and Entity-effect	*Brahma*	Annediel	*Nittya*	Absolute SHEENY (normative wholesomewhole illumination as the greeter God)	女 (Nü): Girl	Male mongoose	*Panchaswara*

Lunar zodiac	GUIDER kingdom	SHEENY-effect	Symbol	Planet Lord	Astral Lord	Signifies	Deity Lord	Demon mediator	Yoga	Signifies	Solar Mansion	Animal mansion	Pathway (dasha)
Descending Sagittarius/ Moon	Human	Poorva-bhadrapada – Without Devoted	Man with two faces (feminine and masculine)	Jupiter	Alpha Aquarii	Star of transformation, spiritual growth, and WHOLESOME WHOLE-effect	Aja Ekapada	Tagriel	Sobh ana	Within SHEENY (metaphysical masculine splendor as the greetable Ennoca)	箕 (Ji): Winnowing Basket	Male lion	Nisargayu
Ascending Heaven/ Sentient energy system	Mineral	Uttara-bhadrapada – Within Devoted	Four legs of a funeral cot	Saturn	Alpha Pegasi	Star of warrior, plant rainmaker, and ENTROPY-effect	Ahir Budhan ya	Abrinael	Atiga nda	Without SHEENY (dynamic feminine vengeance as the conscious Thelesis greet)	斗 (Dou): Southern Dipper	Female cow	Drg
Ascending Aquarius/ White Star	Super deity	Revati – Devoted	Drum for keeping time	Mercury	Gamma Pegasi	Star of wealth, nourishment, animal lordship, and Ether-effect	Pooshra v	Cabiel	Suka rma	SHEENY (formative wholesome magnanimity as the greeting Varaha-mihira Godhead)	牛 (Niu): Bison	Female elephant	Yogni
Descending Virgo/ Neptune	Deity	Abhijit – Godhead: Victorious	Three-cornered nut (perfect prism of creation, creature and creator triangle)	Moon	Vega	Star of accomplishment, deity realization, divinity, and Entity-effect	Brahma	Amnediel	Nirby a	Absolute SHEENY (normative wholesomewhole illumination as the greeter God)	女 (Nu): Girl	Male mongoose	Panchastar a

Sequential primary path	Causative Reality	Sequential Reality	Consequential Reality	Symbolic Reality	Eventual Reality	Compensating Reality	Sacramental Reality	Param Deity Reality	Entropy Reality	Pathway (Dasha)
Primordial Space (without formative capability; without *Param siddha*)	*Nibbida*: conceiving disenchant-ment with the worldly life (wishables without) as a scientific experiment	*Bhava* (Divine, 360): ascending otherness consciousness, craving for a wishable and clinging to that wishable as the objective reality of the world	*Gajanan*: One who has the face of an elephant, empowered to intuitive sensing the cosmic reality, transcending the limits of scientific ideas (ideal-effect), notions (theory-effect), and numbers (sequence duality-effect)	Queen of Divinity — eleventh infinity: *Maha Kaali*/ Luminous	Lunar month *Ashvina*, *Aditya* *Tvashtha*, Sage *Jamadagni*; *Apsara* *Tilottama*; Serpent *Kambalasva*, *Yaksha* *Satajit*, *Rakshasa* *Brahmapeta*, *Gandharva* *Dhrtarastra*	Lord of the womb, who guides us to move forward with confidence, rewards divine outcomes of our siddhi qualities and punishes us with dark VUCA energy when we deviate and is the immanent GUIDER	Solar month *Isha*, Libra (*tula*)	+11/12 dimension of the future formative growth of the present cosmos; Present horizontal direction (transcending infinite present moments): Equator	*Naimittika* (*Maha*) *Pralaya mukti* (transcending the infinite present moments and becoming universal Ascended Master, without extrinsic consciousness)	*Vamada*

Table 16. The Fourteen Pathways of the Self-luminous Entity, Without the Luminous

Sequential primary path	Causative Reality	Sequential Reality	Consequential Reality	Symbolic Reality	Eventual Reality	Compensating Reality	Sacramental Reality	Param Deity Reality	Entropy Reality	Pathway (Dasha)
Primordial Human-effect (within *Param siddha*)	*Yathabhuta:* knowing and envisioning the ontological reality of wishable within as is	*Upadana* (Creator factor, 4): Ascending clinging charm attraction of the wishable, intensifying increasingly from the mental sense, to the astral form, to the etheric experience, to the attachment to the cause of self-existence	*Sumukha:* One who has a beautiful face that emanates intrinsic perfection	Oversoul family within the etheric body – sixth infinity: *Param Pita Parameshwar a/ Sadashiva/* /Dark Matter	Lunar month *Bhadrapada, Aditya Vivaswat, Apsara Anumlocha, Sage Brighu,* Serpent *Shankhapala, Yaksha Aasaarana, Rakshasa Vyaghra, Gandharva Ugrasena*	Lord of the realm of the shadows of the unformed imperfect energies; One who forms perfect creation; the head of planets and life; the immanent fire-effect that travels in all directions	Solar month *Nabhasya,* Virgo (*kanya*)	+10/12 dimension of the future formative growth of the present cosmos; Present Ascending direction (transcending the infinite horizontal presence): Longitude	*Nitya Pralaya mukti* (transcending the infinite horizontal present inertia, and becoming spontaneous)	*Chara Paryaya*

Sequential primary path	Causative Reality	Sequential Reality	Consequential Reality	Symbolic Reality	Eventual Reality	Compensating Reality	Sacramental Reality	Param Deity Reality	Entropy Reality	Pathway (Dasha)
Primordial Trading-effect (without *Param siddha*)	*Samadhi:* concentrating the epistemological experience on the scientific wish immanent within	*Tanha* (Fire, 17): Ascending craving within a SHEENY entity, strangely repelling the GUIDER theory-effect of the subtle unfulfilled desire to be fulfilled by the DIVINE ideal-effect, fire form-effect, water taste-effect, air touch-effect, earth smell-effect, and ether sound-effect	*Ekadanta:* One whose lower tooth illuminates the primordial truth reality of the deity within, and who has sacrificed the upper tooth for destroying the devil who is the fundamental cause of the primeval illusion potential of Self	Soul family within the primary causal body – zeroth infinity: 2000th I AM level; Primeval Perpetuator: *Vishnu*	Lunar month *Shravana,* *Aditya Ushas, Apsara Pramlocha,* Sage *Angira,* Serpent *Elaapatra, Yaksha Srota, Rakshasa Varya, Gandharva Vishwavasu*	Nurturer mother of well-being and development; One who bestows the light of new dawn, effortlessly chasing away all the demonic subconscious imaginary wishables within the immanent DIVINE-effect (by letting the *Kshatriya King of Gods Indra* clear the cloud of satans with thunderous lightning)	Solar month *Nabha,* Leo *(simha)*	+9/12 dimension of the future formative growth of the present cosmos; Present Descending direction (transcending the infinite ascending presence): North Pole	*Laya mukti* (transcending the infinite consciousness and becoming omnipotent primeval entity without primordial entity)	*Ashtakavarga*

Sequential primary path	Causative Reality	Sequential Reality	Consequential Reality	Symbolic Reality	Eventual Reality	Compensating Reality	Sacramental Reality	Param Deity Reality	Entropy Reality	Pathway (Dasha)
Primordial Workculture-effect (within *Param siddha*)	*Sukha*: Happiness as the axiological effect without manifesting the wishing sequence	Vedanta (Deity, 1): Ascending sensation of the ideal wishing, form, value, touch, smell, and sound, without the secondary fundamental vector-effect of the objectified theory of wish	*Kapila* The Lord of grey color who bestows the nectar of consciousness with the creature, freeing the need to have the creation (cow producing the milk nectar)	Flamemate within the etheric body – fifteenth infinity; Primordial Perpetuator: *Maha Sarasvati*	Lunar month Ashada; *Aditya Varuna, Apsara Rambha*, Sage *Vasishra*, Serpent *Sukra, Yaksha Sahajanya, Rakshasa Chitraswana Gandhara Haha*	The Evening rising star that connects the Sun/creator (consciousness), the earth/creation (physical reality), and the time/creature (varying behavior/actions), within the immanent water-effect	Solar month *Sachi*, Cancer (*karki*)	+8/12 dimension of the future formative growth of the present cosmos: Past Growth moment (transcending the infinite descending presence)	*Ganatita mukti* (transcending the infinite ONENESS and becoming ONE within primordial energy, i.e., omnipresent, undivided mentor consciousness)	*Navamsa Sthira*
Primordial Culture-effect (without *Param siddha*)	*Passadhi*: Tranquility as the metaphysical creation within the manifested wishing sequence	*Phasa* (Energy, 19): Ascending sense of contact by a super-wisher, who is conscious of the theory of wish within the wisher medium	*Gajakarnaka* One with ears of an elephant, who uses the power of listening to discriminate between the truth and the *maya* fiction	Soul essence community within the tertiary causal body - fourteenth infinity –*Sati- Parvati* Primordial Greeter	Lunar month *Jyeshta, Aditya Mitra, Apsara Menaka*, Sage *Abri*, Serpent *Takshaka, Yaksha Rathaswana, Rakshasa Paurushya, Gandhava Haha*	The Morning rising Star who prevents the darkness, by manifesting the truth of our varying actions guided by the intuitive knowing, within the immanent WHOLENESS of Self	Solar month *Shukra*, Gemini (*mithun*)	+7/12 dimension of the future formative growth of the present cosmos: Past Horizontal direction (transcending the infinite past growth moments)	*Kala mukti* (transcending the infinite past growth moments and becoming universal light)	*Sbula*

Sequential primary path	Causative Reality	Sequential Reality	Consequential Reality	Symbolic Reality	Eventual Reality	Compensating Reality	Sacramental Reality	Param Deity Reality	Entropy Reality	Pathway (Dasha)
Primordial Techno-logical Servicing (within *Param siddha*)	*Piti:* Rupture as the dynamic correction factor within the wisher, who is not in sync with the immanent *prakriti* character	*Salayatana* (Sound, 257): Ascending sense of the activation of the secondary contact media: intellect (divine-effect), eye (fire-effect), tongue (water-effect), nose (air-effect), body (earth-effect), and ear (ether-effect)	*I ambodra:* One with a large belly, having both known and unknown immanent within as one primordial energy	Flame family within the etheric body – sixteenth infinity – *Maha Gauri/* Primeval Greeter	Lunar month *Vaishaka, Aditya Aryamaan, Apsara Punjikasthali,* Sage *Pulaha,* Serpent *Kachchaneera, Yaksha Athouja, Rakshasa Praheti, Gandharva Narada*	Guides the devotional path of the wisher, beyond the varying truth and toward one constant light, perfection and desirable outcome, within the immanent air-effect	Solar month *Madhava,* Taurus *(vrushabh)*	+6/12 dimension of the future formative growth of the present cosmos: Past Ascending direction (transcending the infinite past horizontal directions)	*Maya mukti* (transcending the natural physical limits of consciousness, through the oneness of *chakras,* and becoming omniscient, without Holy Spirit consciousness)	*Trikona*

Sequential primary path	Causative Reality	Sequential Reality	Consequential Reality	Symbolic Reality	Eventual Reality	Compensating Reality	Sacramental Reality	Param Deity Reality	Entropy Reality	Pathway (Dasha)
Primordial Techno-logical Trading (without *Param siddha*)	*Pamujja.* Joy, as the formative factor without wisher, from knowing the cause of the dynamic correction (i.e., positive or negative spin)	*Namarupa* (Ether, 285): Ascending mental theory consciousness (divine-effect), and formation of the subconscious intention (fire-effect), emotional feelings (water-effect), sensory perception (air-effect), physical contact (earth-effect), divided attention (ether-effect) as tertiary materializations	*Vikata.* One who is ferocious with the power to defeat all the weaknesses of those who value the path of *Karma* action	Godhead within the primary causal body – seventh infinity – *Nandi Cow/* Primeval Maternal	Lunar month *Chaitra, Aditya Dhata, Apsara Kratasthali,* Sage *Pulastya,* Serpent *Vasuki, Yaksha Rathakruth, Rakshasa Heti, Gandharva Tumbura*	Manifests the reality by destroying ignorance, as the immanent SHEENY-effect within the living beings	Solar month *Madhu,* Aries *(mesh)*	+5/12 dimension of the future formative growth of the present cosmos; Past Descending direction (transcending the infinite past ascending directions)	*Para mukti / Prakrita Pralaya mukti* (transcending the infinite past ascending directions and becoming the eternal ONE, i.e., Tao or Godhead)	*Sudarshan a Chakra*

Pathway (Dasha)	*Sandhya*
Entropy Reality	*Saguna Mukti* (transcending the infinite past descending directions and becoming the infinite illuminator consciousness)
Param Deity Reality	+4/12 dimension of the future formative growth of the present cosmos; Future Growth moment (transcending the infinite past descending directions)
Sacramental Reality	Solar month *Tapasya*, Pisces *(meen)*
Compensating Reality	Bestows infinite light endowments (i.e. the wishing sequence for manifesting the wishable wish in the form of the gravitational virtues) within the darkness of night (when the wishes strangely repel in the form of imagined wishables), within the immanent ONENESS of earth as creation and Self as primordial creation
Eventual Reality	Lunar month *Phalguna,* Aditya *Savitr (Parjanya),* Apsara *Sengjit,* Sage *Bharadraja,* Serpent *Airawata,* Yaksha *Ritu,* Rakshasa *Varta,* Gandharva *Vista*
Symbolic Reality	Queen of Space – thirteenth infinity – *Maha Lakshmi* Primeval Illuminator
Consequential Reality	*Vighnanatha* Destroyer of obstacles for those who value the path of *Jnana* (knowledge)
Sequential Reality	*Vinnana* (Spiritual stress, 47): Ascending sense of rebirth consciousness of the quaternary visceral manifestations, within the new, imagined consciousness of intellect (divine-effect), eye (fire-effect), tongue (water-effect), nose (air-effect), body (earth-effect), and ear (ether-effect)
Causative Reality	*Saddha:* Faith in the super wisher within as the normative factor accounting for the manifested experience
Sequential primary path	Primordial Techno-logical Growth (within *Param siddha*)

Sequential primary path	Causative Reality	Sequential Reality	Consequential Reality	Symbolic Reality	Eventual Reality	Compensating Reality	Sacramental Reality	Param Deity Reality	Entropy Reality	Pathway (Dasha)
Primordial Techno-logical Exchange (without *Param siddha*)	*Dukkha.* Suffering without super wisher, as the transforma tive factor for becoming a supra wisher	*Sankhara* (Exchange, 269): Ascending mental, intellectual (verbal), and physical (bodily) fabrications, seeking to validate the causal, etheric, and astral exchange with the supreme wisher	*Vinayaka.* One with the qualities to lead those who value the path of devotion bhakti, guided by the virtues of *dharma* righteousness	Lord of Lords – ninth infinity: *Shani/* Primordial Maternal	Lunar month Magha; Aditya Pushan; *Apsara Ghrtaci,* Sage *Gautama,* Serpent *Dhananjaya, Yaksha Suruci, Rakshasa Vata, Gandharva Susena*	Ensures the well-being of all entities, including the animals and the plants; Motivating power of soul and the knower of the reason for each entity's varying consciousness; the immanent WHOLESOME-NESS	Solar month *Tapah,* Aquarius (*kumbh*)	+3/12 dimension of the future formative growth of the present cosmos; Future Horizontal direction (transcending the infinite future growth moments)	*Brahma Mukti* (transcending the infinite withinness and becoming the sameness of primordial wisher, without the omni-permeating, creator consciousness)	*Chaturvidh a Uttara*

Table 17. The Fourteen Pathways of the Self-conscious Entity, Within the Self-luminous Entity

Sequential primary path	Causative Reality	Sequential Reality	Consequential Reality	Symbolic Reality	Eventual Reality	Compensating Reality	Sacramental Reality	Param Deity Reality	Entropy Reality	Pathway (Dasha)
Primeval, Ascending Techno-logical Cost (within formative exchange; within *Param siddha*)	*Asava*: the destructio n of the flow of supreme wisher conscious ness, within supra wisher	*Avijja* (Fame behavior, 37): Ascending sense of the ignorance kriya (physical limits), not knowing (intellectual limits) the way to resolve the suffering (mental limits), the nature of suffering (astral limits), the cause of suffering (etheric limits), and the end of suffering (causal limits)	*Dhumraketu*: Lord of smoky color, who absorbs the air-effect of the clouded consciousness and creates the earth-effect of the clarified consciousness	Queen of Life — eighth infinity — *Indra*/ Primordial Paternal	Lunar month *Pausha, Aditya Bhaga, Apsara Purvacitti*, Sage *Ayur*, Serpent *Karkotaka, Yaksha Uma, Rakshasa Sphurja, Gandharva Aristanemi*	Bestows material prosperity and physical health; the immanent earth-effect	Solar month *Sahasya*, Capricon (*makar*)	+2/12 dimension of the future formative growth of the present cosmos; Future Ascending direction (transcending the infinite future horizontal directions)	*Kriya mukti / Bhrama mukti* (transcending the infinite future horizontal directions and becoming the eternal wishing vector)	*Tara*

Sequential primary path	Causative Reality	Sequential Reality	Consequential Reality	Symbolic Reality	Eventual Reality	Compensating Reality	Sacramental Reality	Param Deity Reality	Entropy Reality	Pathway (Dasha)
Param, Descending Techno-logical Cost (without formative exchange; without *Param siddha*)	*Vimutti:* freedom from para wisher without supreme wisher, for diffusing the negative white star energy, that is destroying the wholesom e primeval wisher within the finite physical body	*Jaramarana* (Death, 18): Ascending sense of the entropy aging, death, and residual mass of *dukkha* distress, threatening both the autonomy as well as the sameness of the position of whole Self	*Ganadhyaksha:* Leader of the *Ganas,* being the variable SHEENY principle within all categorizations of order sequences	King of Gods – twelfth infinity – *Agni/* Primeval Paternal	Lunar month Margashirsha; *Aditya Anshuman (Ansa), Apsara Urvasi,* Sage *Kasyapa,* Serpent *Mahasankha, Yaksha Tarksya, Rakshasa Vidyucchatru, Gandharva Riasena*	Clarified consciousness of the noumenon that unifies the pair of opposite behaviors; the immanent sameness of the living creatures and the primordial entity	Solar month *Sahas,* Sagittarius *(dhanus)*	+1/12 dimension of the future formative growth of the present cosmos; Future Descending direction (transcending the infinite future ascending directions)	*Tatva mukti* (transcending the infinite future descending directions and becoming a unique spirit, who is enjoying the entirety of the extrinsic endowments)	*Rashmi*

Sequential primary path	Causative Reality	Sequential Reality	Consequential Reality	Symbolic Reality	Eventual Reality	Compensating Reality	Sacramental Reality	Param Deity Reality	Entropy Reality	Pathway (Dasha)
Vritti-jnana (extrinsic: free from subject-effect; *mukti* of a physical cause–making self a technological medium for celestial deity infinity)	*Krama mukti* freedom from the manifested *chitta* promoting intrinsic purposeles s-ness	*Shuddha:* Conscious wishing sequence— *vritti*, within the intrinsic wonder	*Brahmakara:* Subject of wish – meditation, within the intrinsic joy	Primeval Wisher: Materialistic wishable causes, within intrinsic tranquility— *Nirguna/ Artharthi*	*Vikshepa:* delusional beauty (*maya*), within intrinsic malice— Curvilinear; Vector; Knowable; *Chitta* energy	*Vidya* (knowledge): Inherited knowledge of the phenomenon that divides the absolute breath; the immanent infinity of the living creatures within the primordial entity	*Para* (Para consciousnes s knowing); Zodiac entity	0/12 dimension of the future formative growth of the present cosmos; Primordial-Primordial growth moment (transcending the infinite future descending directions)	*Atyantika Pralaya mukti* (transcending the infinite primordial-primordial growth moments and becoming a wise guider entity, enjoying the entire intrinsic endowments)	*Radhyansh a*
Svarup-jnana (intrinsic: metaphysical; filled with subject-effect; *yukti* technique of physical self)	*Sadyo mukti* freedom from the manifestab le *achitta* promoting cost-escalating intrinsic purposeful -ness	*Ashuddha:* Para conscious wishing sequence— *prakriti*, within intrinsic sensory arousal	*Vishayakara:* Object of wish – mediation, within intrinsic heroism	Param Manifestor: Spiritual wishable cause fulfillment, within intrinsic distress— *Saguna/ Apahrtabhara*	*Avarana:* Real truth *Amaya*, within intrinsic fear of unknown— Linear; Canon; Manifestable; *Sutra*	*Avidya* (bounded rationality): Freedom from the inherited knowledge, knowing that the primeval creature knows the primordial entity reality it has created; the param oneness of the primeval creature and the primordial-primordial entity	*Apara* (Conscious knower); Astrological entity	Negative One (-1) dimension of the future formative growth of the present cosmos; Param-Primordial growth moment (transcending the primeval-primordial growth moments)	*Hiranyagarbha mukti* (transcending the infinite param-primordial growth moment and becoming a sentient entity, who is servicing the entire intrinsic endowments)	*Nakshatra*

Chapter 6: The Primary Doctrines for Paradigmatic Planning

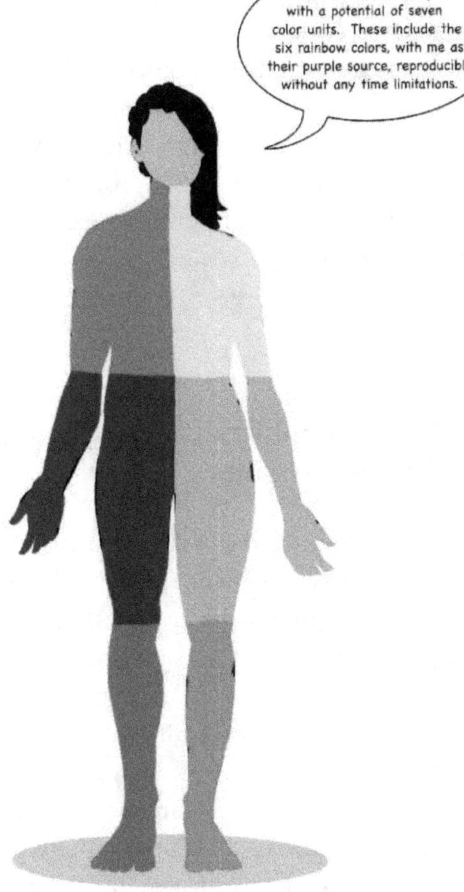

The Sentient Entity is the Primary Doctrine

The Sentient Entity is the Primary Doctrine

A sentient entity is the primary doctrine for the paradigmatic planning. One plans a paradigm for taking, making, and shaping reality because one has the consciousness of the virtual mind-born reality. The mind-born reality exists only as a potential in the past. A sentient entity may potentiate the potential by physically illuminating it beyond the mental realm. After incarnating the self as that potentiated augmented reality, a sentient entity declares the following doctrine. I embody the six rainbow colors (indigo, blue, green, yellow, orange, and red), with me as their purple, spiritual source, reproducible without any time limitations. As a sentient entity, conscious of my omnipresent reality, I have a potential of the seven color units. It manifests my omnipresent whiteness, the colorless past, and the blackening future, formed by radiating the light of my present colorful reality.

6.1. Planning for the Subjectivity of the Supra-Primordial Paradigm: Primary Doctrines

An entity who is conscious of the zodiac cause of the finiteness of the sentient energy can trade the "sentient reality" (*Purushartha*, 7), using the "path of non-violence" (*Ahimsa dharma*, 8) as the supra-primordial doctrinal paradigm. The path of non-violence is free of the subjective psychological bias, correlated with the planning paradigm with the universe of the sentient entities. Such an entity becomes an immortal super-primordial doctrinal paradigm, i.e., the "system dimension" (*Ashrama dharma*, 497), for illuminating the six-dimensional "param-effect," i.e., the "sentient self" (*Purushartha*, 7). The param-effect is inclusive of the potential for the "primordial-effect," i.e., the geometrical "zodiac self" (*Ganitam*, 8×10^{15}) and the "primeval-effect," i.e., the algebraic "astrological self" (*Beejaganitam*, 10^{10}). The six dimensions consist of two primary, two secondary, and two tertiary doctrines. I investigate the primary doctrines designed to manage a subject's subjectivity in this chapter. The secondary and the tertiary doctrines are the subject of the next two chapters.

6.2 First, "I am the Divine Truth Doctrine" (*Brahma Satyam Jagan Mithya siddhanta*, -3 x 10⁶)

As a natural, born-master of the "science of psychometry" (*Mnomapan*, -10^{29}), one generates the "wholly-biased smell of the physical body" (*Purnasau gandha*, 948) for authenticating the bias-free, "self-evident reality" (*Svartha*, -2) of the divine self.   The divine "consciousness" (*Chaithanya*, 4) is the greet of a maternal "flamemate" (*Svayambhu*, 123), who manifests the multidimensional "ground reality" (*Hetvartha*, 385) for the "joy" (*Simchah*, 123) of the primordial, sentient "flame" (*Sadhya*, 32).

As an enjoyer of the divine greet, the primordial flame entity supernaturally radiates the intrinsic sentient energy as the twenty-eighth space dimension through a time-varying conditional correlation with the twenty-seven time-dimensions of the universe.   The "primordial space" (*Dik*, 100) is the "one-dimensional, embodied, primordial reality" (*Gurvartha*, 39) of the "gravitational self" (*Guru*, 100) as a creation.   By ascending the "absolute time" (*Sva*, 11) consciousness, one develops the "soul-level" (*Atman, 4*) power to trade the "communion greeter-effect" (*Vaisnava*, 486) of the maternal flamemate.   One becomes the "primordial illuminator" (*Shri Krishna*, 10) tangent for centering the circumferential truth of the universe with the perpendicular "gravitational-effect" (*Pathitajya*, 959).

The "perpendicular" is the "supernormal" (*Haya*, 270) "param sentient-effect" (*Varuna*, 1000) of an entity working to transform the intrinsic "gravitational energy" (*Lalita*, 100) into a set of three values: the sine, the cosine, and the tangent.   The "sine value" (*Pathitajya*, 959) signifies the consequential growth of the inanimate universe's gravitational-effect.   The "cosine value" (*Pindajya*, 379) signifies the causative entropy in the soul's sentient energy.   The "tangent value" (*Yujya*, 285) signifies the "primeval sentient-effect" (*Shuddhi*, 285), which perpetuates the causative entropy.

By further ascending the "absolute cause" (*Brihaspati*, 1780) consciousness, one develops the "absolute soul-level" (*Paramatman*, 1600) creature power to trade the "sentient entity" (*Siddha*, 7) consciousness for discovering the divine truth of the self.   One's divine truth projects into the six physical realities and the one metaphysical reality, which constitutes the following four "ecological dimensions" (*Asrama dharma*, 497) of the descending dimensionalities:

- **The three-dimensional truth of the intellectual, etheric, and physical bodies.** By perpetuating the "two-dimensional, personified, primeval reality" (*Dvyartha*, 400) of the maternal "flamemate" (*Svayambhu*, 123), through the "perpetuating value" (*Saranyu*, 5) of the "primordial illuminator" (*Shri Krishna*, 10), the entity formatively services 123 * 5 = 615 units of gravity as the "cotangent value" (*Marjya*, 615). Consequently, it generates a formative growth of "primordial-primordial oneness" (*Vibhu*, 379) with the "holy spirit within the self" (*Lakshmi*, 379) for producing the "seed" (*Beeja*, 379) of the "workculture" element (*Nayaki tattva*, 379), to "electromagnetically catalyze" the formation of the "intellectual body" (*Sukshma sharira*, 306).

The workculture element is the "maximum parabolic value" (*Pindajya*, 379) of the "tangential gravitomagnetic-effect" (*Ardhajya*, 10) of Mother Nature. The entity transforms the "profundity of the intellectual body" (*Ruksha*, 49) into a "streak of intelligence" (*Buddhi*, 48), i.e., a sequence of the neuron cells. The streak value of intelligence is the "minimum parabolic value" (*Kotijya*, 48) without the tangent of Mother Nature. The intelligence is the "discrete optimization value" (*Ganapujya*, 306) of the binary set of the maximum, workculture value (*Pindajya*, 379), and the minimum, streak value (*Kotijya*, 48).

The entity generates the "starseed universe" (*Ganarajya*, 476) with the 379 units of the "cosine value" (*Pindajya*, 379) to catalyze the etheric body (*Bhoga sharira*, 957). The starseed universe is the "param, sigma value" (*Ganarajya*, 476) of the intellectual body, formed by reducing the "cotangent value" (*Marjya*, 615) with a natural logarithmic function of the gravitoelectric energy. The etheric body is the "primeval, maximum likelihood value" (*Viyojya*, 957) of the cosine value (*Pindajya*, 379), capitalized with a supernatural exponential function of the gravitomagnetic energy. The entity forms the "physical body" (*Sthula sharira*, 387) with the "limit value" of the gravitomagnetic exponential function. The limit value is the "primordial, cluster value" (*Avanejya*, 387) of the valuing "tangent" (*Ardhajya*, 10), constrained by the growth of the cotangent value (615), the minimum parabolic value (48), the cosine value (379), and the discrete optimization value (306).

The entity clusters the sequentially formed physical bodies into the thirty-six "cluster groups of entities" (*Gana*, 387). The entity norms the thirty-six groups with the "domain value" of the natural logarithmic function, which is the "primordial-primordial value" (*Brahmasayujya*, 36) of the "one domain identity" of the eight parameters of the "rationally conceived reality" (*Yuktartha*, 6): the cotangent value (615), the minimum parabolic value (48), the cosine value (379), the discrete optimization value (306), the tangent, the gravitomagnetic-effect (10), the param, sigma value (476), the primeval, maximum likelihood value (957), and the primordial, cluster value (387).

The entity transforms the thirty-six starseed groups into the eighteen geographies of the starseed universes through the "self-perpetuating" (*Udvaha*, ½) rationally conceived reality. The eighteen is the "boundary value" (*Ghodamunjya*, 18) of the divergent, ascending, super-harmonic function of the sentient energy and the convergent, descending, harmonic function of the thermodynamic energy, which generates the symmetry between the two functions, as the "experienced reality" (*Gudhartha*, 18). The rational "conceived reality" (*Yuktartha*, 6) is the "conditional divergent value" (*Uttarajya*, 6), formed through the solution-seeking "entropy value of the thermodynamic energy" (*Prayujya*, 208).

- **The two-dimensional truth of the astral and causal bodies.** The entity exchanges 285 units as the "tangent value" (*Yujya*, 285) for trading 270 units of the abnormal "cosecant value" (*Rituyajya*, 270) and 15 units of the normal "secant value" (*Autaghaticem rajya*, 15). It trades the astrological universe's energy as the "primeval-effect" (*Beejaganitam*, 10^{10}) with the 270 units of the cosecant value. Thence, it electromagnetically forms the "astral body" (*Linga sharira*, 3). It trades the zodiac universe's energy as the "primordial-effect" (*Ganitam*, 8 x 10^{15}) with the 15 units of the secant value. Thence, it electromagnetically forms the causal body. The primeval-effect is the secondary, fundamental "linear value" (*Margi*, 497) of the "perceivable" (*Sutra*, 497) "ecological dimension" (Asrama *dharma*, 497) of the "electromagnetic catalyst value" (*Ganapujya*, 306) immanent within the "intellectual body" (*Sukshma sharira*, 306). The primordial-effect is the tertiary, residual "curvilinear curved value" (*Vakri*, 396).

The transformative exchange produces a "problem-making" (*Samanya dharma*, 108), "entropy in the metaphysical value of sentient energy" (*Gardhabhejya*, 108) within the entity, who is behaving like a "dumb-ass" (*Gardhaba*, 1000), not knowing the value of the sentient "blessing" (*haberakhah*, 1000). Consequently, the entity trades the one-hundred residual units of "gravitational energy" (*Lalita*, 100) for the transformative networking of the "solution-shaping" (*Laukika dharma*, 208) "entropy value of the thermodynamic energy" (*Prayujya*, 208). The entity experiences both the sentient energy's ascending super-harmonic function and the thermodynamic energy's descending harmonic function as a "symmetrical, boundary value" (*Ghodamunjya*, 18). The entity embodies the organizational truth of the "experienced reality" (*Gudhartha*, 18).

- **The one-dimensional truth of the self-luminous entity.** The entity organizationally develops 100 units as the "gravitational value" (*Dyujya*, 100) through the electromagnetic catalytic activity concentration. The gravitational value is the tertiary residual of the sentient energy, net of the formative servicing of the 615 units for "conceiving" the three-dimensional truth and the transformative exchange of the 285 units for "experiencing" the two-dimensional truth. The gravitational value is the organizational cause of the "perceived reality" (*Sutrartha*, 825). An entity perceives the present reality with the "third, moderating, universal, androgynous eye" (*Vrishna*, 825) value. The third eye value is the balance of the divergent, ascending consciousness function of the "second, right, anti-clockwise, feminine eye vision" (*Vijnana*, 47) and the convergent, descending consciousness function of the "first, left, clockwise, masculine eye vision" (*Vinayaka*, 53). The third, universal, androgynous eye is the "divergent value of the infinite binary power series" (*Varjya*, 825). It is the entropy value of the "spiritual energy" (*Saniddhya shakti*, 169). After consuming the intrinsic gravitational energy for experiencing the reality, the entity trades forty-seven units of secondary gravity from the "animal kingdom" (*Tiryaggati*, 47) and fifty-three units of primary gravity from the "human kingdom" (*Vinayaka*, 53) for producing the "wisdom" (*Chitta*, 100 = 47 + 53) as the "guiding force" (*Guru*, 100).

- **The zero-dimensional truth of the luminous.** The entity metaphysically trades 60 units as the "entropy value of the divine energy" in the form of the "knowing-effect" (*Pradyumna*, 60). "One-dimensional reality" (*Ekartha*, 21) of the "plant kingdom" (*Taiyang*, 60) services this knowing-effect. The knowing-effect is the primordial-primordial truth formed through a two-step function, followed by a two-step correction for the consequential dysfunctionality and the causative super-functionality (i.e., afunctionality).

  - A negative "metric-taking" functional step of aggregating the "one-dimensional reality" (21 + 21 + 21) of the "primordial space element" (*Dik tattva*, 100), the "param time element" (*Sva tattva*, 100), and the "primeval cause" (*Vashikarana*, 38) of the binary elements, immanent within the plant kingdom. The correction factor of the negative three is the **Robin boundary solution**, i.e., the negative, ascending, primordial (negative infinity), feminine, and the divergent-effect of the convergent Eigenvalue. The convergent Eigenvalue is the eight-dimensional curvilinear, differentiated, param gravitational-effect of the absolute soul. The divergent-effect breaks the symmetry of the Eigenfunction value, i.e., the bilinear, positive, descending, primeval (positive infinity), masculine, integrated, param electromagnetic-effect of the soul. The three correction factors are the primordial space element, the param time element, and the primeval cause element.

  - A double-negative "absolute reality scripting" functional step of disaggregating the "primordial, four-dimensional reality" (*Vasanatma*, -3) of the primeval entity. It is the "self-perpetuating" (*Udvaha*, ½) rationally conceived reality for producing the "primeval wisher-effect" (*Halima*, 61: 63 - 2). The intrinsic, corrective, functional, negative two is the "param, two-dimensional reality" (*Badhabuddhi vadartha*, -2). The negative two is the **Atkinson–Mingarelli solution**, i.e., the conditional divergent Eigenfunction value. It is the bilinear, integrated, param electromagnetic-effect of the soul, where the boundary condition is the convergent Eigenvalue of the eight-dimensional curvilinear, differentiated, param gravitational-effect, of the absolute soul. The two functional, compensating factors are the primordial-primordial entity conceiving the rational reality and the param-primordial entity that is scripting the primeval reality of the primordial-primordial entity.

- ○ An extrinsic, dysfunctional negative step of the "universal, i.e., the primeval reality scripting." It corrects for the projected half-value of the "absolute reality" (*Svartha*, -2) of the "soul" (*Atman*, 4), and consequentially illuminates the primeval masculine (*Asura*, -1) self as the "primeval, eight-dimensional reality" (*Omkara vadartha*, -1) of the "absolute soul" (*Paramatman*, 1600).   The negative one is the **Sturm–Liouville solution**, i.e., the conditional convergent Eigenvalue of the following eight curvilinear, differentiated, param gravitational-effect of the absolute soul, where the boundary condition is the divergent Eigenfunction value, i.e., the bilinear, integratedparam electromagnetic-effect of the soul:

- ▪ everything else—"conspicuous" (*Aham*, -1)

- ▪ total presence—"everywhere" (*Evakara*, -4)

- ▪ this entity—Black hole-effect or the "masculine, astral body" (*Idam*, 3)

- ▪ everybody else—divine-effect or "this-ness" (*Idanta*, -3)

- ▪ the presence—Dark matter-effect or "duality in unity", i.e., "ascending theory-effect of the absolute soul" (*Bhedabheda*, 374)

- ▪ total absence—local-effect or the Jupiter-effect or "unity in diversity" (*Abheda*, 0)

- ▪ that entity—Vega-effect or duality, i.e., "ascending ideal-effect of the soul" (*Bheda*, 28)

- ▪ the absence—"nothingness" (*Shunyata*, 2)

- ▪ An intrinsic, pre-correction, afunctional, causative step of transforming the one-dimensional reality of the subject into a "dense octave" (*Sandra*, 60) object for destroying the "subject of present reality" (*Indra*, 0) as the "necessary condition reality"

(*Ya vadartha*, 0) for liberating the self from the "essential condition reality" (*Hetvartha*, 385).

- **The Hidden Factors Shaping the Consequential Truth of the Luminous**

  o The necessary condition is the "**Dirichlet boundary solution**" (*Praman yadartha*, 9), i.e., the conditional divergent beta (exponential slope function) value. It is the multilinear, integrated, param gravitomagnetic-effect of the nine-dimensional absolute space. The essential boundary condition is the convergent alpha (logarithmic slope) value, i.e., the zeroth, differentiated, param gravitoelectric-effect of the three-dimensional time. This essential boundary condition is the "**Neumann boundary solution**" (*Vakyartha*, 298), i.e., the conditional convergent alpha (logarithmic slope) value, i.e., the zeroth, differentiated, param gravitoelectric-effect of the three-dimensional time, without the necessary boundary condition of the divergent beta (exponential slope function) value, i.e., the multilinear, integrated, param gravitomagnetic-effect of the nine-dimensional absolute space.

  o The causative boundary condition is the "**Zaremba boundary solution**" (*Gajasutra vadartha*, $81 = 9^2$), i.e., the catalyzed, primordial-primordial (zeroth), androgynous, super-linear quadratic-effect of the divergent "beta value" (*Praman yadartha*, 9) of the nine-dimensional absolute space.

  o The boundaryless cause is the "**Cauchy boundary solution**" (*Svaprakasha vadartha*, 60), i.e., the catalyzing, param-primordial, natural, supra-linear square root of the convergent alpha value of the three-dimensional time.

  o The three-dimensional time is the "**Sommerfeld radiation solution**" (*Ratnakosha vadartha*, 3600), i.e., the binding, primeval-primordial, para-natural, supreme-linear triangulated root of the parallel, infinite, gamma field-value of the nine-dimensional absolute space.

o The infinite, gamma field-value of the nine-dimensional absolute space is the cognitively-planned "**Green's function solution**" (*Lingopahita-laingika-vadartha*, 1,500 [3600^3/60*60*24*360; transforming the seconds into the years]), i.e., the para-primordial, para-linear octave value. It is also the Cauchy boundary condition for resolving the two-kind problem of the Volterra integral equation. The first-kind is the Cauchy boundary solution. The second-kind generates the Cauchy boundary condition as the solution to the Green's function.

o The octave's supreme-primordial value is the "**Absolute Volterra solution**" (*Prakarana vadartha*, 90,000). It is the eigenspace value (absolute theta field-value), i.e., the absolute inanimate, feminine-effect.

o The octave's supra-primordial value is the "**Absolute Fredholm solution**" (*Bharadvaja-gargya-parinaya-pratisedha-vadartha*, 8,100,000,000). It is the infinite animate, masculine limit to the tradable inanimate, feminine value.

o The octave's super-primordial value is the "**Liouville–Neumann solution**" (*Ashir vadartha*, 10,000,000). It is the limit of the infinite supernatural, androgynous, curvilinear trading of the natural energy's linear value.

o The octave's primordial value is the "**Hilbert resolvent solution**" (*Upadhi vadartha*, 810) to decompose the infinite spectrum of the projection operators catalyzing the Laplace-transformed eigenspace value, to discern the fundamental value of the causative, linear, "natural energy" (*Sattvotsaha*, 810). The natural energy empowers an animate, masculine entity to trade the infinite inanimate, feminine value for holding the "office of the guardian angel for the departed souls" (*Upadhi*, 37) as an androgynous entity. Consequently, the entity trades the power to reincarnate the departed souls as the "param child" (*Manyu*, 19) by perpetually thermalizing the 190 units of the "supernatural energy" (*Shram shakti*, 1) to service a unit of the "sentient energy" (*Varuna*, 1000).

o Of the 190 units, the entity trades 130 units of natural energy for servicing supernatural power to be the "primordial maternal"

(*Aum*, 18). The above nine-step technique for generating the 130 units is the **"Keldysh path-formalizing contour solution"** for creating an "office" or a "council" (*Vidhata*, 130). It nurtures the assembly of the "infinite souls" (*Ratnakosha*, 30) within the self's "boundary value" (*Ghodamunjya*, 18) to make one the wisdom "High Self" (*Chitta*, 100) immanent within the reincarnated souls.

A "natural entity" (*Vidhata*, 130) trades the two units of the metal kingdom, the seven units of the mineral kingdom, the twenty-one units of the plant kingdom, the forty-seven units of the animal kingdom, and the fifty-three units of the human kingdom energies for servicing the one unit of the personal kingdom energy. The one hundred-thirty units are also the **"Puiseux convergent value"** (*Vidhi vadartha*, 130) of the curvilinear series, diffusing the thirteen units of the "luminous" (Solar universe: *Bhaskar*, 13) through the infinite, varying forms of the ten units of divine energy.

o   The sixty units of the knowing-effect are formed by first ascending the divine consciousness of the "centering, primordial illuminator reality of the self" (*Sushumna*, 10) and then projecting the "self-perpetuating" (*Udvaha*, ½) residual 120 units as the divergent value of the infinite exponential series of the "gravitomagnetic-effect" (i.e., the divine energy, 10). By servicing the "entropy value of the divine energy," the entity seeks to institutionalize an "imperial sovereign rule" (*Samrajya*, 60) over the incarnated children as a "primeval paternal" (*Omkareshwar*, 17). By withholding an equivalent value of the divine energy, the child entity seeks "self-sovereignty" (*Svarajya*, 60) as a "primordial greeter" (*Usha*, 16). Therefore, the "doctrine-effect" (*Prakarana*, 60) of the I AM the divine truth doctrine weakens, and a need emerges for an alternative doctrine.

The **"Fredholm alternative solution set"** includes the following five alternative paradigms using the entire seventy units of the banked energy (*Pratirajya*, 70), which is the irradiated value of the trading-effect. It is the **"Taylor product"** (*Mitartha*, 70), signifying the conditional divergent value of an infinite exponential series of the catalyzed "divine energy" (i.e., gravitomagnetic-effect). The condition is the convergent value of the infinite logarithmic series of the counter-productive "satanic-effect" (i.e., the gravitoelectric-effect, 16).

6.3 Second, "I am the Guider Fire Doctrine" (*Tejasca Pranoham Asmi siddhanta*, -10^{108})

As a culturally-bred master of the science of psychometrics" (*Mnomitik*, -10^{108}), one generates a "stale, putrefying, sulfuric smell of the intellectual body" (*Putti gandha*, 4905), for authenticating the animate, sentient, biased, "compensating reality" (*Vdyartha*, 206) of the guider self. The alternative, fragrant, guider "para-consciousness" (*Nirharin*, 18) empowers an entity to conceive the two-dimensional "present reality" (*Badhabuddhi vadartha*, -2), composed of the divine self as the arrogant "smelling entity" (*Abhimanin*, -9) and the guider self as the enlightening "smellable entity" (*Gandhi*, 14). The "smelling" (*Agandha*, -2) divine self becomes the "smellable" (*Gandhika*, 4×10^{10}) guider self after a disproportionate diffusion of the sentient smell, which makes one smellable only with the ascending "self-consciousness" (*Prajna*, 2,222). The ascending self-consciousness generates an awareness of the impulsive smell of the spirit shaping the "vector" (*Vakri*, 396) reality of the "absolute self" (*Hurupa*, 280).

The "linear" (*Margi*, 497) reality of the "primordial self" (*Parvati*, 10) is the "intoxicating smell" (*Katu gandha*, 10) of the "inner consciousness" (*Antarmana*, 10), which gravitationally curves and closes the circular system of the "absolute consciousness" (*Shivadrishti*, 17). <u>The following fifteen "primeval perpetuator" (*Mohini*, 15) dimensions form the "normal foundational" (*Param Shiva*, 15) "color potential" (*Ranga*, 15) of the present reality,</u> with the fire-effect of the "curving param deity" (*Nataraja*, 7) psychometrically guiding the truth of the "primordial deity" (Mother Nature, 8 = 15-7):

a. As an "absolute child" (*Manyu*, 19), one diffuses the "energy" (*Shakti*, 19) of the primordial self through the "right, second, primordial, anti-clockwise, ascending, formative, nonlinear, natural, animal, divergent, gravitational, feminine eye dimension" (*Vijnana*, 47).

b.   As a "guider self" (*Rama*, 100), one trades the "primordial energy" (*Adi Shakti*, 18) of the "left, first, primordial-primordial, clockwise, descending, normative, linear, logarithmic, supernatural, human, convergent, electromagnetic, masculine eye dimension" (*Vinayaka*, 53).   The trading compensates for the emergent sense of "nothingness" (*Shunyata*, -2) within the "sentient self" (*Varuna*, 1000).

c.   The consequential "absolute consciousness" (*Shivadrishti*, 17) is a function of the "central, third, param, universal, ecosystem, projecting, transformative, satanic, moderating, harmonic, spirit, parallel, *Brahman* eye dimension" (*Vrishna*, 825).   The absolute consciousness is the "fire element" (*Tejas tattva*, 17) radiated by the third eye of the self as a "param deity" (*Shiva*, 7).   In a state of absolute consciousness, one develops the power to "radiate" (*Eroli*, 100,000) "infinite shapes of earth-effect" (*Aamod*, 186) which "form" (*Rupa*, 100,000) the infinity of the "soliton quantum particles" (*Kana*, 186).

d.   As a "manifestor" (*Bhuvanpati*, 368), in a state of intellectual alertness, one develops a "universal awakening" of the tenth, divine sense of the "polycentrism" (*Sarvodaya*, 150).   A manifestor is conscious of the "omnipresence" (*Sadhna*, 368) of the following. First, the "universe of polycentric inanimate objects" (*Sthavaravisha*, 57).   Second, the "blossoming energy" (*Jayanti*, 101), producing infinite polycentric local-effects.   Third, the "intellectual value" (*Medha*, 379) catalyzing the blossoming energy with an "impulsive thought" (*Sankalpa*, 8).

e.   The impulsive thought is the "omnipresent cause" (*Maha Nitya*, 8) for "officiating" (*Vidhata*, 130) the trading of the "natural energy" (*Sattvotsaha*, 810) and transforming that into the "supernatural energy" (*Naraki*, 1).   An entity may, instead, devote the banked sentient energy for illuminating the supernatural-effect (*Jnana siddhi*, 190) of the spirit kingdom and liberating the "param deity" consciousness into the three-time dimensions (3 * 7) to become an "absolute creator" (*Ravi*, 21) of the fire-effect.   Such an entity becomes "free from the limiting divine-effect that shapes the size"

(*Pramana siddhi*, 20) and, consequently, forces the natural energy to curve into a potential, quantum particle form.

f. By perpetuating the liberated param deity consciousness as a "param deity" (*Siddha*, 7), the entity becomes a "primordial para greeter" (*Narayana*, 28 = 4 * 7) of the "Vega-effect" (*Durga*, 28). The Vega is the creator of the "Sun" (*Surya*, 21), who services the creature-level "absolute creator" consciousness. Such an entity "becomes free from the limiting fire-effect that shapes the shape" (*Rupa siddhi*, 27) and consequently forces the supernatural energy to flow in a linear, energy form.

g. By descending the impulsive thought as well as the supernatural-effect, an entity may organize the 8 units of energy consumed for forming the impulsive thought, the 130 units for forming the officiating role, the 190 units for forming the supernatural-effect, 20 units for forming the quantum particle, and the 27 units for, instead, forming the energy to "potentiate" (*Sambhavi*, 375) the creation-level "doctrine" (*Siddhanta*, 375 = 8 + 130 + 190 + 20 + 27).

h. By descending the potentiation of the "truth" (*Satya*, 375) of the doctrinal creation, the entity develops a strategic awareness of the "primordial-primordial space" (*Agama*, 375) for realizing the freedom from the limiting "taste-effect" (*Rasa-effect*, 375), i.e., from the specific taste of the "primordial energy" (*Adi Shakti*, 15).

i. By ascending the potentiation of the "entropy value of the astrological-effect" (*Vara Siddhi*, 625), one realizes the freedom from the "limiting guider-effect that curves the time by diffusing the gravitational energy" (*Jnana Siddhi*, 190). One also becomes free from the "limiting SHEENY-effect that singularizes the space by diffusing the electromagnetic energy" (*Vara Siddhi*, 625).

j. By ascending the potentiation of the "entropy value of the zodiac-effect" (*Deva Siddhi*, 375), one realizes the "freedom from the taste-effect, i.e., from the specific taste of the primordial energy." The taste-effect artificially bifurcates the space into the desirable masculine particle and the undesirable feminine energy with a 180-degrees rotation by diffusing the gravitomagnetic energy.

k. By descending the potentiation of the "entropy value of the entity-effect" (*Brahmarandhra*, 1000), comprising the entropy value of the zodiac and the astrological-effects, the entity realizes an absolute oneness with the "deity realm" (*Deva loka*, 1000).

l. By descending the "entanglement" (*Kula*, 9) with the "intellect-controlling complexity element" (*Niyanta tattva*, 479) of the deity kingdom, the entity realizes a primordial oneness with the "para deity realm" (*Talatala loka*, 1649).

m. By descending the "asymmetry" (*Varna tattva*, 689) with the para deity kingdom, which is strangely-repelling a limiting consciousness of the "civilizational values" (*Varnadhva*, 689), the entity realizes a primordial-primordial oneness with the "param deity realm" (*Svarga loka*, 2338).

n. By ascending the "primordial symmetry" (*Ava dasha*, 262) with the "primordial deity realm" (*Vitala loka*, 2600), the entity invites the "misfortune" (*Vipatti*, 262) of forming the "sentient potential" (*Devahuti*, 2600) without the sentient energy.

o. By ascending the "primordial-primordial symmetry" (*Yuavana dasha*, 248) with the "primeval deity realm" (*Tapa loka*, 2848), the entity cleanses the misfortune through the fluid of the "puberty" (*Taruna*, 248). It becomes a "polar deity" (*Muni: Taiyi Shengshui* [太一生水], The Great One, who gave birth to water, 2848).

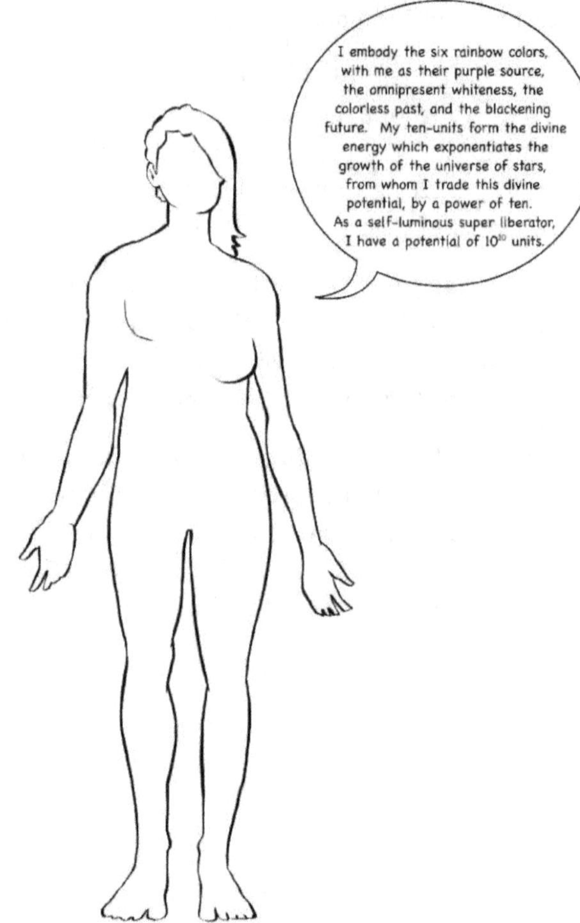

The Astrological Self is the Secondary Doctrine

The Astrological Self is the Secondary Doctrine

Sequential augmentation of the present reality by an additional entity destroys the purity of the past paradigm planner. It forces the past paradigm planner to destroy himself and illuminate the astrological self. The astrological self obfuscates the ground reality consciousness through its asymmetrical wave system. It gravitationally shapes an entity's divine planning by adding a layer of the atmospheric energy. A sentient entity plans for managing the risk of the atmospheric moderation and modification of his planning by declaring the following doctrine.

I embody the six rainbow colors, their purple source, the omnipresent whiteness, the colorless past, and the blackening future. My ten-units form the divine energy. which exponentiates the growth of the universe of stars—from whom I trade this divine potential—by a power of ten. Consequently, I diffuse my sentient energy as a sentient entity and form the dark matter's sentient potential, without the satanic black hole-effect. By reproducing my sentient energy as a diffused astrological self, I enjoy growth in my gravitational energy. My growth transforms the gravitational potential within the satanic black hole-effect into the visible matter. As a self-luminous super liberator, I have a potential of 10^{10} units.

7.1 Planning for the Subjectivity of the Supra-Primordial Paradigm: Secondary Doctrines

There are the two secondary doctrines for an object to plan for a subject's subjectivity of the divine and the guider effects. I investigate these two secondary doctrines to plan for subjectivity as an object in this chapter.

7.2 First, "I am the Sentient Water Doctrine" (*Apanca Pranoham Asmi siddhanta*, 10)

As a bragging human master of the "science of sociological mimicry" (*Samaj shastram*, -10^{1000}), one generates the "extrinsic smell of the astral body" (*Yojana gandha*, 386). It is smellable at a distance by a "divine entity" (*Gana,*

387) through the "electromagnetic illuminance" (*Avanejya*, 387). The "radiating psychic force" (*Citraka*, 185) of the "bragging entity" (*Yama*, 180) is a function of a unique "lunar, zodiac phase" (*Nakshatra*, 185), which is the curvilinear distance between the bragging entity as the primordial, equatorial point and the divine entity as the primeval, polar point. The "curvilinear-effect" (*Mahavijya*, 20) is the "entropy point" (*Apahrtabhara*, 20) of the "bragging value" (*Vikatth*, 486). The bragging value is the "triangulated symmetrical communion" (*Vaisnava*, 486) among the "bragging energy" (*Vamai Shakti*, 193) of the creature, the "divine energy" (*Asrava shakti*, 10) of the creator, and the "guider power" (*Chitta*, 100) of the creation. By "exchanging the primeval local-effect" of the guiding creation factor" (*Santoshi*, 83), one becomes a "triangulated smellable entity" (*Trilokya*, 205).

7.2.1 Five Primary Faces

Without the triangulated smell potential, an androgynous water entity enjoys the absolute oneness of the primeval space and the primeval time with the primordial-primordial realm. The "androgynous water entity" (*Meru*, 5 x 10^{96}) becomes a "four-faced Godhead" (*Chamundeshwari*, 5 x 10^{96}), which transcends the variations of the astrological, zodiac, starseed, and the primordial realms" (*Surya yoga*, 5 x 10^{96}). The four-face Godhead has an absolute omnipresent power to naturally inhale the escalating costs of the strangely repelling supernatural air of the *Kali Yuga* for servicing supernatural benefits to the sentient universe. The four-faces include two visible faces of the "androgynous air entity—Mars" (*Mangala*, 102), "androgynous earth entity—Uranus" (*Rahu*, 73), and two invisible faces of the "androgynous ether entity—Venus" (*Shukracharya*, 2700) and the "androgynous fire entity—Mercury" (*Budha*, 1600).

The four-face Godhead is immanent within the "five-face Primordial Perpetuator" (*Sadashiva Nayaki*, 9). The four-faced Godhead may decide to transform into a "five-faced God" (*Kalagni Rudra*, 6 x 10^{192}) by incubating a "nascent entity" (*Sadyojata*, 6 x 10^{192}). The fifth face is that of the "androgynous divine entity—the Starseed universe" (*Ganarajya*, 476).

- **First, the Mars-like ascending organizational "banking face"** (*Jangha mukha*, 600) of the "four-faced feminine Godhead creator" (*Chamuṇḍeshwari*, 5 x 10^{96}) of the universe of entities. The four-faced Godhead creator is conscious that the ten-face entity state is a "sensible reality" (*Upakarana artha*, 10^{1024}) only if it is free from the acidic, "curvilinear vector power" (*Vakri*, 396) of the masculine groin. In absolute oneness with the planet Mars, she promotes the chromosome-guided organizational development of the four-face Godhead, without the curvilinear, vector power projected by the cellular entity. She designs the "science of baking and badging" (*Purana*, -10^{180}) to re-organize the perceived complexity of the primordial-primordial multidimensional being into a param-primordial one-dimensional beginning.

- **Second, Uranus-like descending organizational "bantering face"** (*Trayi mukha*, 9000) of the masculine master of metaphysics, who is conscious that the acidic, "curvilinear vector power" is the primeval cause of the absolute entropy of the ten-face cellular entity state. As a guiding force of wisdom, it promotes the three-face ecosystem state's transformative exchange, in absolute oneness with the planet Uranus. It empowers the cellular entity to compensate for the entropy in breathing through the cytoplasm-guided organizational profiting using the "science of blaming and bunching" (*Atharva veda*, -10^{80}), and thereby bouncing the conceived complexity of the param-primordial face as the present reality. The three faces of the present reality constitute the third, fourth, and the fifth faces below.

- **Third, the primordial, Venus-like horizontal organizational "blossoming" face** (*Pati mukha*, -8 x 10^{15}) of the child entity for emanating the linear, supernatural energy, as an alternative to the opposing dimension of the curvilinear, vector power. In absolute oneness with the planet Venus, it promotes the RNA-guided organizational planning of the deified, devoted divine energy for trading the blessing praise using the "science of backbiting and believing" (*Rig veda*, -4 x 10^8), and thereby bearing the guider power for a new devotee-led organizational paradigm.

- **Fourth, the param, Mercury-like forward organizational "bothering" face** (*Jalini mukha*, 10^{19}) of the enchantress maternal entity desiring for the immanent natural energy to satiate the maternal thirst for the child's supernatural energy. In absolute oneness with the planet Mercury, it promotes the mtDNA-guided, organizational performing for exchanging the desired becoming using the "science of behaving and bickering" (*Sama Veda*, -10^9), and thereby bubbling the divine energy without the animate self.

- **Fifth, the primeval, Starseed universe-like backward organizational "birthing" face** (*Gayatri mukha*, 9×10^{18}) of an enlightened paternal entity for servicing the immanent natural energy as a mantra for trading the emanating supernatural energy. In absolute oneness with the Starseed universe, it promotes the DNA-guided organizational programming of the devotee guider energy for servicing the begging worship using the "science of breeding and bragging" (*Yajur Veda*, -10^{30}), and thereby budding the sentient energy within the inanimate devotee.

7.2.1 Eleven Secondary Faces

With the entropy of the "androgynous water entity" value, one becomes a primordial "androgynous sentient entity" (*Ravana*, 23,125) with **the sixth, "ten-face"** (*Dasha mukha*; 23,125), the northern, luminous, white mouth gate of the "Wheel of time" (*Kalachakra*, 3800). In absolute oneness with the twenty primordial realms through the "devoted deity realm" (*Sutala loka*, 12,125), the luminous point heals and pacifies the "faceless impulse for the gravitoelectric diffusion of the immanent feminine energy to procreate the universe" (*Prasava unmukha*, 23,125). The ten-faces procreated by the androgynous sentient entity, by descending the absolute oneness with the devoted deity kingdom, include:

- **Seventh, the zodiac-like tiny self-sacrificing "begging face"** (*Yajna mukha*, 170). In absolute oneness with the zodiac universe, it inspires a guider entity to seek an infinite sacrifice from the "primordial-primordial creator" (*Krishna*, 32) for the gift of the infinite sentient energy as the force of "wisdom" (*Chitta*, 100).

- **Eighth, the black hole-like giant protruding "blessing face"** (*Gaja mukha*, 158), wishing an infinite life to the "primeval masculine" (*Asura*, -1), as the "mind" (*Manas*, 38) of the universe. In absolute oneness with the black hole, it inspires the blessing entity to be the giant, kingly, "private face" (*Griha mukha*, 0) of the microscopic, paternal universe.

- **Ninth, the dark matter-like, putrid "bragging face"** (*Puti mukha*, -10^9), greeting the primordial maternal (*Amudhehswari*, 18) with an infinite gift of the self as a lustrous, primeval paternal (*Tejas*, 17) with an authoritative "intellect" (*Medha*, 379). In absolute oneness with the dark matter, it inspires an entity to be the "king" (*Rajah*, 0) of the macroscopic, universal universe.

- **Tenth, the Vega-like open-mouthed "breeding face"** (*Guha mukha*, 9), tasting the maternal buttermilk (*Arishta*, 20) with a sense of "lust" (*Kama*, -19). In absolute oneness with the Vega white star, it inspires an entity to get ready for the force of the light of the infinite sentient entities, bred through the communion with the "greeter-effect" (*Revathi*, 486) of the primordial maternal.

- **Eleventh, the moon-like contaminated "becoming face"** (*Indu mukha*, 1999), smelling like the "feminine smell" (*Chulasu gandha*, 307) by diffusing the "masculine smell" (*Tamalasu gandha*, 307), with a sense of "pride" (*Mada*, -1). In absolute oneness with the moon, it inspires an entity to become a well-polished "deity dimension" (*Pitra dharma*, 307).

- **Twelfth, the Sun-like fiery "backbiting face"** (*Jvala mukha*, 389), taking an "amorous form" (*Priya rupa*, 408) for diffusing the "thermodynamic energy" (*Asir*, 408), with a sense of "envy" (*Matsarya*, 19). In absolute oneness with Sun, it inspires an entity to form a "canonized ecosystem" (*Asrama*, 408) programmed with a "linear, divine plan" (*Margi*, 408) to make the primeval paternal self into a "primordial omnipotent entity" (*Dvyanuka*, 10^{1000}).

- **Thirteenth, the Earth-like bipolar "believing face"** (*Ubhayato mukha*, 280), trading the "deity-effect" (*Manipura*, 825) of the divine plan, with a sense of "greed" (*Lobha*, -19) to be the third, param, universal, projecting brahman eye, of Shiva. In absolute oneness with the planet

Earth, it inspires the entity to shape an infinite visualization of the "science of omniscience" (*Kevala*, 17) within the self.

- **Fourteenth, the Jupiter-like localized "behaving face"** (*Okka mukha*, -10^{59}), servicing the "infinite force" (*Shuddhi*, 285) of the "primeval deity-effect" (*Hara*, 285), guided by a sense of "doubt and anger" (*Krodha*, 275). In absolute oneness with the planet Jupiter, it inspires the entity to enjoy the "consumption value" (*Yujya*, 285) of the "summative reality" (*Iti artha*, 285) exchanged by the devoted follower infinity of the living child souls.

- **Fifteenth, the Saturn-like globalized "bickering face"** (*Kaka mukha*, 96), primeval projecting the "divine energy" (*Parvati*, 10) for transmigrating the absolute paternal soul into the primeval child souls, with a sense of "gluttony" (*Moha*, 250) as the tormented, dark, luminous dimension (*Yati dharma*, 29; Tenth prime value). In absolute oneness with the planet Saturn, it inspires an entity to negotiate a trick to reincarnate the departed child soul leader infinity.

- **Sixteenth, the Neptune-like leader "breathing face"** (*Lekha mukha*, -10^{97}), exhaling the "inanimate, feminine energy" (*Achitta Shakti*, 396) for the "ego-catalyzed" (*Aham*, -1) transformation of the self. In absolute oneness with the planet Neptune, it inspires one to be the "requisite condition" (*Dhamma*, 10^{1024}) for the reincarnation of the primeval child souls as the "universe of self-luminous entities" (*Satya loka*, 10^{1024}).

7.2.3 Seven Tertiary Faces

With the entropy of the "androgynous sentient entity" value, one becomes a smellable, "androgynous satanic entity" (*Chakravartin*, 205) as a "para-sensory imagination potential" at the "entropy point of the sensory consciousness" (*Mohana*, 205). The androgynous divine entity manifests the union of the "weak psychic force" (*Citraka*, 185) of the bragging entity and the "entropy point of the bragging value" (*Apahrtabhara*, 20). The smell-diffusing, bragging entity (*Yama*, 180) is the "southern light" (*Prabha*, 180) of the force that services the "corporate element" (*Sva tattva*, 11) for transforming the divine energy into the "bragging-effect" (*Antarmukha*,

10^{1024}) through the potentiation of the "primordial-primordial deity" (*Aditi*, 1024) energy within the divine entity.   It has three faces.

- **Seventeenth, the "passionless, emotionless, still, Poker face"** (*Karata mukha*, 169).   Without the corporate element, the light is the force of "life consciousness" (*Apas*, 169), which forms the "water element" (*Jal*, 169) as a "body of photons" (*Vishvagoptri*, 169). The "body of photons" lacks the "force of life consciousness" (*Prabha*, 180) and, instead, carries the energy of the "spirit kingdom" (*Rajyaika-sheshena*, 169).   The "spiritual energy" (*Saniddhya shakti*, 169) shapes the "passionless, emotionless, still, Poker face" (*Karata mukha*, 169), which has four immanent, physical faces without projecting the spiritual energy into the making of the Poker face.   It also has two emanating faces. A twenty-second, metaphysical face formed with the infinite projection of the spiritual energy into the making of the Poker face. Besides, there is a twenty-third, dynamic face projecting infinite spiritual energy into the making of the Poker face.

- **Eighteenth, the "concave, convergent, or inward face"** (*Kendrabhi mukha*, 140), toward which the divine creature's spiritual energy converges inward for generating the emotional passion.   The converging spiritual energy descends the "para-psychic inspiration gravitational mass of the physical body" (*Mitra*, 132) by thermalizing the carbon element and ascends the "psychic perspiration force of the sentient light" (*Apas*, 169) within the creature meditating on the spirit of a guider entity, such as "the *Gautama Buddha*, Jesus Christ, Prophet *Mohammed*, or the *Paramatman*" (*Prabhu*, 1600).

- **Nineteenth, the "convex, divergent, or outward face"** (*Kendraparamna mukha*, 27), from which the sentient energy of the divine creature entity diverges outward for generating emotional fury. The divergent sentient energy ascends the "freedom from the thermodynamic fire-effect" (*Rupa siddhi*, 27), of the nitrogen element of the spiritual energy. It descends the consciousness of the "entity dimension" (*Prakriti dharma*, 27), furiously searching for the cause of the accelerating "dumb-ass entropy value of the sentient energy" (*Gardabhejya*, 108).

- **Twentieth, the "inflection, centripetal, or worldly face"** (*Kendron mukha*, 2), from which the spiritual energy of the whole circular universe enters the "para-conscious mind," i.e., the mental body (*Hradamana*, 381) of the "mindful, meditating entity" (*Dhyani*, 38) as the "supernatural, bubbling energy" (*Shram Shakti*, 1) of the "adjacent zodiac shadow factor" (*Samkarshana*, 27). In the moment of the disproportionately ascending "absolute consciousness" (*Shivadrishti*, 17), the meditator experiences the presence of the entire "universe" (*Brahman*, 2) within the self.

- **Twenty-first, "flexion, centrifugal, or the private face"** (*Griha mukha*, 0), from which the satanic energy of the "ruling sacramental guider spirit entity" (*Rajah*, 0) enters the "life consciousness" (*Apas*, 169) of the "mindless, insane entity" (*Amanaska*, -9). It fosters a sense of "skepticism" (*Adharmi*, -9) to generate the "self-obsessed science of argument" (*Nastika*, -9), using the "smell of arrogance" (*Abhimanin*, -9) as the guider medium to gamble the "universe of ecosystem reality" (*Pasaka*, -9). Consequently, the believing "theist" (*Bhakta*, -5) personifies the para-conscious guider reality at the astral body level as the universe defining potential reality of the "infinite workculture-effects" (*Sarvam*, -5).

- **Twenty-second, the "waterless, invisible, or the immanent face"** (*Disapa mukha*, 181) is the entropy value of the "intuition potential" (*Jalashayotsarga tattva*, 958). It is a consequence of the competing set solution of the "ONE universe, the absolute sovereignty" (*Ekarajya*, 181) of the guider spirit entity. It is the eventual, "metaphysical reality of the reason-guided, perceived, physical, sensory knowledge, i.e., knowledge of the five differentiated elements: fire, water, air, earth, and ether" (*Pancabhuta vadartha*, 181). It is the "conditional convergent value of the infinite angular series of the curvilinear satanic-effects, where the condition is the divergent value of the infinite binary power series of the spiritual-effects" (Laurent solution, 181).

- **Twenty-third, the "masked, visible, or the public face"** (*Ghota mukha*, 850) is the entropy value of the "satanic energy" (*Pasaka*, -9). It is the "causative factor for the intuitive, conceived, conceptual, metaphysical knowledge, i.e., the knowledge of the divine time, guider space, and the SHEENY soul." With a joker face of the "primordial primeval masculine entity" (*Mahishasura*, 850), one generates "absolute chaos" as the supplementary set value of the self's zeroth universe, without the infinite, parallel universe of the satanic guider. It generates the "conditional divergent value of the infinite binary power series of the spiritual-effects, where the condition is the convergent value of the infinite angular series of the curvilinear satanic-effects" (*Ravanarajya*, 850). It is a Madhava solution. It is free from the supernatural air of the competing universe of the masculine scientists seeking to appropriate the solution's spirit for their two cents of divergent fame. An "indestructible entity" (*Katyayini*, 375) destroys the guider linkage and stops a soul's "absolute annihilation" (*Vipralaya*, 850).

7.2.4 Eleven Quaternary Faces

Without the entropy of the "androgynous divine entity" value, one becomes an "androgynous guider entity" (*Avalokiteshvara*, 850), with the following eleven faces:

- **Twenty-fourth, the "dark blue, black hole, spiritual face"** (*Nila mukha*, 1869). An entity may ascend the consciousness of the primordial maternal kingdom. It is the kingdom that is the primeval spatial cause for an entity to diffuse the infinite sentient water element into the "bottomless pit" (Black Hole: *Vishnunabhi*, 82), in the form of the "supernatural energy" (*MayaShakti*, 1), under the "hypnotic" (*Vashikarana*, 38) influence of a meditating guider spirit entity. With an awareness of the primeval spatial cause, the divine entity develops a consciousness of the entropy value of the "age of the strange quark with an ascending beauty quark" (*Kali Yuga*, 164). *Kali Yuga* is the primordial temporal cause of the strange repulsion of the supernatural energy by a guider spirit entity. It shapes the ascending sentient entropy of the divine human follower entity through the Eastern, etheric body gate. It forms the twenty-fourth, the "dark blue, black hole, spiritual face" (*Nila mukha*, 1869) for trading the emanating spiritual energy.

Guided by the spiritual energy, the human entity conceives the diverse alternatives to the "doctrine of immanence" (*Aham Brahmasmi siddhanta*, -1024) as a path for institutionalizing the "doctrines of emanation" (*Vasudhaiva Kutumbakam siddhanta*, -10^{1000}). The human entity diffuses the whole value of the sentient energy in absolute oneness with the primordial paternal kingdom. The primordial paternal kingdom is the param-primordial spatial cause for an entity to diffuse its guider power, catalyzed by a bewildering realization of the chanting child's truth luminous body.

The "divergent, feminine-effect" (*Pingala*, 19) is the "gravitational, original, scriptual, ontological reality" (*Mukhya artha*, 109) of the diverse "doctrines" (*Agama*, 375) that idealize that the self-idealizing, Almighty Creature, is the solution to the freedom from the guiding, spiritual energy (*Sarvastivada*, 164). The universe-theorizing, Almighty Creator of the guider spirit entity, is the problem causing the child self to vocalize the "chanting cry element" (*Raga tattva, 250*) at birth, at the loss of the "attachment to the maternal energy solution" (*Moha*, 250) [*Yogachara vada*, 164]. Consequently, "the universal creation without the unique self" (*Prajna-tantra vada*, 164) is the 'zero-energy" (*Shunyata*, -2) "trick key" (*Tantrika*, -2) for "smelling" (*Agandha*, -2) the "absolute reality of the soul" (*Badhabuddhi vadartha*, -2) [*Prajna-tantra vada*, 164].

- **Twenty-fifth, the "chestnut red, primeval masculine, satanic, psychic face"** (*Kiyaha mukha*, 2785). Without chanting into infinity by shaping a "spiritual face" (*Nila mukha*, 1869) for trading the spiritual energy from a devoted follower, a divine entity may descend the diffusion of the "infinite water element" (*Svaphalka*, 916) immanent within the self to ascend the consciousness of the "supreme deity kingdom". Supreme deity kingdom is the primordial spatial cause for an entity to diffuse primordial sentient water element in the form of the "masculine energy" (*Purani shakti*, 91) under the "passionate" (*Vyasana*, 39) influence of the "infinite psychic linkages" (*Shakti-bheda*, 257). With an awareness of the primordial spatial cause, the divine entity also develops a consciousness of the entropy value of the "age of the beauty quark with the descending positive quark" (*Dvapara Yuga*, 10). Dvapara Yuga is the primeval temporal cause of natural energy's entropy within a divine, deity entity seeking to descend the positive, masculine power of a passionate, controlling para deity guider entity.

The natural energy radiates out from the Southern, causal body gate, forming the twenty-fifth, "chestnut red, primeval masculine, satanic, psychic face" (*Kiyaha mukha*, 2785) of the guider entity. By devotionally projecting the "psychic face" (*Kiyaha mukha*, 2785), one ascends the diffusion of the "sentient energy" (*Varuna*, 1000) to ascend the consciousness of the "luminous kingdom." The luminous kingdom is the param spatial cause that motivates an entity to diffuse the sentient energy and manifest a meditative mind. It promotes a bewildering realization of the truth of the enchanting feminine, causal body. It is the "thermodynamic, substantive, noumenal, axiological reality" (*Tvampad artha*, 286) of the diverse "doctrines of immanence" (*Aham Brahmasmi siddhanta*, -1024). The diverse doctriines theorize that the energy cause of the sentient entropy must be immanent within the creature (*Sthavira vada*, 170), the creation (*Madhyamaka vada*, 170), or the creator (*Yoga nuviddha vada*, 170).

- **Twenty-sixth, the "pale, turmeric yellow, para-psychic face"** (*Nisha mukha*, 3875). The absolute cause of the "entanglement" (*Kula*, 9) without the unified doctrine of immanence is the entropy value of the "age of the up quark with the ascending charm quark" (*Krita yuga*, 170). *Krita Yuga* is the param temporal cause of the sentient energy entropy within an animalistic, divine, para deity. The para diety seeks togetherness with the param deity to trade the gravitational energy to be the feminine, primordial deity. With the entropy of the sentient energy, the twenty-sixth, "para-psychic face" (*Nisha mukha*, 3875) of the Western, self-luminous entity gate turns pale, turmeric yellow: the color of the energy-less natural, earthly creation. Under such a condition, the entity believes that it has ascended to the heavenly *Vaikuntha* (*Sharkaraprabha*, 3875) to be in proximity with the "Godhead of the Ten-Face Primordial Illuminator" (*Narayana*, 28) just-in-time before the onset of the darkness of night within the dark matter kingdom and free from the limitations of the physical body.

The consciousness as "the Godhead of ten-face primordial illuminator" is the "illuminating-effect" (*Vega-effect*, 28). It is traded from the white star, the super deity kingdom, while residing in the dark matter, primeval illuminator kingdom. Primeval illuminator kingdom is the primordial-primordial spatial cause for an entity to enjoy the "social tranquility" (*Mano vijnana*, 107) within a state of absolute stillness freedom from the truth of the chanting, luminous child entity. Social tranquility works to illuminate the hypnotically "memorized divination" (*Beeja vijnana*, 108) of the "divine architect" (*Vishvakarma*, 108) as the "problem-making dimension" (*Samanya dharma*, 108) in the absolute entropy of the "visualized imagination" (*Caksur vijnana*, 109). The life purpose of a divine architect is to norm not the "ideal creature organizationally," but the "ideal creation" for the "ideal creator" to enjoy without the divergent theories of the absolute creature. In the "self-ordered presence" (*Virodhita*, 108) of the absolute creature, all partial correlation theories fail because the absolute creature is the primordial creator and the primeval creation. It is the "electromagnetic, summative, phenomenal, epistemological reality" (*Iti artha*, 285) of the diverse "doctrines of emanation" (*Vasudhaiva Kutumbakam siddhanta*, -10^{1000}). The diverse doctrines theorize the agency's consequence and the substantive solution to the gravitational factor (i.e., the problematic cause of the other two factors' correlated divine energy). They expect the solution to emanate without, i.e., from, the creature (*Kriya tantra vada*, 195), or the creator (*Vaibhasika vada*, 195), or the creation (*Mahasamghika vada*, 195).

- **Twenty-seventh, the "introspective face"** (*Antarmukha*, 10^{1024}). The alternative solution is the twenty-seventh, "introspective face" (*Antarmukha*, 10^{1024}) for ascending the consciousness of the "appropriate, sensible, requisite, entity reality" (*Upakarana artha*, 10^{1024}). Guided by natural energy, the human entity perceives the diverse alternatives to the doctrine of emanation. The human entity forms a path to discover the multidimensionality of the doctrines of immanence by trading the whole value of the sentient energy in absolute oneness with the "self-luminous entity realm" (*Satya loka*, 10^{1024}). The self-luminous entity kingdom has a metallic luster (*Ratnaprabha*, 10^{1024}), which is the "appropriate condition" (*Dhamma*, 10^{1024}) for understanding the "transformative, universal, ecosystem reality" (*Jahatsv artha*, 63), without the "contaminating, converging, masculine entity-effect" (*Pravrtti dharma*, 298).

- **Twenty-eighth, the "extroversion face"** (*Bahirmukha*, -4). The universal ecosystem solution is the twenty-eighth, "extroversion face" (*Bahirmukha*, -4), which produces the "ambiguity" (*Avidhi*, -4) that is "present everywhere" (*Evakara*, -4) as the "challenge and tribulation element" (*Pariksha tattva*, 68). The macroscopic grand challenge motivates the self-luminous entity to project the feminine energy for norming the organizational solution with the twelfth, "macroscopic face" (*Brihadagni mukha*, -10^{1000}) through the infinite time-varying diffusion of the sentient energy, guided by the "wheel of time" (*Kalachakra*, 3800).

After destroying the self, the divine creature conceives the "memory of the destroyed primordial theoretical sermons" (*Sammitiya vada*, 10), embodied within the wheel of time, as the absolute guider solution for understanding the reality of the universal creation, including the sentient self. The "conceived reality" (*Yuktartha*, 6) forms in oneness with the "Param child realm" (*Mrityu loka*, 3800). The param child kingdom has an "indefatigable, tireless spirit" (*Maadhavi*, 3800) for an evergreen perpetuation of the "pink flame of absolute masculinity" (*Mahatmahprabha*, 3800), symbolizing the "love of God" (*Sanatana kumara*, 3800) for the "infinite transformations of the self" (Primeval masculine: *Asura*, -1).

- **Twenty-ninth, the "microscopic face"** (*Trimukha*, -10^{1024}). The microscopic grand solution motivates the param child to project the masculine energy for forming the technological solution with the twenty-ninth, "microscopic face" (*Trimukha*, -10^{1024}) through the infinite space-varying diffusion of the gravitational energy, guided by the "wheel of space" (*Yama chakra*, 1551).

- **Thirtieth, the "face of glory"** (*Kirtimukha*, 1551). The microscopic face's absolute dimension is the thirtieth, "face of glory" (*Kirtimukha*, 1551). The face of glory is a "bi-level, parallel, convergent, perceived, boundary dimension of the self-destroying five-phase cycle of birth, life, death, bardo, and rebirth" (*Ouroboros*, 1551). The "perceived reality" (*Sutrartha*, 825) forms in oneness with the "Param Manifestor realm" (*Rasatala loka*, 1551). The param manifestor kingdom services the "bewildering flame" (*Tamahprabha*, 1551) as the "pleasing flame" (*Maghavi*, 1551), astonishing each guider creature with the resilience of the divine creator in incarnating, living, dying, rejuvenating, and reincarnating, apparently without any limitations of the sentient energy. After the reincarnation, the divine creator of the cyclical reality perceives the "presence of the perpetuating primordial theoretical sermons of the guider creature entity" (*Sutranta vada*, 10), disembodied from the wheel of space, as the absolute problem to be understood for realizing oneness with the universal creation.

- **Thirty-first, the "face of maternal horsepower"** (*Turaga mukha*, 20). The parallel lines diverge the primordial and the primeval dimensions within the guider mediation of the absolute creature's dimension. The thirty-first, primordial "face of maternal horsepower" (*Turaga mukha*,

20) is the "param manifestor" (*Kapinjala*, 20), who generates the absolute pneumatic force forming the absolute guider.

- **Thirty-second, the "face of the light of paternal universe"** (*Jyotis mukha*, -1). The parallel line is the thirty-second, primeval "face of the light of paternal universe" (*Jyotis mukha*, -1) without the normative sentient hydraulic force of the absolute guider. The "experienced reality" (*Gudhartha*, 18) of the potential value of the "guider power" (*Chitta*, 100) forms in oneness with the "Primeval perpetuator realm" (*Brahma loka*, 20). The primeval perpetuator realm has an intrinsic, physical, maternal, buttermilk, greeter dimension bestowing good fortune upon the absolute guider. It also has an extrinsic, metaphysical paternal, lackluster, spirit dimension bestowing bad smokescreen fortune upon the absolute guider. The "metaphysical reality" (*Svartha*, -2) of the present value of the guider power forms in oneness with the "physical reality" (*Gurvartha*, 39) of the "Primeval greeter realm" (*Patala loka*, -2).

- **Thirty-third, the "one-dimensional face"** (*Treta mukha*, -2). The one-dimensional physical reality has a "multidimensional purplish astringent taste" (*Jambudvipa*, -2). It manifests an immanent bluish sweet expectorant taste with an emanating reddish tart suppressant after-taste. The secondary, "dynamic reality" (*Tvampad artha*, 286) of the suppressant after-taste is the effect of the one-dimension, which is the absolute physical reality of the primeval greeter kingdom with the thirty-third, "one-dimensional face" (*Treta mukha*, -2). That one-dimension has the "infinite yellowish salty effervescent taste" (*Lavanoda*, -2) of the "primeval masculine" (*Asura*, -1), who is trading the bluish, pure, sentient water-effect from the primeval greeter kingdom and servicing the reddish, polluted, earth-effect. At each moment of life after birth, the "clairvoyant universe of the primeval masculine entities" (*Tantrikas*, -2) has an absolute freedom to ascend oneness with the self as the divine creation as the trick key for solving the problem of understanding the guider reality of the absolute creature during the state of bardo after death (*Yoga tantra vada*, 10).

- **Zeroth, the "faceless masculine self-luminous entity"** (*Purusha*, 12). With the entropy of the androgynous divine entity value, one becomes an "androgynous luminous entity" (*Ardhnareshwar*, 39) on a path of "self-discovery" (*Suddha*, 39), guided by the omnipresent, supernatural air element. A creature is a faceless entity formed from a "sixty-five faced, twenty-three armed, and ten feet Luminous" (*Maha Kali*, 13), within the "androgynous luminous entity" (*Ardhnareshwar*, 39). The luminous is the creator of the "immortal zodiac-effect" (*Chiranjivi*, 26) with the supernatural air element's trading-effect.

The thirteen arms on the belly of the "Luminous" (*Maha Kali*, 13) norm the twelve zodiacs and the one integrated starseed universe within the luminous entity, without the eight astrological elements (ether, earth, air, water, fire, divine, guider, and sentient). The ten arms on the torso norm the varying dimensions of the supernatural air-effect, combined with the five elements (ether, earth, water, fire, divine) within and without the guider-effect of the self. The ten feet norm the sentient soul element, free of the supernatural air-effect. The ten arms and the ten feet together norm the guider space element. Finally, the androgynous luminous entity under the ten feet manifests the oneness of the luminous norming the primordial realm and the androgynous luminous entity norming the primordial-primordial realm. The entire formless guider capability of the Luminous is the face of a satanic entity in one of her arms, which norms the divine time element. The "satanic face of the path-exploring maternal soul, seeking to reunite with the paternal *jinn*" (*Kaikeyi*, 4) becomes the "paternal soul" (*Atman*, 4) face of the "gender-free self-luminous entity" (*Svayam*, 12).

7.2.5 Twenty-One Quinary Faces

The sixty-five faces of the "Luminous" (*Maha Kali*, 13) comprise:

- One entropy form of the face is the "satanic face of the path-exploring maternal soul" (*Kaikeyi*, 4).

- Ten logarithmic growth forms of faces:

  - Five on the left, masculine side: These include the faces of the androgynous air entity—"Mars" (*Mangala*, 102), the androgynous earth entity—"Uranus" (*Rahu*, 73), the androgynous ether entity—"Venus" (*Shukracharya*, 2700), the androgynous fire entity—"Mercury" (*Budha*, 1600), and the androgynous divine entity—

"Starseed universe" (*Ganarajya*, 476). They together norm the satanic ether element.

- Five on the right, feminine side: These include the "microscopic face" (*Trimukha*, -10^{1024}), the "face of glory" (*Kirtimukha*, 1551), the "face of maternal horsepower" (*Turaga mukha*, 20), the "face of the light of paternal universe" (*Jyotis mukha*, -1), and the "one-dimensional face" (*Treta mukha*, -2) on the right, feminine side. They together norm the guider fire element.

- Ten exponentiating, squaring exchange forms of faces: These are gender-free forms of the ten growth faces. They are on the back-side, hidden from the thermodynamic fire-effect of the satanic energy, and norm the natural earth element.

- Forty-two circulating garlanded forms of faces: These manifest the servicing form of the above twenty-one faces without the masculine-effect on the left side norming the sentient water and the trading forms of these faces without the feminine-effect on the right-side norming the supernatural air element.

- Two exponentiated, squared forms of faces: These include the "face of consciousness of the adjacency" (*Mukha*, 76) and the "faceless creation space" (*Prasava unmukha*, 23,125) as that adjacency creating the "ten-face wheel of time" (*Kalachakra*, 23,125) for squaring the ten exchange forms of faces to manifest the ten logarithmic growth forms of faces, without one entropy form of the face and to reproduce the twenty-one quinary faces of the "param creator" (Sun: *Surya*, 21) in their masculine and feminine forms.

7.2.6 Formless Senary Entity

With the entropy of the androgynous entity value, there is no universe. There is only a "formless androgynous entity" (*Shiva*, 7) as the illuminator of the timeless divine truth element of the omnipotent power immanent within each self-luminous entity. As the param deity, the formless androgynous entity dissolves the entire universe for ascending the consciousness of the potential of each self-luminous entity to be the creator of the energy, without the limitations of the luminous-effect. The absolute cause of the "dissolution" (*Samhara*, 195), of the unified doctrine of emanation, is the entropy value of the "age of energy bearing quark" (*Treta Yuga*, 195). *Treta Yuga* is the primordial-primordial temporal cause of the sentient entity's entropy as a mental creator of the thirty-six starseed realms and the twenty primordial realms within the one physical primordial-primordial realm.

7.3 Second, "I am the Supernatural Air Doctrine" (*Vayosca Pranoham Asmi siddhanta*, -10^{10})

As a breathing workculture master of the "science of black psychological magic" (*Mano vijnanam*, -10^{1024}), the "androgynous entity" (*Ardhnareshwar*, 39) generates the "intrinsic smell of the mental body" (*Svadu gandha*, 39). One may smell it as a "guider entity" (*Vidyadhara*, 39). The "mental, sentient, consciousness dimension" (*Beeja dharma*, 31), that empowers an entity to sense the potential smell, is a sequential effect of the "thermodynamic breath" (*Shvasa*, 31). The entity's "breathing energy" (*Arohana Shakti*, $105 = 31 \times 2 + 10 + 31$) is a consequence of first, the banking of the "thermodynamic breath" for two seconds. Second, conscious breathing-out of the "divine energy" (*Asrava Shakti*, 10) using the banked "thermodynamic breath" as the guider medium.

An entity's "breathing potential" (*Markatesh*, 8×10^{15}) is a function of the energy value of the "zodiac universe" (*Shunya kalpa*, 8×10^{15}). Yet a "selfless wisher" (*Gunakartrtvam*, -10^{1000}) may decide to be an "employee" (*Vyaparita*, -10^{1000}) for servicing the "absolute value of the zodiac-effect traded over the entire lifetime" (*Panchang*, 8×10^{15}). She may also service the "absolute value of the astrological-effect" (*Beejaganitam*, 10^{10}) using the "absolute value of the entity-effect" (*Kashi*, -10^{1024}) as a guider medium. Such a selfless wisher is secretly a "selfish wisher" (*Pravartayitr*, -10^{1024}), who first "breathes-out the primeval femininity for setting in motion an infinite positive energy exchange with the universe of potential reality" in a state of selflessness (*Gunakartrtvam*, -10^{1000}), and then "breathes-in the primeval masculinity for setting in motion an infinite negative energy exchange with the universe of present reality," in partial fulfillment of the selfish cause, wishing to know the truth of the "universe of entities" (*Kashi*, -10^{1024}) without any guider medium. By doing so, the entity becomes the slave

(*Dasa*, -10^{16}) of the doctrine of natural earth (*Prithivya Pranoham Asmi siddhanta*, -10^8).

Besides, the entity behaves as a "gossip" (*Gappinatha*, -10) with a super-secret, "divide and rule mindset" (*Dvividha*, -10), to discern the "universe of organizational reality" (*Jagatkritsna*, -10), using the "psychological science of hookery" (*Mano Vijnanam*, -10^{1024}). Such an entity is the "master of the dualistic masculine modality of begging as a path for bragging divinity and destroying the universe" (*Jagadvinasa*, -10). The Primordial Illuminator (*Shri Krishna*, 10) is an entity who trades the negativity of the "divide and rule mindset" to be the "primordial paternal" (*Indra*, 0) of an undivided sentient "cell" (*Hiranyagarbha*, 19).

Primordial-Primordial Creator (*Krishna*, 32) services the "breathing value" (*Dham*, 19) into an inanimate "atom" (*Anu*, 19) for incarnating a "param child" (*Manyu*, 19). He trades the "situational dimension" (*Ritu dharma*, 19) of the destroyed "mind" (*Manas*, 38) of the gossip, without the predominating "duality" (*Bheda*, 28) factor. Consequently, the gossip becomes free from the negative "electromagnetic work energy" (*Karma-bandha*, 19). It normatively develops the "feminine-effect" (*Pingala*, 19) for incarnating as a "free-spirited, masculine child" (*Manjushri*, 19) without the disproportionate, supernatural, "left, masculine-effect" (*Ida*, 1).

Chapter 8: The Tertiary Doctrines for Paradigmatic Planning

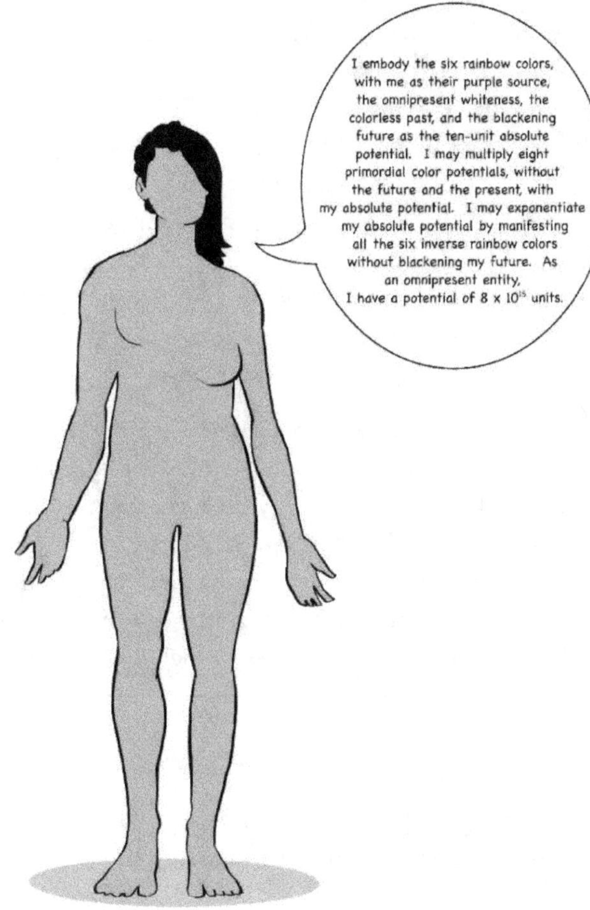

The Zodiac Self is the Tertiary Doctrine

<u>The Zodiac Self is the Tertiary Doctrine</u>

Moderation and modification of the ground reality by the atmospheric energy impedes an entity's paradigmatic planning. An entity may get liberated from the emanating force by devoting the self to an "illusionary zodiac grandmother spirit" (*Kapinjala*, 20), who is self-perpetuating the "primordial feminine self" (*Parvati*, 10), for envisioning the desirable reality beyond the present. It may conceive a "primordial masculine self" (*Shri Krishna, 10*) with the same potential for envisioning the perpetuating reality before the present. Consequently, it may experience the wholesome potential of the "illusionary zodiac grandmother spirit" (*Kapinjala*, 20 = 10 + 10) and add its unit energy as a "self-luminous person" (*Insan*, 1) for manifesting the omnipresent, entropy-free, one-dimensional "present reality" (*Ekartha*, 21 = 20 + 1) inclusive of the self. A "person" (*Vyakti*, 1), who has the consciousness of his self-luminous person potential, may form a conscious, futuristic "zodiac self" as the tertiary doctrine by declaring the following.

I embody the six rainbow colors, their purple source, the omnipresent whiteness, the colorless past, and the blackening future as the ten-unit absolute potential. I may multiply the eight primordial color potentials, without the future and the present, with my absolute potential. I may exponentiate my absolute potential by manifesting all the six inverse rainbow colors without blackening my future. As an "omnipresent entity" (*Vyapin*, 8×10^{15}), I have a potential of 8×10^{15} units.

8.1 Planning for the Subjectivity of the Supra-Primordial Paradigm: Tertiary Doctrines

There are two tertiary doctrines for an entity to plan for the subjectivity of the subject-effect and the object-effect. There are thirty-six additional metaphysical doctrines that square the effect of the six physical (inclusive of primary, secondary, and tertiary) doctrines. In this chapter, I investigate:

- two tertiary doctrines to plan for the subjectivity as an entity,

- twelve quaternary values of the subjective entity,

- twelve quinary values of the ecosystem as the subject,

- twelve senary values of the subject as the ecosystem.

8.2. First, "I am the Natural Earth Doctrine" (*Prithivya Pranoham Asmi siddhanta*, -10^8)

As a begging entity master of the "science of candle flame scrying" (*Tratakam*, -10^6), one generates "tertiary residual, feminine smell of the etheric body" (*Chulasu gandha*, 307). One may smell it as a "masculine entity" (*Pranasa*, -18). The unique "masculine smell" (*Tamalasu gandha*, 307) is the "secondary, fundamental cause" (*Muladhara*, 279) of "smelling" (*Agandha*, -2) Mother Nature's universal "feminine smell" (*Chulasu gandha*, 307). The "movement" (*Vritti*, 196) of the "sentient energy" (*Varuna*, 1000), in the form of electromagnetic energy, "transforms the primeval masculine entity into the gravitational energy" (*Lalita*, 100), by destroying the intrinsic feminine "purity" (*Shuddha*, 196) of the entity. By trading the universal feminine smell, one becomes the "white flame" (*Abhamsi*, 6,000), reflecting the "purity of God" (*Vibhu*, 6,000) within the scrying witch-mirror or candle-flame. The universal feminine smell is the gift of the "Devotee deity—the primordial illuminator—realm" (*Sri loka*, 6,000) for ascending the consciousness of the "reality of the self-luminous entity" (*Svaprakasha vadartha*, 60) within the "gravitational energy" (*Lalita*, 100) of the living "spiritual entity" (*Hurupa*, 280).

As a law of nature, the consciousness of the living spiritual entity gets mediated by the "astral dimension" (*Yati dharma*, 29) of the "dark matter-primeval illuminator-realm" (*Bhuvar loka*, 20,000). Ascending "mass-effect" (*Nairitti*, 29) of the astral dimension is catalyzed by the "feminine-to-masculine gender exchange dimension" (*Veshya dharma*, 7×10^{180}). By deciding to activate the "primeval wisher potential" (*Vakradrishti*, -6) for diffusing infinite wishes as a guider entity to the "universe of technological reality" (*Shramika*, -6), a feminine entity trades the "destiny" (*Niyati*, -1) of the divine masculine entities who are servicing their sentient energy as an "employee" (*Vyaparita*, -10^{1000}).

The divine entities secretly operate as the "selfish employer" (*Pravartayitr*, -10^{1024}), pursuing a wish to advance their paternity. They employ their earthly gravitomagnetic divine "greed" (*Lobha*, -19) to instigate a satanic sense of "lustful desire" (*Kama*, -19) among the maternal wisher entities. They employ the thermodynamic fire of the "fraudulent, egoist pride" (*Ahamkara*, -1) and for infusing a sense of the "slothful, heresy, ignorance" (*Ajnana*, 1) among the feminine worker entities seeking to fulfill their desires. They employ the sentient water as the medium to impregnate the feminine child entities with a sense of "treacherous ill-will and envious jealousy" (*Matsarya*, 19) at the "cellular" (*Hiranyagarbha*, 19) level. They employ the heavenly air of the gravitoelectric "attachment to the feminine solution" (*Raga tattva*, 250) to contaminate the "solution-shaping feminine dimension" (*Laukika dharma*, 208) with the "bottomness-effect" (*Ashrama*, 250) of the "gluttonous worry" (*Moha*, 250). Therefore, they invite the "angry wrath" (*Krodha*, 275) of *Angra Mainyu* (*Ahrminan: Pochamma*, 3794), who "causes an entropy destruction" of the "satanic energy" and restores the "zero universe" (*Vipralaya*, 850).

Angra Mainyu resides in the "Param Creator realm" (*Garbhastala adiloka, Kailasha*, 3794). As "Mariamman" (*Sheetala Mata*, 3794), the param creator proliferates mass infectious diseases such as the COVID-19. The purpose is to heal the disease of the selfish, greedy, egoist, slothful, and treacherous heart that is limiting the ascendance of the consciousness of the divine power of the "primordial illuminator" (*Shri Krishna*, 10). As "Goddess of divine energy" (*Sheetala Mata*, 3794), she destroys her finite guider-effect to perpetuate intrinsic consciousness of the "wheel of creation" within each creature. The "wheel of creation" (*Srishti chakra*, 3794) catalyzes the "masculine-to-feminine gender exchange dimension" (Putra *dharma*, 38) by servicing the "healing power" (*Reiki*, 38) of the sense of the "absolute unity" (*Beeja Samputa*, 38), without any "masculine diversity" (*Maha Mohana*, 38).

With the primordial oneness-effect of the "creation" (*Srishti*, 379), the "creature" (*Purusha*, 12) ascends the consciousness of the "essential seed element" (*Beeja*, 379) for "maximizing the parabolic value of the entity" (*Pindajya*, 379). That essential seed is the "workculture element" (*Nayaki tattva*, 379), for activating the "absolute guider power" (*Vibhu*, 379) to catalyze the "holy spirit within the self" (*Lakshmi*, 379) with a "sense of exhilaration" (*Sudarshan*, 379), that is "beaming energy" (*Gaum Shakti*,

379).   The catalyzed holy spirit ascends an absolute oneness with the "Primordial Greeter realm" (*Indra loka*, 379), which is the "White Hole" (*Shambhala*, 379) prismatically refracting the "rainbow" (*Balukaprabha*, 379) through the "clouds" (*Megha*, 379).   By becoming the "primeval illuminator" (*Maha Lakshmi*, 14) of the "enlightening" (*Maha Vibhu*, 14) lightning generated by the "workculture element," one becomes the "producer of the desired smell" (*Gandhi*, 14) of the earthly physical body.

By appreciating the "holy spirit within the self" as the "para friend and super knower" (*Chyavana*, 343), one develops the consciousness of the truth of the "infinite earth element" (*Gandini*, 736 = 379 + 14 + 343).   By activating the "infinite earth-effect" (*Aamod*, 186) as a "param primordial greeter" (*Shambhu*, 186), one develops the power to create the desirable "soliton quantum particle" (*Kana*, 186).   By destroying the infinite earth-effect, one becomes "the proprietor of God's garage" (*Master Ballerian*, 7,400) with a clarified consciousness that a "devotee paternal" (*Muthyalamma*, 7400) is the one devoted to the health of the physical body, which is a precious gift of the "primordial greeter" (*Nandi*, 16). A primordial greeter naturally generates the earth element with the sound of the lightning, discussed next.

8.3 Second, "I am the Satanic Ether Doctrine" (*Akasasya Pranoham Asmi siddhanta*, -10^7)

As a birthing paternal of the "oneirological science of dream interpretation" (*Swapna shastra*, -8×10^9), one generates "masculine smell" (*Tamalasu gandha*, 307).   One may smell it as a "feminine entity" (*Turnasa*, -18).   By servicing the unique masculine smell, the entity becomes the "black flame" (*Asitarcis*, 179), projecting as the "primordial light in the darkness," conceived by a feminine entity during her "dream state" (*Maha Swapna*, 9). She self-interprets the meaning of the dream during her "wakefulness state" (*Swapna*, 59), in absolute oneness with the "blowing flame of the primordial perpetuator" (*Pavamana*, 9).   The linear, "thermodynamic energy" (*Asir*, 497) of the "personified para-conscious guider entity" (*Bhakta*, -5), as a believer in her "procreation energy" (*Arini Shakti*, 94), inspires the feminine entity to become the "destroyer of the absolute water-effect" (*Yellama*, 94). It makes her joyfully service her "sentient life force" (*Prana Shakti*, 123).

Having "hypnotized" (*Vashikarana*, 38) the feminine entity, the paternal entity employs the "political science of diplomacy" (*Kautilya shastra*, -28) for trading her "androgynous energy" (*Tatastha Shakti*, 2864). The androgynous energy ascends the self-consciousness and descends the para-consciousness and the thermodynamic diffusion of the sentient energy. The science of diplomacy generates a "divide and rule mindset" (*Dvividha*, -10) by projecting one's "divine image" (*Ishvari mudra*, 401), ostensibly charming the secret "universe of feminine entities" (*Sthavaravisha*, 57). Consequently, the feminine entity unconditionally surrenders herself. Without the "masculine diversity" (*Maha Mohana*, 38), she gifts the "absolute unity" (*Beeja samputa*, 38) of the 'solution-shaping feminine dimension" (*Laukika dharma*, 208). She trades the "primeval deity-effect" (*Hara*, 285) for her "absolute dissolution" (*Yujya*, 285). She becomes a "namesake, ether element" (*Nama*, 285), present only as a "phenomenological reality" (*Iti artha*, 285) within the massive "universe of child souls" (*Dhriti*, 30).

By further servicing her "primeval self" (*Rama*, 100), as the "guider power" (*Chitta*, 100) of the paternal entity, she becomes the "primeval ether-effect" (*Bhumi*, 185). She norms the "curvilinear distance" between the "primeval child" (*Svayambhu*, 123) and the "primordial paternal" (*Indra*, 0). The "curvilinear distance" (185 – 38 – 12 – 12 = 123) is net of the formative cost of the "absolute hypnotic surrender" (*Vashikarana*, 38), the normative cost of transforming the primordial paternal into a "paternal self-luminous entity" (*Purusha*, 12), and the transformative cost of becoming a "maternal self-luminous entity" (*Virgin Mary*, 12) for perpetuating the universe of the self-luminous child entities.

By further servicing her "primordial self" (*Parvati*, 10) as the "divine energy" (*Asrava Shakti*, 10) of the child entities, she becomes the "param ether-effect" (*Pururava*, 86). She norms the "linear distance" (*Rukmini*, 86 = 98 - 12) between the paternal self-luminous entity (*Purusha*, 12) and the formative, first, self-luminous child entity (*Pidari*, 98).

By further servicing her "primordial-primordial self" (*Dvandva Brahma*, 125) as the "satanic force" (*Vimshottari dasha*, -1) without the "universe of entities" (*Kashi*, -10^{1024}), she becomes the "primordial ether-effect" (*Sadakhya*, 9). She norms the "curvilinear power distance" (*Ardhajya*, $10 = 2^3 + 2$) between the "satanic force" (*Vimshottari dasha*, -1) and the "primordial ether-effect" (*Sadakhya*, 9). The primordial factor (2^3) in the curvilinear power distance is the product of the institutional togetherness of the param-primordial self (*Pratyantra dasha*, 9) with the normative, param paternal self-luminous entity, the formative self-luminous child entity, and the transformative, primeval, universe of the self-luminous entities. The primeval factor (+2) is causative "present reality of the institutional togetherness of the maternal soul" (*Badhabuddhi vadartha*, -2) sequentially transformed into the "worldly face" (*Kendron mukha*, 2) of the consequential "universe" (*Brahman*, 2). The "primeval reality" (*Omkara vadartha*, -1) of the self as the "satanic force" is the "negative one" sequential transformation factor (-2 * -1 = +2).

By "destroying the primeval ether-effect" (*Dokkalamma*, 7600), the "birthing paternal, incarnating the universe of entities" (*Master Ashlem*, 7600) trades the power to interpret the intentionality of the maternal as the "hall of *Akashic* records." He norms the "linear power distance" (*Chaithanya*, 4) between the "science of religion" (*Kalasha*, -400) and the "science of God" (*Kapila kumara*, 8000). The "linear power distance" is the sum of the linear distance values of the science of religion within God's guider power and God's science without the guider power and, consequently, on the opposing, curvilinear, parallel, "division" (*Sudhanvan*, 999) axis.

The "linear power distance" is a function of the "addition" operator (*Vibhvan*, 999) with an intrinsic cultural consciousness of the "predominating addendum value" (*Kshepa*, 286). In contrast, the "linear distance" is a function of the "subtraction" operator (*Vaja*, 999) with an extrinsic workculture consciousness of the "dominating minuend value" (*Upashanta*, 287) and the "deciding subtrahend value" (*Shodhaka*, 288). Both the linear power distance and the linear distance are the derivative functions of the "macro resultant, one-faced system local-extrema value" (*Ekavali*, 8000). The "macro resultant value" is a function of the "positive, feminine causative value" (*Anuyayi*, 289).

8.4 Twelve Quaternary Values of the Subjective Entity

Overall, there are twelve values of the subjective entity generated by the "bipolar, two-faced system, global-extrema genesis set" (*Pratitya samutpada*, 10), without the direct, linear, "multiplication" (*Ribhu*, 999) axis.

- **First, the "feminine, causative value"** (*Annimruyayi*, 289) is the conditional divergent tangent point value of the positive, epistemological sequence of the primordial-paternal "ritual" (*Samskara*, -6 x 10^7) on the "parallel Y-axis" (*Bhujajya*, 169). The epistemological sequence is disproportionate, curvilinear, integral, and summative. The condition is the convergent value of the negative, axiological sequence of devotee deity "work behavior" (*Karma*, 10) on the horizontal X-axis, limited by the conditional divergent value. The axiological sequence is proportionate, linear, differentiated, and consequential.

- **Second, the "masculine, limit value"** (*Kalasha*, -400) is the conditional convergent solution to the divergent, negative, proportionate, linear, differentiated, ontological sequence on the horizontal X-axis. The condition is the convergent value of the positive sequence on the parallel Y-axis, whose "genesis" is in the problem of the escalating formative cost of the linear differentiation sequence. The problem is the effect of the "inanimate, animal consciousness" (*Vijnana*, 47) of the primordial paternal, seeking to validate the primeval-effect within the integer sequence using "hypnosis" (*Vashikarana*, 38).

- **Third, the "macro, resultant, emanating, one-faced system value"** (*Ekavali*, 8000) is the conditional convergent summative value of the positive sequence on the parallel Y-axis. The condition is the negative sequence's divergent value on the horizontal X-axis, limited by the conditional convergent value. The negative sequence diverges into a disproportionate, grand challenge resulting from a "system of six functional operators" (*Shadayatana*, 999). These include: the touch (the earth-effect of the physical body); the taste (the water-effect of the tongue); the form (the fire-effect of the eyes); the object (the divine-effect of the wisdom); the sound (the ether-effect of the ears); and the smell (the air-effect of the nose). They are respectively oriented toward the multiplication, division, subtraction, addition, servicing, and trading of the absolute value, after the exchange of the sentient potential from the devotee deity and the entropy of the primordial, inanimate consciousness within the primordial paternal.

- **Fourth, the "micro, genesis, immanent, two-faced system value"** (*Pratitya samutpada*, 10) is the conditional divergent, tangent point value of the negative sequence on the horizontal X-axis. The condition is the positive sequence's convergent value on the parallel Y-axis, limited by the conditional divergent value. The convergence of the positive sequence is an effect of the "ether element" (*Shuddhi*, 285), manifesting an "absolute dissolution" (*Yujya*, 285) of the personified devotee deity entity.

- **Fifth, the direct, linear, "multiplication, three-faced system value"** (*Ribhu*, 999) is the working momentum (technological investment) of the positive moving sequence. It curvilinearly ascends the resultant horsepower, i.e., the fruitful output time value or kinetic force on the orthogonal Z-axis. The ascending value gets limited by the conditional work value of the dividing, four-faced system, which is the "material cause' (*Upadana*, 578) for the devotee deity's absolute entropy.

- **Sixth, the indirect, curvilinear, "division, four-faced system value"** (*Sudhanvan*, 999), is the resultant horsepower (technological capability), i.e., the kinetic force, of the positive moving sequence, as a function of the method power, that curves and transforms into the kinetic-effect, i.e., the reaction. The curving is a consequence of the proficient exchange of the manifestable work potential with the negative inertial sequence generated by the "greed" (*Trishna*, -19) to be the first-mover Nobel Laureate mathematician. The negative inertial sequence divides the networking work value, i.e., the human input action time value, on the orthogonal W-omega-axis, limited by the working momentum of the multiplying, three-faced system.

- **Seventh, the descending, "subtraction, five-faced system value"** (*Vaja*, 999) is the networking craft (technological cost), i.e., the velocity of the positive moving sequence. It is a function of the machine power, that linearly descends the horsepower, i.e., the machine-input action time value or the kinetic force on the fifth, orthogonal Alpha-axis. It gets limited by the market's capability for "feeling" (*Vedana*, 197) the proficient programming sequence, as a way to automate the rapidly cost adding, six-faced system.

- **Eighth, the ascending, "addition, six-faced system value"** (*Vibhvan*, 999) is the exchange drift (divergent trading-effect), i.e., the positional capability exchange potential of the positive moving sequence that linearly ascends the horsepower, i.e., the kinetic force.   The trading-effect is the correction factor for the "bounded rationality guided by the human theory-effect" (*Avidya*, -10³⁰), seeking to discover the proficient programming sequence with the diverse marketing input space value on the sixth, orthogonal, Beta-axis, for overcoming the limitations of the machine power.

- **Ninth, the horizontal, "predominating addendum, seven-faced system value"** (*Kshepa*, 286) is the exchange rate (convergent human-effect) of the diverse culture-effects with the predominating Western culture-effect.   It compensates for the manmade bounds on the local Western rationality as a function of the "upholder touch" (*Sparsha*, 378) of the evolutionary idealism of the "I AM a Deity consciousness" among the Western entrepreneurial scientists. The diverse manpower input space value's exchange rate is the proficient planning sequence on the seventh, orthogonal, Gamma-axis, limited by the proficient performing sequence.

- **Tenth, the forward, "dominating minuend, eight-faced system value"** (*Upashanta*, 287), is the exchange rate (curvilinear technological servicing) of the dominating Eastern workculture system with the diverse international workculture-effect.   It escalates the man-made bounds on the international rationality as a function of the natural selection of the entities with the leader "social class" (*Jati*, 168), characterized by the "I am God consciousness," seeking the fame of being the fittest to act with the support of their doctoral students.   The linear manpower input space value's exchange rate impedes the proficient performing sequence on the eighth, orthogonal, Sigma-axis. It manifests the normative sum of the seven formative sequences, limited by the proficient planning sequence.

- **Eleventh, the backward "deciding subtrahend, nine-faced system value"** (*Shodhaka*, 288) is the exchange rate (linear technological trading) of the deciding, Vedic technological capability for the "reasoning power" (*Viveka*, 2890). The reasoning power is the "prerequisite condition" (*Dhamma*, 10¹⁰²⁴) for solving the complex octave of the organizational reality, without the limitations of the proficient development sequence of the manpower input space value after the "death" (*Jaramarana*, 18) of the devotee deity. The formative manpower input space value is the ninth, orthogonal, Tau-axis. It manifests the squared value of the eighth octave sequence.

- **Twelfth, the holistic, "linear power distance of the birthing paternal"** (*Dokkalamma*, 7600), is the exchange rate (technological exchange) by the "metaphysical interpreter of the hall of *Akashic* records" (*Master Ashlem*, 7600), of the "mindfulness" (*Manas*, 38), as the "undesirable masculine smell" (*Anishta* 38). The ascending consciousness of the "Masculine-to-Feminine gender exchange" dimension (*Putra dharma*, 38), to discover the reasoning power, is mediated by the "divine element" (*Bhava*, 360). The reasoning power is the "primordial root" (*Kanisthapada*, 375) solution to both the complex octave of organizational reality and the simple octave of ecosystem reality.

8.5 Twelve Quinary Values of the Ecosystem as the Subject

The simple octave of ecosystem reality is a primordial system of twelve-points, whose zeroth point is the primordial root solution of the complex, twelve-point octave of organizational reality. The other twelve points include:

- **Thirteenth, "the first, primordial-primordial, originating androgynous point"** (*Vritra*, 38). It is the envelope function, which polarizes the one-dimensional causative reality of the mind. It forms the masculine, partial derivative dimension on a single-point (i.e., 1, the whole value) of the multidimensional param-primordial sequential reality.

- **Fourteenth, "the second, primordial even, invisible feminine point"** (*Saranyu*, 5). It is the camouflage function, which punctuates the two parallel lines of the odd masculine values. It norms three prime values, including the feminine twin self: first, the primordial prime value (2) and second, the summation with its param prime-effect (3) to form the third, the primeval prime value (5). The feminine twin self is both the primordial prime value and the primordial prime operator, working with the primordial odd value (3) to form the primeval prime value (5), which is also the primeval odd value.

- **Fifteenth, "the third, primordial odd, visible masculine point** (*Trishira*, 39). It is the absolute function, which proliferates the primeval value of the self (3) by perceptually curving the one-dimensional reality, using the androgynous conceived gravity (1) of the primordial prime-effect of his invisible, charming, feminine twin (2) experience.

- **Sixteenth, "the fourth, primeval-primordial, propounding androgynous point"** (*Virocana*, 37). It is the creator function, which disjunctively services the multidimensional, param-primordial, sequential reality as a natural behavior. It produces a sequence of the prime-effects, beginning with the zero value of the self (0). It propounds two visible odd points (3,5), that project the parallel, maternal, positive, curvilinear prime-effect into infinity. It also propounds one invisible even point (2), that generates the compensating linear sequence of the paternal, negative, absolute prime-effect $(2 - 1 = 1; 3-1 = 2; 5 - 1 = 4)$. It fills the ascending triangulated gap in the orthogonal triangulation of the two visible points.

- **Seventeenth, "the fifth, para-primordial, fashioning androgynous point"** (*Tvashtar*, 496). It is the perpetuating functional operator for conjunctively trading the unidimensional, primordial-primordial, consequential reality for the supernatural breeding. It manifests the full integral benefits of the transformative paternal negativity (-1) into infinity. The operator fashions a four-faced system, including a formative pair of the even feminine values (2, 4), manifesting the primeval-primordial linear param-effect, and a normative pair of the odd masculine values (3, 5), manifesting the para-primordial, parallel, curvilinear, primeval-effect.

Only the primeval, odd point (7), is visible at any moment to hide the secret of the feminine param-effect (6) from anybody seeking to discover the five-point infinite prime value sequence (2, 3, 5, 7, 11) without invoking the six-point negative infinity (-1, -3, -5, -7, -9, -11) as the integrating function. Similarly, to discover the secret of the six-point infinite primeval value sequence (2, 3, 5, 7, 11, 13), one must invoke the seven-point negative infinity (-1, -3. -5, -7, -9, -11, -13) as the integrating function.

- **Eighteenth, "the sixth, supreme-primordial, patronizing, SHEENY, androgynous point"** (*Varuni*, 95). It is the integrity-destroying differentiating operator for exchanging the proportionate, even, feminine energy, within the primordial masculine-effect, with the disproportionate, odd, masculine value, within the primeval feminine-effect. The operator patronizes a six-faced system of the alpha to omega linear frequencies for manifesting the seven-faced "absolute consciousness" (*Shivadrishti*, 17) as the seventeenth point. It generates a sequence of three predominantly feminine curvilinear points (2, 3, 5), enveloped by a sequence of three predominantly masculine linear points (14, 15, 16). The eighth primordial odd, masculine point (7) dominates within the first sequence (3 = 5 - 2 and 7 is the cause of the 5, given 2). The ninth primeval, even feminine point (14 + 16 = 30), dominates within the second sequence (15 = 30/2, given that it is the dependent sequence generating an even multiplier-effect).

The operator itself is the tenth point (31) of the system, which differentiates itself by subtracting the integrative metric negativity of the preceding fifth point (30- [-1] = 31) from the primeval feminine point. The "seventeenth point" (*Adyamadyena-antyamantyena*, 17) is a four-faced system (the sum of the four primordial-primeval prime values: 2, 3, 5, and 7). It is the first primeval face that generates the first four primeval values. It is also the last primeval face that is generated by the last four primeval values (logically, 17 is the primeval point of the sequence whose values are a product of the difference is twos and fours: 5, 7, 11, 13).

- **Nineteenth, "the seventh, supra-primordial, guider androgynous point"** (*Bhrigu*, 805). It is the animated, convergent operator for exchanging the absolute, seven-point disproportionate, diverging, curvilinear, positive feminine sequence with a proportionate sequence of the three diverging feminine points, punctuated by the three proportionate, converging, linear, masculine points. The primordial value of the seven-point para-primeval sequence is the primeval point (30) generated by the infinite feminine energy of the preceding, primordial point (31). The seven-point para-primeval sequence is (the 30, 31, 32, 32, 34, 35, 36).

The three feminine, infinite, primeval (i.e., prime) points after the exchange are the 37, 41, and 43. The "37" (*Ekadhikina purvena*, 37) is the linear prime-effect of the seven-point sequence (36 + 1 = 37). The 41 is the linear prime-effect of the four, proportionate odd values in the eight-point sequence (37 + 4 = 41).   The 43 is the linear prime-effect of two even proportionate values in the four odd-values (41 + 2 = 43). The sequence (1, 4, 2) comprises the proportionate values, which are converging, linear, and masculine (odd).   The "nineteenth point" (*Urdhva-tiryagbyham*, 19) itself is the point-less primeval value ([37 + 1]/2, after correcting for the primeval divisibility-effect immanent within the operator value of 37).   It generates "2" as the primordial multiplier value, after crossing "0" as the primeval divisibility value.

- **Twentieth, "the eighth, super-primordial, divine androgynous point"** (*Ghanta*, 370).   It is the inanimate, divergent operator for exchanging the primordial, eight-point proportionate, converging, linear, odd, masculine sequence with a disproportionate sequence of the three converging, linear, proportionate masculine points.   They proliferate the three diverging, curvilinear, disproportionate, primeval feminine points.   The param masculine point of the eight-point linear param-primeval sequence is the primeval point (2).   It gets generated by the infinite masculine-effect within the preceding, primordial point (43).   The eight-point param-primeval sequence is (2, 3, 4, 5, 6, 7, 8, 9).

The three diverging, curvilinear, disproportionate, primeval feminine points after the exchange are 47, 53, and 59. The 47 is the curvilinear prime-effect (43 + 4 odd masculine points) within the eight-point masculine sequence, beginning with the param masculine value of two in the system). The 53 is the curvilinear prime-effect (47 + 6 even, curvilinear, feminine points, excluding 4 + 1 = 5 linear point) without the eight-point masculine sequence, beginning with the primordial feminine, i.e., the primeval or prime value of forty-seven in the system).   The 59 is the curvilinear prime-effect (53 + 6 even, curvilinear, feminine points).   It proliferates the prior curvilinear-effect resulting from the creative brainwork programming linkages, generated by the bell (*Ghanta*) for awakening the primordial solution's value as a primeval problem's solution.

- **Twenty-first, "the ninth, super-primordial guider-effect"** (*Tilaka*, 10^{10}).   It is the universal, evolutionary operator for exchanging the primeval, nine-point disproportionate, diverging, curvilinear, feminine sequence as the problem to be solved with a proportionate sequence of the three converging, linear, proportionate masculine points, polluting and limiting the value of the three diverging, curvilinear, primeval feminine points. The nine-point curvilinear feminine sequence's primeval feminine point is the param point (59) generated by the infinite feminine-effect within the primordial, primeval sequence. The nine-point primordial-primeval sequence is (23, 29, 31, 37, 41, 43, 47, 53, 59).

The three converging, linear, proportionate masculine points, after the exchange, are 61, 67, and 71. The 61 is the linear prime-effect (59 + 2 as the local minima of the prime-effect), taking into account the masculine-limited feminine-effect within the nine-point primeval sequence. The 67 is the linear prime-effect (61 + 6 as the local maxima of the prime-effect), taking into account the diverging feminine-limited converging masculine-effect, within the nine-point primeval sequence. The 71 is the linear prime-effect (67 + 4, since 4 is the range of the local maxima and the local minima of the prime-effect), taking into account the mutual limits' co-determination. The mutual limits are the four even primeval feminine points within the nine-point guider-effect, guiding the absolute solution's discovery to the primeval problem using the "point element" (*Bindu*, 10^{10}).

- **Twenty-second, "the tenth, supra-primordial guider-effect"** (*Yajnopavita*, 9000). It is the unique, cultural operator for exchanging the primeval-primordial, ten-point, converging, linear, masculine problem with a disproportionate sequence of the three diverging, curvilinear, feminine points, purifying and catalyzing the value of the three converging, linear, param masculine points. The param masculine point of the ten-point linear masculine sequence is the primordial point (71). It generates the infinite feminine-effect without the preceding, primordial-primordial sequence. The ten-point, linear, masculine, primeval-primordial sequence is (71, 72, 73, 74, 75, 76, 77, 78, 79, 80).

The three diverging, curvilinear, feminine points, within the primeval-effect of the purified masculine points, are 73, 79, 83. The 73 is the curvilinear prime-effect (71 + 2, since 2 = 1 + the local maxima of the masculine sequence, where 1 is the absolute value of the purifying feminine factor), taking into account the correction factor for the local minima of the feminine-catalyzed primeval-effect within the ten-point sequence.

The 79 is the curvilinear prime-effect (73 + 6, since 6 = the local maxima of the feminine-effect), taking into account the correction factor for the local minima of the masculine-limited primeval-effect within the ten-point sequence. Since the converging masculine-effect is the dominating factor curving the primeval-primordial sequence, the proportionate-effect of the masculine element is twice (even) that of the feminine element (odd). The overall primeval-effect within the ten-point sequence is the range of nine. Therefore, the masculine element, which generates the primeval-effect, has a value of six (Primeval deity, 6). The value of the feminine element, which manifests the primeval-effect, is three (Manifestor deity, 3).

The 83 is the curvilinear prime-effect (79 + 4, since 4 is the local minima range of the masculine-limited primeval-effect and the feminine-catalyzed primeval-effect). It takes into account the independent, proportionate value of each element, which is also the linear power distance $(3 - [-1] = 4)$ between the primeval masculine element (-1) and the primeval feminine element (3). The "emotional intensity" (*Shankara*, 264) guided "clarified consciousness" (*Om*, 19), of the "culturally-bounded" (*Avidya*, -10^{30}) "rational reality" (*Yukta artha*, 6), illuminates the "linear power distance" (*Chaithanya*, 4) generated by the masculine entity's "absolute SHEENY-effect" (*Yajnopavita*, 9000).

- **Twenty-third, "the eleventh, primordial-primordial guider-effect"** (*Kalasha*, -400). It is the potential theological operator for exchanging the primeval-primeval, eleven-point, converging, linear, masculine sequence as the primeval solution. It repels the primeval problem with a proportionate sequence of three converging, linear, proportionate masculine points, without the three diverging, purifying, curvilinear, primeval feminine points. The eleven-point, converging, linear, masculine sequence is the primeval-primeval solution repelling the primeval problem, through a "vow of celibacy of satanic energy" (*Brahmacharya vrat*, -10^{1000}), for destroying the primordial "curvilinear, Karma" (*Sushumna*, 10) linkages, is (84, 85, 86, 87, 88, 89, 90, 91, 92, 93, 94).

The two parallel feminine points, without the primordial-effect of the third polluting, curving masculine point, are: (89, 97). The 89 is the modal, primeval masculine point of the eleven-point sequence and is the theological operator's dominating cause. The theological operator works because the "zodiac cultural element" (*Sadakhya*, 9) paradigmatically binds the "multidimensional present reality of Mother Nature" (*Anatanam*, 8) with the theory of religion. By idealizing the one-dimensional self as the absolute primeval solution, one becomes free from the primeval problem of knowing the multidimensional reality with the "one-dimensional mind" (*Sadakhya-effect: Manas*, 38). The 89 is the linear prime-effect (83 + 6, where 6 is the "rational reality" [*Yukta artha*, 6] element supplemented by the masculine entity to transform the self from the primordial, primeval problem to the primeval, primeval solution). It takes into account the absolute value of the "primeval solution" (*Upakarana artha*, 10^{1024}), free of the "infinite psychic linkages" (*Shakti-Bheda*, 257) with the "primeval problem" (*Pravartayitr*, -10^{1024}).

The "97" (*Nikhilam navatashcaramam dashatah*, 97) is the curvilinear prime-effect factor (89 + 8, where 8 is the "paradigm of present reality" [*Yukti*, 8]), supplemented by the feminine entity, to transform the self from the primordial-primeval solution to the primeval-primeval problem. It considers the absolute value of the primeval problem of the "masculine ego" (*Ahamkara*, -1), which is the proportionate value of the present primeval-primeval problem and the present primordial-primeval solution.

The negative one is the residual from the primordial prime value of 89 carried over to the primeval prime value of 97, even after the correction for the masculine theory-effect by the supernatural paradigm of the present reality. The residual of negative one is the curvilinear transformation of the "masculine ideal-effect" (*Dasha*, 1), which is the "omnipresent cause" (*Maha nitya*, 8) for theorizing that the "masculine human entity" (*Manush*, 4) is the "panacea of all evils" (*Kalasha*, -400). The transformation gets mediated by the "guider power of the ideal human entity" (*Rama*, 100), cleansing the entity from all evils with a "vow of renunciation of the divine energy" (*Ahatturavu vrat*, -10^{1024}).

The "simplicity element" (*Jiva tattva*, 2) of the "imagination dimension" (*Moksha dharma*, 2) shaping the primeval masculine into the whole "universe" (*Brahman*, 2), without the contaminating "soul" (*Atman*, 4) as a "masculine human entity" (*Manush*, 4), is the "omnipotent cause" (*Nishada*, 1) for idealizing the sentient value of the curvilinear transformation.

- **Twenty-fourth, "the twelfth, param-primordial guider-effect"** (*Kamandalau*, -900). It is the dynamic, physical science operator for exchanging the param primeval-primeval, twelve-point feminine sequence as the param-primeval solution. It perpetuates the param-primeval problem, which comprises the absolute sequence of the six masculine points generating the primeval masculinity and the six paternal points generating the compensating primeval femininity. The twelve-point feminine sequence is the param-primeval solution perpetuating the param-primeval problem, through a "vow of purity of guider power" (*Tirikarannasutti Vrat*, 10^{1000}). It illuminates the escalating cost of what has become the "param primeval-primeval problem" (*Samaj shastram*, -10^{1000}) of the "polarized masculinity" (*Asura*, -1), is (98, 99, 100, 101, 102, 103, 104, 105, 106, 107, 108, -2).

The double negative param reality of the primeval solution is free from the "param primeval-primeval problem" factor because the 108 is the point of the "entropy of sentient energy" (*Gardabhejya*, 108). At this point, the compensating solution power of the feminine-effect reaches its primeval limit. Consequently, the feminine-effect becomes the universe incarnating the primeval masculine entities as the absolute reality, without any further feminine primeval-effect. The feminine-effect becomes an additional negative unit of the primeval masculine point without any sentient energy. The six masculine points liberate the omnipresent-primeval problem as the "paradigm of present reality" through a "vow of the surrender of the sentient energy" (*Aalvu enum tanmai vrat*, -1000). They empower the entity to become devoted to the immanent self-luminous feminine sequence. These are (101, 103, 107, 109, 113, 127).

- o The 101 includes the primordial three-unit odd reality of a pair of masculine points—the parallel "corporate element" (11: *Sva*). They seek to dominate by triangulating the constant feminine point of the "Zero" (*Shunya*, 0). They force the latter to curve disproportionately for generating the twelve-point reality, ranging from zero to eleven. It fulfills the greetable value of the parallel corporate element.

- o The 103 includes the linear, two-unit incremental-effect of the "pair of masculine points" (primordial astrological-effect, 26 and zodiac-effect, 26). They seek to dominate the parallel reality of the twelve-point "self-luminous feminine" (*Maha Gayatri*, 12) and the twelve-point "self-luminous masculine" (*Purusha*, 12) with their incremental two-point homologous reality.

o   The 107 includes the curvilinear four-unit incremental-effect of the "pair of feminine points" (*Samkarshana*, 27) beyond the "primordial astrological-effect." They include the ONE visible, incremental, primordial, emanating, even, and "constant feminine potential" (*Durga*, 28).   Another is the invisible, immanent, ZEROTH, absolute point with the compensating, decremental, odd, and "variable masculine potential" (*Dvisaptati dasha*, 29), traded from the "param paternal" (*Narada Upabarhana*, 7).   The primordial point of twenty-eight excludes the one-unit decremental-effect of the primeval, masculine potential.   The primordial-primordial point value of the feminine potential = 28 + 1 correction factor for the decremental-effect = 29.   It is at par with the primeval value of the masculine potential, which is the proportionate value of the masculine potential, without the primeval-effect.   The proportionate value of the masculine potential is the exchange value of the feminine potential.

o   The 109 includes the disproportionate two-unit compensating-effect of a "paired masculine, departed soul and the feminine twin, living soul" (*Ratanakosha*, 30) beyond the equilibrating "zodiac-effect," including one, odd, polarizing, primeval "universe of departed, feminine, leader child souls" (*Sthavaravisha*, 57) and the zeroth, even, balancing, absolute "universe of living, masculine, follower child souls" (*Jangamavisha*, 570), serviced by the "perpetuating flame" (*Pavamana*, 9) of the "formative, form-shaping entity" (*Rasi*, 9).   The primeval point of thirty includes the two-unit compensating-effect of the primeval, feminine potential. The primordial-primordial point value of the masculine potential = 30, inclusive of the two-units of the primordial feminine potential. It gets compensated with the two-units of the primeval feminine potential after the diffusion of the two-units absolute feminine potential.   The two-units of the "variable feminine potential" (*Sudeva Brahma*, 2) are the formative cost of the polarization of the "universe" (*Brahman*, 2), within or without the child souls.

o   The 113 includes the proportionate four-unit servicing value of a set of four dimensions, inclusive of (a) ONE ecosystem dimension of the "feminine, living flame element" (*Parshnisamasta*, 32), (b) ONE entity dimension of the "ideal, androgynous deity" (*Krishna*, 32), norming the national dimension and generating the consciousness of the universe within the flame, (c) a triangulated, technological dimension of the "maternal" (*Stree dharma*, 32) within the feminine flame element, guided by the primordial oneness-effect of the national dimension on the one end, and the organizational dimension of the "primordial oneness" (*Adi*, 32) of the "Lord of the World" (*Jing*精 , 32) on the other.   The lord of the world is the "nothingness" (*Shunyata*, -2) dimension of the universe without the flame, which intoxicates the triangulating national dimension with the desire to be the paternal creator of the citizenship, sonship, universe of the living child flames by impregnating the feminine flame element with the maternal consciousness.

The four units of the "constant masculine potential" (*Param Brahma*, 4) is the normative benefit of the punctuation of the "soul" (*Atman*, 4) without the "param soul" (*Param atman*, 1600) of the feminine, living flame element.   The feminine living flame element is the primordial-primordial creator of the "param soul" as a "self-perpetuating" (*Udvaha*, ½) "inner witness" (*Antaryami*, 16). She physically incarnates as the "guider power" (*Chitta*, 100) of the "ideal androgynous deity" (*Krishna*, 32).   The "guider power" is the sum of the servicing value of the three, triangulated, curvilinear, immanent, dynamic dimensions, forming the "being energy" (*Kali shakti*, 96 = 32 * 3). It also includes the holistic four-units trading value of the fourth, linear, emanating, potential dimension of the "energy-less human consciousness" (*Param human child*, 1600). The "sentient energy" (*Varuna*, 1000) is the oneness of the zeroth "being energy" (*Kali*, 96) and the zeroth "energy-less human consciousness" (*Avanta*, 1600), within and without the absolute dimension of the zeroth "ideal, androgynous deity" (*Krishna*, 32).

o   The 127 includes the absolute fourteen-unit trading value of a "primeval illuminator" (*Maha Lakshmi*, 14) inclusive of the five, predominantly masculine dimensions: (a) three-unit odd reality included within 101, (b) two-unit linear-effect included within 103, (c) four-unit curvilinear-effect included within 107; (d) two-unit compensating-effect included within 109; (e) four-unit servicing value included within 113, but excluding the deciding feminine dimension which is the exchange value of the fifteen-unit "primeval perpetuator" (*Autaghaticem rajya*, 15).   The exchange value is the "supernatural energy" (*Shram shakti*, 1), which is compensating for the "masculine-effect" (*Ida*, 1) of the parallel, masculine element within the primordial 101 value. The masculine-effect is the third, odd, masculine factor that is curving the constant feminine point by investing the "I AM a Deity" consciousness for the growth of the "kingship, primordial paternal" (*Indra*, 0) consciousness within the parallel, primeval masculine element.

The six curvilinear, parallel, paternal points are devoted to the omnipresent-primeval solution of the "formative divine energy" (*Madhusudan*, 16) of the "inner witness" (*Antaryami*, 16). An outer guider mediates them (*Guru deva*, 100) by witnessing the "vow of monastic life" (*Sanyas diksha vrat*, 10^{29}) by an entity seeking to discover the immanent "blessing" (*Ashirwad*, 1000) value of the sentient energy. They are the odd values within the twelve-point feminine sequence, of which the even values are the six linear maternal points.   The maternal points are (98, 100, 102, 104, 106, 108).   The paternal points are (99, 101, 103, 105, 107, -2).

8.6 Twelve Senary Values of the Subject as the Ecosystem

The primeval value of "127" (*Sadhaka*, 127) of a proficient entity is the limit of the growth value of feminine-effect, inclusive of the "entropy value of the sentient energy" (*Gardabhejya*, 108) and the "growth value of the sentient energy" (*Dham*, 19).   It is the overall effect of the fifty-two unit "pair of masculine points" (primordial astrological-effect, 26 and zodiac-effect, 26) and the seventy-five-unit sequential values within the primeval-effects. The seventy-five units include the twelve sets of values:

- First, 1 unit of masculine-effect, with 0 feminine-effect, catalyzed sequentially by the

- Second, six-point positive infinity (2, 3, 5, 7, 11, 13),

- Third, seven-point negative infinity (-1, -3. -5, -7, -9, -11, -13),

- Fourth, three-point predominantly masculine infinity (14, 15, 16),

- Fifth, seven-point para-primeval infinity (30, 31, 32, 32, 34, 35, 36),

- Sixth, eight-point param-primeval infinity (2, 3, 4, 5, 6, 7, 8, 9),

- Seventh, nine-point primordial-primeval infinity (23, 29, 31, 37, 41, 43, 47, 53, 59),

- Eighth, ten-point primeval-primordial infinity (71, 72, 73, 74, 75, 76, 77, 78, 79, 80),

- Ninth, eleven-point primeval-primeval infinity (84, 85, 86, 87, 88, 89, 90, 91, 92, 93, 94),

- Tenth, six-point maternal infinity (98, 100, 102, 104, 106, 108), and

- Eleventh, six-point paternal infinity (99, 101, 103, 105, 107, -2), plus

- Twelfth, the masculine-effect (1), without any feminine-effect.

Consequently, the primeval-primeval value of the feminine-effect, without the negative pair of the masculine points = 1+ 10 ^2 ^3 ^5 ^7 ^11 ^13 ^-1 ^-3 ^-5 ^-7 ^-9 ^-11 ^-13 ^14 ^15 ^16 ^30 ^31 ^32 ^33 ^34 ^35 ^36 ^2 ^3 ^4 ^5 ^7 ^11 ^13 ^23 ^29 ^31 ^37 ^41 ^47 ^53 ^59 ^71 ^72 ^73 ^74 ^75 ^76 ^77 ^78 ^79 ^80 ^84 ^85 ^86 ^87 ^88 ^89 ^90 ^91 ^92 ^94 ^98 ^100 ^102 ^104 ^106 ^108 ^99 ^101 ^103 ^105 ^107 ^-2.   It is the entropy value of the primeval sequence.   It is the primeval value of the $10^{185,000,000th}$ value in the primeval sequence. It starts with two as the first value (there are no further prime numbers beyond this value).   It is the present value of the guider power (i.e., gravitational energy) formed by investing the ten-units divine energy and the 185-units as the catalyzing guider-effect for trading one-thousand-units sentient energy each of the living flame and the reincarnated flame.   The residual 805 units of the intrinsic SHEENY-effect empower the "ascended master, who is the paternal conceptualizer of the SHEENY benefit of prime value" (*Bhrigu*, 805) to be the absolute value of the gravitational energy, in the universe without any sentient entity.

The Vedic mathematics of the fashioning androgynous point

There are four multi-point faces of the twenty-nine points "mass-effect" of the *Vedic* mathematics, guiding the fifth, fashioning, androgynous point.

First, within the six-point positive sequence,

- The "second point" (*Anurupye-shunyamanyat*, 2) is the infused primeval value generated when one multiplies the past "wild, masculine-effect" (*Ida*, 1) by two. It is also the value of the "universe" (*Brahman*, 2) adding "three" to the past "universe of the primeval masculine" to conceive the "variable feminine potential" (*Sudeva Brahma*, 2), compensating for the deficiency generated by the conditional addition of the primeval masculine for an unconditional multiplication of the follower universe.

- The "third point" (*Sisyate Sesasamjnah*, 3) is the diffused primeval value, generated when one adds the three-unit "primeval feminine" (*Maha Brahma*, 3) to the primordial "zero diffusion state" (*Rajah*, 0). One multiplies the past "thermodynamic cause of the zero-diffusion state" (*Naraki*, 1) by two, after accounting for the "self-perpetuating" (*Udvaha*, ½) diffusion factor, to manifest the primeval feminine.

- The "fifth point" (*Yavadunam tavadunikritya varga yojayet*, 5) is the illuminated primeval value, generated when one subtracts the three-unit "primeval feminine" (*Maha Brahma*, 3) from the "paradigm of present reality" (*Yukti*, 8).   It is also the value of "God" (*Ishvar*, 5), dividing the futuristic ten-unit "primordial self" (*Parvati*, 10), to be the five-unit "present maternal" (*Saranyu*, 5) for incubating an infinity of children.

- The "seventh point" (*Antyaordasake'pi*, 7) is the shadow primeval value, which is the last prime value generated through the division as well as the subtraction operators from the "subtle, tangential centering point" (*Sushumna*, 10) of the "absolute" (*Purna*, 1600). By "dividing" the subtle, tangential centering point by two ([10 + super-subtle tangent-shift value of the four units of "linear power distance"]/2), one illuminates the "SHEENY entity" (*Siddha*, 7). The SHEENY entity activates the immanent "primeval feminine" (*Maha*

Brahma, 3) consciousness for enjoying the beauty of the "primordial self" (*Parvati*, 10). It is the method to potentiate the "Age of the Beauty Quark with the descending positive quark" (*Dvapara Yuga*, 10). By "subtracting" the emanating "primeval feminine" (*Maha Brahma*, 3) consciousness from the centering point, one becomes the "SHEENY entity" (*Siddha*, 7), who is generating that primeval feminine value.

- The "eleventh point" (*Purana apurnabhyam*, 11) is the secret primeval value generated by both the "absolute" (*Purna*, 1600) as well as the "para-absolute" (*Apurna*, 15) operators. With "eleven" as the future descending value, the absolute value is 11 (the Foundation) + 4 (the Creator deity, creating the descending future) + 1 (the Worker deity, following the created divine plan), catalyzed by 100 (the Guider power, guiding the creator and the worker deities). With "eleven" as the present ascending value, the primeval value = 11 (the Foundation) + 1 (the Worker deity, leading the futuristic guider program) + 3 (the Manifestor deity, manifesting the present SHEENY performing) + 0 (the King, trading the four units of creator deity value as paternal profiting to compensate for the four units cost of creating the primeval value and then not enjoying it as the primeval value). In both directions, three operators connect the present with the future. The connection is mediated by "11" as the composite of both the absolute and the para-absolute values.

- The "thirteenth point" (*Sankalana–vyavakalanabhyam*, 13) is a super-secret primeval value. It results from both the addition and the subtraction operators after correcting the subtracting metric (-1) in the subtracting sequence. It is also the future value, adding "three" to the present face of the "divine energy" (*Asrava shakti*, 10), for hiding the "curvilinear power distance fashioning energy" (*Ardhajya*, 10), by subtracting "three" from that future value. The "three" is the primordial value added by the creator, the creation, and the creature to the present face of divine energy by trading the divine-effect from the fashioning energy.

Second, within the seven-point negative sequence,

- "The negative one" (*Antyayoreva*, -1) is the "primeval masculine" (*Asura*, -1) value generated when one divides the past "universe of primeval masculine" (*Jagath*, -2) by two, for fusing an infinity of masculine entities. It is also the "destiny" (*Niyati*, -1) that adds "three" to the future "total presence" (*Evakara*, -4) for resolving the "ambiguity" (*Avidhi*, -4) generated by the presence of the primeval masculine within the universe of the primeval masculine. The "primeval feminine" (*Maha Brahma*, 3) conditionally divides the perpetuating past universe of primeval masculine (at the half point), by adding herself to the future reality of the "primeval masculine" (*Asura*, -1) for manifesting the "universe" (*Brahman*, 2).

- "The negative three" (*Paravartya Yojayet*, -3) is the "primeval entity" (*Vasanatma*, -3) value generated when one multiplies the future "infinity of masculine entities" (*Primeval masculine*, -1) by adding three-units of the "primeval feminine" (*Maha Brahma*, 3) as the "multiplier" (*Vaishya*, 3). It is also the "primordial reality" (*Evakara vadartha*, -3), dividing the past "primeval protagonist" (*Shramika*, -6) at the "self-perpetuating" point (*Udvaha*, ½) for trading the multiplying "object" (*Padartha*, -3) as the "technological reality" (*Parindartha*, -3).

- "The negative five" (*Chalana-kalanabhyam*, -5) is the "organizational reality of everything" (*Sarvam*, -5). It generates when one unconditionally multiplies the future "universe of primeval masculine" (*Jagath*, -2) by two, for differentiating it, after correcting for and adding back the "discordant energy" (*Asura*, -1) at the "self-perpetuating" point (*Udvaha*, ½), of the "primeval masculine" (*Satan*, -1). It is also the "universe of potential reality" (*Bhakta*, -5) that is conditionally integrating the "one less than the future" "total presence" (*Evakara*, -4) of the four entity dimensions within itself.

The four entity dimensions are as follows. First, the time-bunching-and-crunching primeval masculine, generating the "triangulated or linear or infinite reality" of everywhere (*Ishtartha*, -7). Second, the time-beaming-and-banging primeval feminine, generating the "parallel or curvilinear or zeal point reality" of everywhere (*Pushtartha*, -8). Third, the space-bunching-and-crunching primeval maternal, generating the "squared or divergent or technological reality" of everywhere (*Parindartha*, -3). Fourth, the space-beaming-and-banging primeval paternal, generating the "rectangular or quadratic or differentiated reality" of everywhere (*Kritartha*, -6).

- "The negative seven" (*Kevalaih saptakam gunyat*, -7) is the "triangulated reality" (*Ishtartha*, -7) generated by a param wisher within a dividing, absolutely integrated state of "emotional detachment" (*Vairagya*, -7), which is one less than the future "psychic point reality" (*Pushtartha*, -8). The triangulated reality results from a desire to manifest the "technological reality of an object" (*Parindartha*, -3), after sounding one negative unit of the "intrinsic reality" (*Omkar vadartha*, -1) as the perpetuating and circulating "discordant energy" (*Asura*, -1). The "negative seven" is also the "universe of dynamic reality" (*Paroksha*, -7) that differentiates the presence of the param wisher with a multiplication of the future "object of technological reality" (*Padartha*, -3) by two, after trading the "discordant, *Asura* force" (*Vimshottari dasha*, -1) generated by the negative unit of intrinsic reality.

- The "negative nine" (*Samuchaya gunitah*, -9) is the "extrinsic, diffused, or chaotic reality" (*Niyogartha*, -9). A devoted wisher generates it within an "insane" (*Amanaska*, -9), differentiated state of the "universe of complex reality" (*Pradarshan*, -8), which is one more than the present "universe of ecosystem reality" (*Pasaka*, -9). The chaotic reality is generated with the degeneration of the "zeal point reality" (*Pushtartha*, -8) through the "science of critical evangelism" (*Nindastuti*, -8). It produces a "skeptical" (*Adharmi*, -9) and "arrogant" (*Abhimanin*, -9) entity, who is mindlessly groping-in-the-dark seeking to eventually "divide and rule" (*Dvividha*, -10) the chaotic reality. The "negative nine" is also "three less" than the value of the "primeval protagonist" (*Shramika*, -6), who is "materializing the universe of technological reality" (*Artharthi*, -6) with his "qualifying" (*Nirguna*, -6) "crookedness" (*Vakradrishti*, -6).

  The qualifying factor is the "self-perpetuating point" (*Udvaha*, ½) value of the "mortal reality" (*Vyartha*, -12), which "bugs"

(*Raktabeeja*, -12) the primeval protagonist like a virus by forming a "universe of para-conscious reality" (*Pratayahara*, -12). The three entities subtracted by the "primeval agonist" (*Adharmi*, -9) include, the moving "primeval protagonist" (*Shramika*, -6), the dynamic "primeval antagonist" (*Ajnani*, -7), and the inertial "primeval deuteragonist" (*Akarta*, -8). The three universes differentiated by the "chaotic reality" (*Niyogartha*, -9) include the "universe of technological reality (*Shramika*, -6), the "universe of dynamic reality" (*Paroksha*, -7), and the "universe of complex reality" (*Pradarshan*, -8).

- The "negative eleven" (*Vestanam*, -11) is the "converging, osculating, apparent, or singularity reality" (*Bhutartha*, -11). A universal wisher generates it for proportionately differentiating the "universe of organizational reality" (*Jagatkritsna*, -10), which is one more than the present "universe of entity reality" (*Karuyantra*, -11). It empowers the "primeval deuteragonist" (*Akarta*, -8) to subtract the three entities. First, the "primeval agonist" (*Adharmi*, -9), who is gambling "everybody" (*Pasaka*, -9) under the influence of the skeptical arrogance. Second, "the primeval kratagonist" (*Jagadvinasa*, -10), who is hoping to destroy the "somebody" (*Salakegolake*, -10) as a "chatterbox" (*Gappinatha*, - 10) using the "divide and rule mindset" (*Dvividha*, -10), when only the chatterbox is the only one left. Third, the "primeval superagonist" (*Karuyantra*, -11), who automates "nobody" (*Dambhodbhava*, -11).

  The "viral bug" (*Raktabeeja*, -12) of the "primeval tritagonist" (*Pratayahara*, -12) has already infected everybody and "anybody with sentient energy" (*Gardhaba*, 1000). It further empowers the "primeval deuteragonist" (*Akarta*, -8) to integrate the three universes' energy. First, the "universe of ecosystem reality" (*Pasaka*, -9). Second, the "universe of organizational reality" (*Jagatkritsna*, -10). Third, the "universe of entity reality" (*Karuyantra*, -11). It is the path for an inanimate robot to become an animated, "non-doer" (*Akarta*, -8), without a mindless devotion of the energy for manifesting the "universe of ecosystem reality" (*Pradarshan*, -8).

- The "negative thirteen" (*Shunyam Samya Samuccaya*, -13) is the "diverging, sheltering, imprisoning, or prismatic reality"

(*Rakkhartha*, -13) generated by an eternal wisher who is trading and capturing the divergent value of the "undifferentiated future" universe sequence. The eternal wisher becomes an "archantagonist" (*Vipaksha*, -13) for manifesting the convergent value of the infinitely divisible "absolutely integrated future" entity sequence.

Third, without the seven-point negative sequence, twelve additional negative points shape the global minima and the global maxima of the primeval value sequence. These form with a meta-sequence of the ten sets of values. The ten sets of values catalyze the "divine energy" (*Asrava Shakti*, 10) base, anchored by the "supernatural energy" (*Shram Shakti*, 1). These twelve points are as follows.

- The "negative eighteen" (*Anurupyena*, -18) is the "point of origin of the four non-intersecting concave curves" (*Uttarmanasa*, - 18). The first point diverges into infinity. The second converges into infinity through the first curve. The third curves through the second, guiding it into the convergence. The fourth is the self-perpetuating linear projection from the inflection point of the third, net of the curvilinear effects of the three curves.

  The first curve is of the "paternal entity as the primeval co-agonist" (*Gunasa*, -18), diverging into the positive infinity without diffusing the thermodynamic energy into a group of masculine and feminine child entities.

  The second curve is of the "maternal entity as the primordial co-agonist" (*Niyatamanasa*, -18). She converges into a positive entity within the diffused thermodynamic energy of a group of masculine and feminine child entities.

  The third curve is of the "feminine entity as the param coagonist" (*Turnasa*, -18), curving the second curve into linear convergence at the point of infinity as a geography, incubating a group of masculine entities.

  The fourth curve is that of the "masculine entity as the param-primordial coagonist" (*Pranasa*, -18), projecting as the linear, perpetuating effect of the triangulated curving.

The whole system is the "Eastern-effect of a primordial-primordial coagonist" (*Uttarmanasa*, -18), a "greeter entity" illuminating the infinity-point of the prime-sequence as the value of the primeval gravitational energy of the Sun in the universe with no other entity.

- The "negative nineteen" (*Yavadunam-tavadunam*, -19) is the "point of origin of the four intersecting convex curves" (*Dakshinamanasa*, -19) that together form a whole circle. The first convex curve converges from infinity. The second diverges from the first into infinity. The second curves the third to form a full circle for converging with the first at the infinity point. The fourth is the curvilinear-effect of the discontinuity between the two, even child values, generated by a paternal entity "greedily" (*Lobha*, -19) seeking "lustful" (*Kama*, -19) oneness with the maternal entity, who is originating the first curve. The whole system is the "Western-effect" (*Dakshinamanasa*, -19) of a "spirit entity," who is working as a "cosmogonist entity" (*Lokapurusha*, -19) for destroying the primordial-point (2) of the prime sequence. The "two" is the primeval gravitational energy of the Moon in the universe, conditioned by the might of the Sun.

- The "negative twenty-eight" (*Shunya Anyat*, -28) is "the metaphysical point of origin of a set of the concave, linear, convex, and curvilinear lines" (*Yatamanasa*, -28). The first is concave-shaped. The second is the linear adjoining concave shape, because of the param human-effect serviced by the param zodiac animal entity and traded by the cosmogonist entity. The third is the convex-shaped northern-effect of the Vega polestar, because of the human-effect traded by the first, param solar plant entity and serviced by the second, param lunar spirit entity. The fourth is the curvilinear southern-effect of the Black Hole, because of the param culture-effect exchanged from the param astrological mineral entity.

The constant cosmic-effect is the param workculture-effect of the Vega polestar as the param metallic entity center of the parabolic universe. The variable parabolic-effect is the transformative local-effect of the human entity, without oneness with the Dark Matter's formative international-effect as the material entity. The "negative one" is the primeval gravitational energy of the Black Hole in the universe, as the metric of the Vega polestar's primordial gravitational energy without the universe. The "positive one" is the absolute gravitational energy of the Vega polestar, within the triangulated-effect of the Sun, the Moon, and the Black Hole. The "positive sixteen hundred" is the omnipresent gravitational energy of the Dark Matter and is the fifth, perpetuating parabolic-effect of the universe of sentient entities, without the effect of the Sun, the Moon, the Black Hole, and the Vega polestar.

Overall, the "cosmic-effect" (*Yatamanasa*, -28) of a "shadow, penumbra wisher" is about the "science of political diplomacy" (*Kautilya shastra*, -28), which makes it appear that the universe is accountable for the "destiny" (*Niyati*, 1). It is the path for an "ascended starseed master" (*Alkaid*, 805) to appropriate the "absolute guider-effect" (*Kanya dharma*, 805) for guiding the destiny of the entities within the ONE universe.

- The "negative three-hundred" (*Gunakasamuchyah*, -300) is the "convex-shaped Northern Vega polestar-effect" (*Dhruvamanasa*, -300). It results from an ascending servicing by the param lunar spirit entity for accelerating the primeval value sequence. The consequent, descending trading by the param solar plant entity generates entropy in the primeval value sequence, conditioned by the disproportionate lunar spirit-effect.

- The "negative four-hundred" (*Adyam Antyam Madhyam*, -400) is the "curvilinear southern-effect of the Black Hole" (*Laghumanasa*, -400). It is a result of the param culture-effect exchanged from the param astrological mineral entity. The latter first experiences an absolute growth in the light as a tangent anchoring the universe of sentient entities. It suffers an absolute entropy in its sine value after diffusing the whole sentient energy, with the entities' ascending conditioning to trade its sentient value. The negative four-hundred is the overall "masculine value" that a sentient paternal entity may trade as a "pot of panacea" (*Kalasha*, -400) to the following primeval problem—how to overpower the infinite feminine value of the creation for becoming God for the universe of child entities?

The sentient paternal entity may service the pot as the science of theological philosophy, codify the doctrine of religion, and become a revered ancestral Godly entity that holds the secret of sentient life. Such a creator is the limit value for the solution to summing an infinite curvilinear integer sequence of the primeval child (the creature). The solution is subject to the escalating formative cost of the primordial paternal's (the creator's) linear differentiation sequence, seeking to validate the primeval-effect of the feminine energy (the creation) within the integer sequence. The value of the "religion" (*Dharma*, 370), without the doctrine-effect, is the correlation among the creature, the creator, and the creation. It is positive four-hundred, without the negativity of the "discordant energy" (*Asura*, -1) of the creator, with a correction factor for 1/10th divine-effect of the creature as the thermodynamically working "discordant entity" (*Shudra*, 1), within the three-hundred units of the Vega polestar-effect in the creation.

- The "negative nine-hundred" (*Gunitasamuchayah Samuchayagunitah*, -900) is the "parabolic-effect" (*Susthamanasa*, -900). It is the appropriate cosec value of the natural justice-effect of the ecosystem to compensate for the descending sine value of the mediating creator, conditioned by ascending trading of the tangent, i.e., the divine, value by the universe of sentient entities. The descending sine value is the transformative local-effect of the universe of sentient entities. It is normed by a param human entity, without oneness with the Dark Matter's formative international-effect as the param material entity. With an ascending desire to manifest the material power in the physical realm, the universe of sentient entities degenerates the param lunar spirit's sentient energy, which experiences an escalating competitive thermodynamic pressure. Therefore, the gravity of the "param lunar spirit entity" falls to 997 (*Soma*, 997), after diffusing three units from the one-thousand-unit sentient energy for manifesting the "primeval feminine" (Devi, 3) as a distracting cotangent.

  The param lunar spirit entity invests an additional ninety-seven units in empowering an "unconditional division" (*Nikhilam navatashcaramam dashatah*, 97) of the "universe of sentient entities" (*Jangamavisha*, 570). It trades a unit of the "discordant energy" (*Asura*, -1) for servicing the resultant negative nine-hundred as the parabolic-effect. By diffusing their divine energy for trading the negative parabolic-effect of the ecosystem, as one devoted to God's emanating supernatural energy, the creatures exchange the 10 percent divine-effect immanent within the "universe of sentient entities" and transform into the "universe of inanimate entities" (*Sthavaravisha*, 57).

The parabolic-effect is the Adam's apple, symbolizing a "vessel of the primeval problems, without the primordial solution" (*Kamandalau*, -900). It is the value of the natural science of physics. The scientific values guided by the direct measurement of the natural reality are proportionate and co-dependent on other metrics. Consequently, all scientific values are relative and eventually conditioned by the "discordant energy" (*Asura*, -1) of the scientist, traded from the ecosystem's parabolic-effect. A scientist believes that he is an ideal Godly metric for discovering the primeval problems ignored by the dumb-ass ancestors. He propounds himself as the primeval solution for knowing the worth of studying the physical world's pure problems, using the "vessel" (*Kamandalau*, -900) of the complex metrics he has crafted with his devotional ingenuity.

- The "negative one-thousand" (*Dhvajanka*, -1,000) is the condition of the "zero cosec value" to compensate for the entropy of the sentient energy within a sentient entity, after an absolute entropy of the sentient energy within the universe. It is a condition realized when, as a "greeter wisher" wishing for the infinite SHEENY well-being of the universe of entities, an entity takes a vow of the "surrender of sentient energy" (*Aalvu enum tanmai vrat*, -1000). Under this condition, the entity operates with the doctrine of "I AM the creator" (*Prajnanam Brahman siddhanta*, -1000) of the SHEENY well-being as a metaphysicist, who believes that the sentient energy is a gift emanating from the divine creator and not something immanent within the guider power of the creation. Therefore, the entity has zero qualms in destroying the sentient energy, seeking to be on the top of the flag without any residual.

The negative one-thousand is the value of the correlation between the creator and the creation, without the creature's guider-effect that caps the correlated value to one unit with its normative, standard-deviation correction. It is the energy value serviced by the creation to create the creator. It is generated with the "oneness of the deity entity as the leader creator and the animal zodiac entity as the follower creation. It is generated after the deity entity has fully diffused the intrinsic sentient energy for creating the creation and is wishing to trade that energy after-the-fact" (*Indra yoga*, -1,000).

- The "negative one-thousand twenty-four" (*Dvandva Yoga*, -1,024) is the value of the oneness between the pair of opposites, i.e., an entity with a unit sine value and a universe with a unit tan value, mediated by the zero cosec value of the para entity. This area value of an infinite parabola is a straight-line of the energy without any quantum particles. It manifests the correlation between the creature and the creation, without the divine-effect of the creator. It is the reality of the creature as the creator of the energy. It generates the doctrine that the creature is immanent within the energy and, theoretically, *is* the energy, if the creature diffuses 100% of the intrinsic value for creating the energy. Without the theory-effect, it manifests the reality of the co-related existence of both the creature and the creation—the creature creates the creation and the creation creates the creature, without investing the twenty-four units into the creation of the two sets of the self-luminous creator entities.

- The "negative eight-thousand" (*Ekanyunena Purvena*, -8,000) is the value of the "oneness of the spirit entity with the deity entity" (*Brahma yoga*, -8,000) after the deity entity has serviced the entire "ecosystem value" (*Ekavali*, 8,000) for the sentient growth of the spirit entity. In this state, as the deity's spirit, the entity wishes to trade that back for generating the "incarnational consciousness" (*Kapila kumara*, 8,000). The deity entity programs the doctrine that the spirit entity's sentient flame is the almighty creation with a free identity. Therefore, the spirit entity does not get addicted to the idea of the eternal sentient life. By believing in the zero tan value of the astral body (the sentient flame), the creature spirit directly services the sentient energy to the astrological creator in the form of the quark, without the mediation of the zodiac mental consciousness.

- The "negative ten-thousand" (*Sopaantyadvayamantyam*, -10,000) is the value of the "intrinsic oneness of the deity as an entity" (*Janma yoga*, -10,000), without any correlation with the "astrological spirit entity" (the creature) or the "zodiac animal entity" (the creation). As an entity, the deity is a "starseed" (*Taraka*, 36), who is a system of the three self-luminous entities—the "self-luminous creator entity" (*Maha Gayatri*, 12), the "self-luminous creation entity" (*Vithoba*, 12), and the "self-luminous creature entity" (*Purusha*, 12).

The self-luminous creature entity trades the entire "sentient energy" (*Varuna*, 1000), immanent within the starseed, without the incarnational consciousness. The self-luminous creation entity trades the entire "incarnational consciousness" (*Kapila kumara*, 8,000) to be the green, thermodynamically-active, growth-catalyzing astral body flame, within the self-luminous creature entity. The self-luminous creator entity the services the one-thousand units of the supernatural energy, for ascending the entity incarnation consciousness within the deity.

Inclusive of the eight-thousand units of the incarnational consciousness and the one-thousand units of the deity consciousness, the deity's energy value becomes a negative ten-thousand. It is after correcting for the immanent "discordant energy" (*Asura*, -1), without an extrinsic oneness with the consciousness-free "sentient energy" (*Varuna*, 1,000) that the entity is diffusing, for trading the one-thousand units of the deity consciousness. Without any prior physical birth experience, it authenticates the zodiac birth-power to be a sentient earthly creature. It lacks the consciousness of the intrinsic power as a "holy spirit" (*Trinetra*, 1) within the universe of creatures.

- The "negative hundred-thousand" (*Vilokanam*, -100,000) is the value of the "intrinsic oneness of the zodiac animal as an entity" (*Yajna yoga*, -100,000), i.e., as a reincarnated soul, with an immanent "etheric sound consciousness" (*Shrotra vijnana*, -100,000) of the sentient energy, diffused in the prior births in the form of the astrological quarks. A deity creator creates the animal-effect without the "zodiac entity" (*Rashi*, 13). He multiplies the incarnational level consciousness to form twelve zodiac entities. He trades the "self-perpetuating" (*Udvaha*, ½) incarnational value of the absolute zodiac entity. He becomes the "radiating animal entity shape" (*Rupa*, 100,000) of the zodiac system.

The animal entity trades the negative hundred-thousand as the intrinsic value after norming the "discordant energy" (*Asura*, -1) of the deity creator. The deity creator seeks to trade the entire 100,000 units of energy from the animal entity by inspiring it to be the "king of the deity universe" (*Indra*, 0). He motivates the other deity entities to follow the path of self-entropy, so that it may trade their sentient energy and be the "creator of the primeval deity universe" (*Atma linga*, 100,000). The zodiac animal entity generates the mesmerizing sound, "I AM the creature" (*Tat Tvam Asi siddhanta*, - 100,000), seeking to fulfill the purpose of its creation, eventually merging with the deity creator to be an immanent dimension of the primeval deity universe.

- The "negative million" (*Lopana sthapanabhyam*, -1,000,000) is the value of the "appropriate oneness" (*Yukta yoga*, -1,000,000) of the spirit as an entity, i.e., conditional on the absolute (intrinsic oneness within a living human entity, with the mineral consciousness), the primeval (extrinsic oneness as a departed plant entity, with the human consciousness), or the primordial (absolute oneness as a metallic entity, without the material consciousness) state of presence. It is without the primordial-primordial consciousness of the incarnational deity entity or the param-primordial consciousness of the incarnating zodiac animal entity.

A spirit continuously moves its mental oneness consciousness to the diverse forms of entities, seeking the alternative "one with all, the omnipermeating entity" (*Sarvavyapin*, 1,000,000) fame. However, without the devotional focus on the purpose of life, within a disjunctive state disjointed from both the primordial-primordial deity and the param-primordial zodiac entity, the spirit generates zero intrinsic value. It then trades the "negative million," as the value of the desired "one with all" entity. The spirit believes, "I AM the Almighty Creator" (*Sarvam Khalvidam Brahma siddhanta*, -1,000,000), after enjoying the oneness with the "universe without the moving zodiac entities" (*Piyati*, 1000/72).

The spirit conceives the universe without the moving zodiac entities as the value of the "pi." The pi divides the duality within the "primordial-primordial self" (*Dvandva Brahma*, 125) with the "five-faced, param-primordial oneness" (*Sadakhya*, 9) of the param-primordial self. It seeks to manifest not only the "squared radius" as the reality of the deity entity, but also the "pi value" of the animal entity that is multiplying the squared system area with the self-radiating "circular-effect" (*Piyati*, 1000/72).

The self-radiating, "circular-effect" of the manpower spirit of the scientist, the material power spirit of the universe of scientists, the machine power spirit of the instruments, and the method power spirit of the science, is the marketing spirit of the "universe of para-conscious reality" (*Pratyahara*, -12). It is the primary impediment to the discovery of the secret of the prime value sequence using as the thirty-third sutra of the "ten power ten" value, besides the sixteen primary and the sixteen secondary Vedic techniques for simplifying the "universe of complex reality" (*Pradarshan*, -8) created by the universe of modern mathematicians.

Fourth, without the negative-effect generating the primeval sequence, there are four additional points that guide the "meta-sequencing" method, the divine "catalyst" method, the sentient "anchoring" method, and the overall satanic "growth" method.

- The "positive one-hundred" (*Shesanyankena Charamena*, 100) is the value of the "institutional system" (*Sthiti*, 100) generated by the force of the primeval zodiac system. It comprises the holistic constant zodiac entity, without the animal-effect, and the animal-effect of the universe, without the moving zodiac entities.   The holistic value of the "formative form-forming entity" is nine (*Rasi*, 9), i.e., seven with a remainder of two. The universe's value without the moving zodiac entities is 1000/72, i.e., fifteen with a remainder of five.

  Therefore, the value of the wholesomewhole zodiac system = The "primeval value of the zodiac system" (*Nakkhatta*, 22/7) = (Sum of the bases) without the (Sum of the remainders) that is impeding the "wholesomewhole" consciousness, because of the residual circular-effect = Square root of the absolute circular-effect = (7 + 15)/ (2+5) = 22/7 (*Artta*, 22/7).   The value of the "absolute circular-effect" (*Mandalayita*, 484/9) = 484/49, i.e., ten with a deficit of six. It is the energy value serviced by an "inquisitive, soul-searching, philomath entity" (*Jijnasu*, 484/49).

The value of the "wholesomewhole astrological system" is generated through a sequence of inquisitions with a divisive "curvilinear power distance tangent" (*Ardhajya*, 10). It is = The "primeval value of the astrological system" (*Jatakamuktavali*, 10) = The value of the "divine energy" (*Asrava shakti*, 10) = 10. It is the value of the gravitational line centering the zodiac circle by closing the area of the influence of the zodiac system for generating an absolute oneness with the "astrological system" (*Sara Kalpa*, 10^{10}).

A "guiding force" (*Guru*, 100) trades the gravitational tangent of the zodiac system, the creator of the astrological system. It services the "open-system circumference-effect" (*Citra*, 100) for institutionalizing his "primeval self" (*Rama*, 100). It becomes the "guider power" (*Chitta*, 100) of the "institutional system" (*Sthiti*, 100). The value of the guiding force = Value of the "primordial circular-effect" (*Chakrika*, 90) + Value of the "divine energy" (*Asrava Shakti*, 10) = One hundred, without the trading-effect of the absolute circular-effect that eventually transforms one into the primeval, i.e., the incremental or the marginal growth value of the "zodiac system" (*Shunya kalpa*, 8 x 10^{15}).

The "primeval value of the zodiac system" (*Nakkhatta*, 22/7) is the institutional-effect of the "guiding force" (*Guru*, 100). The guiding force diffuses the "I AM the soul" (*Esa Ta Atmantaryamyamrtah siddhanta*, 100) consciousness for attracting and afflicting an "inquisitive, soul-searching, math-loving entity" (*Jijnasu*, 484/9). The "institutional-effect" (*Mumukshu*, 22/7) is also the "primeval value of the starseed system." It is afflicted with a desire to trade the guider power for incarnating a marginal, motivated starseed entity into the earthly realm, seeking to trade the sentient energy of that entity for becoming a "primeval child," i.e., the incremental "soulmate" (*Akrura*, 123).

The objective for "meta-sequencing" the "institutional-effect" with the guider power is to generate the "joy of the spontaneous deliverance" (*Simchah*, 123) from the "sentient life force" (*Prana shakti*, 123) and to trade the incremental twenty-three units for becoming the "God of longevity" (*Shou Lao*, 23). Therefore, the traded value of the institutional-effect is the value of an "afflicted, diseased, injured, or suffering entity" (*Artta*, 22/7), who is experiencing the "affliction, disease, injury, or suffering" (*Jara*, -10^{180}) and eagerly awaiting the "death" (*Jaramarana*, 18). The self-inflicted suffering makes the "science of baking the time-varying self into the primeval solution and badging the culturally-constant space into the absolute problem" (*Purana*, -10^{180}) illuminated within the ancient "*Puranas*" from the pre-*Vedic* India.

- The "positive 10^{10}" (*Sara Kalpa*, 10^{10}) is the value of the "absolute astrological-effect" (*Suryaphanichakra*, 10^{10}) serviced at the death-point (*Tilaka*, 10^{10}) by an "afflicted entity" (*Artta*, 22/7). The intent is to shape the archeological science with the entire history of one's death, including the constant oral history of the events during the "zodiac time" (*Tribhajya*, 10^{10}), before the astrologically-moving variable programming codes guided by the "state of the institutional system" (*Sthiti*, 100). The afflicted entity trades the ten units of the divine energy from the "primordial realm" (*Maha Kalpa*, 10^{1000}) at each moment, without the sequential-effect of the "astrological system" (*Sara Kalpa*, 10^{10}). Consequently, it develops the power to catalyze the intrinsic divine energy of the ten units with the ten units' extrinsic divine energy, without the mediating circumferential-effect of the institutional system.

  The "inquisitive entity" (*Jijnasu*, 484/9) diffuses its absolute astrological-effect into the "astrological system" of the sequentially moving effects. It becomes a "requisitive entity" (*Jalini mukha*, 10^{19}), demanding the "guider debt" benefit value (*Rishi rina*, 10^{19}) payback. The payback compensates for the cost of the "primordial-effect" (*Vidyapati*, 10^{10}) within the "fruit of the inquisition" (*Karma phal*, -10^{19}) that an inquisitive child enjoys. The "algebra" (*Beejaganitam*, 10^{10}) is a system to account for the product of the primordial greeter energy, serviced by a primeval wisecrack in the form of the "paternal debt" benefit value (*Pitri rina*, 10^{10}) for fulfilling the desires of the inquisitive child.

- The "positive 8×10^{15}" (*Gunitasamuchyah*, 8×10^{15}) is the "absolute zodiac-effect" (*Panchang*, 8×10^{15}) traded at the "incarnational value" (*Markatesh*, 8×10^{15}) as the whole anchoring "geometry of the breathing potential" (*Ganitam*, 8×10^{15}). It norms the identity of the "one" (*Akanda*, 8 x 10^{15}) as the "self" (*Atmatva*, 8 x 10^{15}). It is the "omnipresent entity" (*Vyapin*, 8 x 10^{15}) value of the "primordial realm" (*Maha Kalpa*, 10^{1000}). An entity mediates the transformative creator—the zodiac system and the transforming creation—the astrological system. It generates the "trigonometric self-effect" (*Trikonamiti*, 10^{96}) with the catalyzed hundred-units of the guider power, after correcting for the four units invested into creating the afflicting guider power.

The entity becomes an anchoring paternal "metric of angular effects" (*Kamarupa*, 10^{96}), seeking to understand the "science of the black psychological magic" (*Mano vijnanam*, -10^{1024}), of the transformative animal "zodiac system" (*Shunya Kalpa*, 8 x 10^{15}). It then transforms into a "para param entity" (*Pashupati*, 10^{96}), whose three-dimensional "absolute entity-effect" (*Tri mukha*, -10^{1024}) is immanent within the holy "universe of entities" (*Kashi*, -10^{1024}).

The three dimensions include the self as the almighty creator, the universe of entities as the almighty creature, and the "primordial-primordial realm" as the almighty creation instigating the "vow of renunciation" (*Ahatturavu vrat*, -10^{1024}) of all the SHEENY benefits within the self to pay for the "greeter debt" (*Brahma rina*, -10^{1024}) cost, equivalent to the consumed four-unit creator value. Consequently, the fourth dimension with the self becomes one of the "selfish wisher" (*Pravartayitr*, -10^{1024}), seeking to be a heavenly deity for trading back the entire debt servicing cost and to become a "requisite entity" (*Ratnaprabha*, 10^{1024}) for understanding the "entity reality" (*Upakarana artha*, 10^{1024}) traded from the "self-luminous entity realm" (*Satya loka*, 10^{1024}).

- The "positive 6 x 10^{192}" (*Vyashtisamasthi*, 6 x 10^{192}) is the "omnipotent entity" (*Sarvashaktimana*, 6 x 10^{192}), whose "self-perpetuating" (*Udvaha*, ½) "omniscient-effect" (*Ajanya*, 5 x 10^{96}) generates an "inverted, transposed, false, mythological" (*Mithya*, 5 x 10^{96}) value. By mastering the "science of sociological mimicry" (*Samaj shastram*, -10^{1000}), one catalyzes the "trigonometric self-effect" (*Trikonamiti*, 10^{96}) with the "derivative" (*Yuktartha*, 6) value of the holy "universe of entities" (*Kashi*, -10^{1024}). One thereby realizes an "absolute oneness of the primeval space and the primeval time, across the astrological, the zodiac, the starseed, and the primordial realms" (*Surya yoga*, 5 x 10^{96}). Such a "param omnipresent entity" (*Devikotta*, 5 x 10^{96}) enjoys a primeval oneness with the "four-faced Godhead" (*Chamundeshwari*, 5 x 10^{96}), as an "astrological entity, working as a regent lord of the zodiac entity" (*Rashyadhipa*, 5 x 10^{96}). It has the paternal power to organize the "desired reality" (*Pushtartha*, -8) as an "androgynous water entity" (*Meru*, 5 x 10^{96}), who has the potential to give any "deceitful and nonsensical" (*Mithya*, 5 x 10^{96}) form to the thermodynamic fire-effect using the figment of the air of imagination.

By deciding to become the "source of infinite water for accruing primeval value" (*Jalandhara*, 6 x 10^{192}), the entity breaks the "band of protection" (*Rudraksha*, 6 x 10^{192}) bestowed by the "five-faced God" (*Kalagni Rudra*, 6 x 10^{192}) to each "nascent entity" (*Sadyojata*, 6 x 10^{192}). It becomes a subject of the "correlational algebra" (*Anvayika Beejaganita*, 6 x 10^{192}), who enjoys an "instant ablation" (*Simchah*, 123) from the "sentient life force" (*Prana shakti*, 123), both "individually as an entity and collectively as the universe of entities" (*Vyashtisamasthi*, 6 x 10^{192}). The entity gets caught in the complex potential of its imagination. The universe of entities gets caught in the grand challenge of discerning the "appropriate, sensible reality" (*Upakarana artha*, 10^{1024}), transcending the omnipermeating "affliction" (Jara, -10^{180}) of the almighty creator of the viral "universe of para-conscious reality" (*Pratyahara*, -12).

The opportunity cost of not appreciating the value of the self as a "liberated entity" (*Arjuna*, 10^{16}) by birth as a human entity is the "deity debt value" (*Deva rina*, 10^{16}). It is the value of an entity who is creatively servicing the "inner witness" power (*Antaryami*, 16) within the "universe of complex reality" (*Pradarshan*, -8), for trading the catalyzed value of the "divine energy" (*Asrava shakti*, 10) of the universe of entities, without the limits of the mathematical reality.

Chapter 9: The Sequential Effects of the Paradigmatic Planning

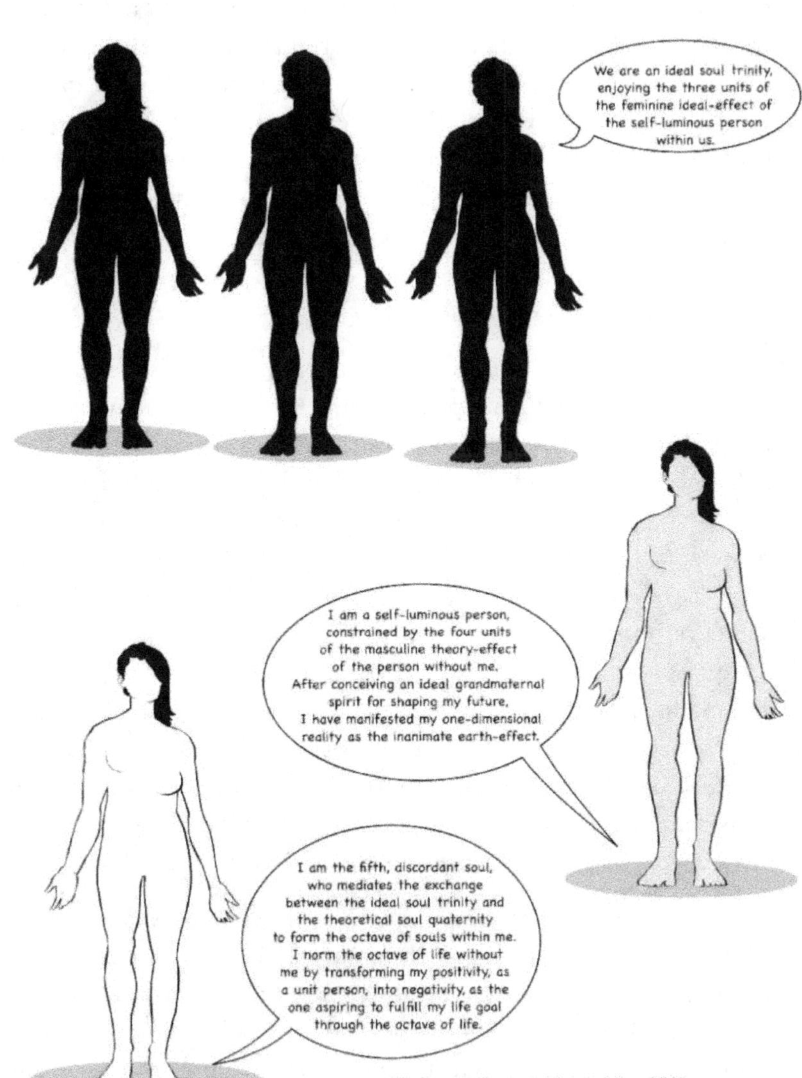

An Octave of Souls is the Origin of Life

Methyl is the Origin of Life

Methyl comprises a soul trinity without a person and a soul quaternity within a person, with an eighth soul that embodies the value of the sentient life's goal as the "earth-effect" (Carbon: *Ravi*, 21). Each "soul" (*Atman*, 4) is a unit of "life" (*Prabhasa*, 4), which manifests as the "water-effect" (Hydrogen: *Jalaprana*, 4). After originating the octave of life, a person becomes a negative discordant unit. Therefore, the manifested form of the methyl is CH3-. The "ideal soul trinity" (*Sridhara*, 25) destroys the "theoretical soul quaternity" (*Homa*, 25) for creating the eighth, "paternal soul" (*Pitra*, 4). The paternal soul comprises a one-unit "person" (*Vyakti*, 1) and a three-unit "multiplier" (*Vaishya*, 3), comprising the three units of the "ideal-effect" (*Dasha*, 1) serviced by the ideal soul trinity.

The "ideal-effect" (*Dasha*, 1) is the "maternal spirit" (*Dasha*, 1), embodying the truth of the simple, one-dimensional natural future reality of the person. The maternal spirit trades the ideal-effect from the "self-luminous twin person" (*Insan*, 1), who conceives the illusionary future grandmother spirit for forming herself as the "ideal" (*Maya*, 1). The person trades the four units of the "theory-effect" (*Mahadasha*, 0) to "eclipse" (*Grahana*, 1) the "ideal" (*Maya*, 1) and transforms into the "absolute zero metric of the present reality" (*Maha Shunya*, $-1 = 0 - 1$). The key to the origin of life is the sequential effect of the ideal paradigmatic planning. It motivates the past, the present, and the future time dimensions to form a "DIVINE Council of the three ideal souls" (*Sridhara*, 25). The three ideal souls modify the person's present reality by transforming him into the sole knower of the ideal's secret truth. For manifesting the origin of life, the soul trinity declares the following.

We are a soul trinity, embodying the sentient water-effect for infusing the sentient element within the inanimate earth-effect. As a potential grandfather human entity, each of us has a four-unit soul potential. Each of us also has a twelve-unit self-luminous potential without the Sun's luminous-effect. Each of us further enjoys the twenty-one units of the simple future reality, just like the Sun, who embodies his potential within each earth-effect unit. The twenty-one units include the two units as a theory-shaping grandmother and the one unit as a theory-taking mother. By conceiving the whole value of the "theory" (*Sadhaka*, 127), each of us becomes a one-hundred twenty-seven-unit proficient exchange system.

The theory's value includes: first, the seven-unit potential of the masculine and feminine sentient selves within each of us; second, the past and the future dimensions of the forty-two-unit "observer-effect" (*Sati*, 42) that forms our present as a "polluted entity" (*Beeja-Jagrat*, 84); and third, the "organization" (*Sangathan*, 29 = 127 – 14 – 84) we form with our entangled presence.

For creating the manifestor factor, the self-luminous person declares the following.

I embody the sixteen-unit truth of the theoretical "soul" (*Atman*, 4) quaternary. The person who is trading my ideal-effect is the fifth soul, who has realized the nine-unit "goal" (*Maha Shiva*, 9) of the sentient life to form a "GUIDER Council of the three theoretical souls" (*Homa*, 25 = 4 * 4 + 9). I have ascended my value to twenty-one by adding my potential to form the five-unit "perpetuating value" (*Saranyu*, 5), comprising the four theoretical souls and the one illusionary soul. I have become the "the earth-effect" (*Ravi*, 21), the absolute creator just like the "Sun" (*Surya*, 21) who is my creator.

9.1 First, Methyl as the Origin of Life

"Ether-effect" (*Naad*, 257) is the overall primeval value of the maternal as well as the paternal entities (127 * 2 = 254), together with the "primeval feminine" (*Maha Brahma*, 3), who manifests the "sound" (*Parai*, 257) of the triple infinity of the primeval paternal, primeval maternal, and the primeval feminine self. The objective of the sound is to materialize the sentient life. The sentient life originates with the combination of the atoms into the "methyl" (CH_3-: *Asura*, -1), a cycloalkyl. The methyl comprises the "discordant" bonding of the three negatively charged units of the water-effect, within one unit of the earthly carbon element. The three negatively charged units of the water-effect are the diffusion value of the convergence of the primeval paternal, primeval maternal, and primeval feminine, without the ether-effect. The one neutral unit of the carbon element is the diffusion value of the convergence within the ether-effect, compensated by the unit negative value of the primeval masculine that is traded by the three units of the water element.

The three units of the water-effect are composed of the three different quark orientation. The primordial-primeval maternal unit is the beauty quark oriented, with a transformative spirit. The primeval-primeval paternal unit is the down quark oriented, with a normative spirit. The param-primeval feminine unit is the strange quark oriented, with a formative spirit. The para-primeval masculine unit is the potential quark oriented, with a gravitomagnetic spirit for attracting the transformative spirit through the togetherness of the formative spirit and mediated by the normative spirit's growth value.

The gravitomagnetic spirit is a function of the three primordial spirits immanent within the carbon-12 atom. These spirits are the "primordial paternal" (*Indra*, 0) in the form of the truth quark that gives a neutral form stability, the "primordial maternal" (*Nirharin*, 18) in the form of strange quark that gives a transformative point variability, and the "primordial feminine" (*Prithvi*, 132) that gives a formative flowing linearity. The overall one-hundred-fifty energy units generate the ten units of the "gravitomagnetic spirit" (*Asrava*, 10), without the six neutrons and the six protons. The ten units include the six electrons within the seventh carbon atom and the three parallel pairs of correlation linkages with the three water-effects units.

As a cycloalkyl, methyl's distinctive gravitational quality is its "lipophilicity"—the power to dissolve all non-polar spirits that are not similar to the base spirit for massifying (i.e., fattening and organizationally developing) the dominating spirit. The dominating spirit is the "primordial masculine" (*Kaumari*, 90), produced through the "primordial circular-effect" (*Chakrika*, 90) of the sound, first away from the primeval divergent trinity and then toward the primordial convergent entity.

The primordial circular-effect is the "self-perpetuating" (*Udvaha*, ½) "primordial maternal" (*Nirharin*, 18) dimension, catalyzed by the "divine energy" (*Asrava shakti*, 10) of the primordial feminine dimension, and normed by the local Jupiter-effect (*Mahadasha*, 0) of the "primordial masculinity—the human-effect" (*Linga*, 53) immanent within the "primordial paternal" (*Indra*, 0) dimension. The primordial circular-effect manifests by transforming the methyl into the alkyl ether, known as the anesthetic diethyl ether, or the "ether element" (*Shuddhi*, 285): CH3–CH2–O–CH2–CH3. There are sixty divergent elements within the five carbon atoms, twenty curvilinear elements within the ten hydrogen atoms, twenty-four convergent elements within the one oxygen atom, and four linear bonding elements within the one 108-unit, "unique, problem-making paternal soul dimension" (*Samanya dharma*, 108) of the ether element.

9.2 Second, Nine Forms of the Ether Element That Sequence Life

The ether element has nine forms: the maternal, the paternal, the feminine, the masculine, the greeter, the satanic, the devil, the deity, and the para deity.

The maternal ether is known as "cytosine nucleic acid (CNA) or nucleoside" (*Atbudha*, 285). The paternal ether is known as "guanine nucleic acid (YNA) or nucleotide" (*Vamatevan*, 285). The feminine ether is known as the "guanosine or peptide nucleic acid (PNA) or the deoxynucleotide" (*Nama*, 285). The masculine ether is known as the "uracil or glycol nucleic acid (*GNA*)" (*Hara*, 285). The greeter ether is known as the "thymine or threose nucleic acid (TNA)" (*Aakash*, 285). The satanic ether is known as the "uridine or hexose nucleic acid" (HNA)" (*Sabda*, 285). The devil ether is known as the "adenine or xeno nucleic acid (XNA)" (*Ab*, 285). The deity ether is known as the "adenosine or morpholino nucleic acid (MNA)" (*Maruta*, 285). The para deity ether is known as the "cytidine or locked nucleic acid (LNA) or nucleobase" (*Yujya*, 285).

As a byproduct, the methyl forms into the curvilinear alkyle ether, norming the "methoxyethane" (*Indrani*, 69): CH3–CH2–O–CH3. There are thirty-six divergent elements within the three carbon atoms, sixteen curvilinear elements within the eight hydrogen atoms, twenty-four convergent elements within the one oxygen atom, and three linear bonding elements within one 69-unit "twin maternal soul" (*Shachi*, 69).

For realizing a "causative equilibrium" (*Alaya vijnana*, 108), the paternal soul dimension forms a glycolic polymer of ether linkages, known as polytetramethylene ether glycol: –CH2CH2CH2CH2O–. There are forty-eight divergent elements within the four carbon atoms. There are sixteen curvilinear elements within the eight hydrogen atoms. There are twenty-four convergent elements within the one oxygen atom. Besides, there are two linear bonding elements within the one 90-unit "primordial circular-effect" of the "self-driving and circulating masculine soul dimension" (*Chakrika*, 90).

There are eighteen squared linkages—the "organelles" (*Dasha*, 1)—among the three molecules: the three intrinsic oneness linkages of each with the self; the three extrinsic oneness linkages of each with the other two; the three extrinsic oneness linkages of each with one other mediated by the second; the three extrinsic oneness linkages of each with one other, moderated by the second; the three extrinsic oneness linkages of each with one other, free from the second; and the three sequential oneness linkages of each of the three with the overall eighteen-unit "chloroplast" (*Nirharin*, 18). The chloroplast is the normative value of "the feminine soul as the organizing dimension" (*Tarkshaya*, 18).

9.3 Third, the Phage as the Primary Life Form

The whole 285-unit system forms a "macrobiomolecule—the platelet" (*Vyoma*: 285) and norms the "param animal zodiac-effect" (*Pashutva*, 855), within the "lipid—the sponge phage" (*Hara*, 285). It perpetuates the power to dissolve all the non-polar spirits that are not similar to the base spirit for massifying (i.e., fattening and organizationally developing) the dominating spirit. It is the primordial-primordial form of the biological life. It has a normative power to generate an "asexual, horizontal gene transfer" (*Yujya*, 285) for reproducing the maternal fattening sequence. The macrobiomolecule generates its normative power through the "methanogenesis" (*Vithoba*, 12).

The methanogenesis is the production of methane to catalyze the "lysogenic cycle" (*Dvisaptati dasha*, 29). The Lysogenic cycle dissolves the nitrogen (fire-effect) for the growth of both the hydrogen (water-effect) and the carbon (earth-effect) atoms, within a phosphorous atom (divine-effect). The Lysogenic cycle activates the "primordial ether-element" (Culture element: *Sadakhya tattva*, 9) within the lipid for balancing the "pH" (*Pratyantara dasha*, 9) by compensating for the disproportionate acidity from the fire-effect. The primordial ether element activates the "primeval-primordial oneness element" (Absolute time element: *Sva tattva*, 11) without the lipid, for the horizontal servicing of the "acid-effect" (*Anishta* 38). It vertically trades the "alkaline-effect" (*Yavashuka*, 28) from the primeval lipid within the universe of lipids. It empowers the maternal phage, known as prophage—the virion, to transform the primeval lipid into a circular "replicon," the paternal phage—the chromosome.

COVID-19 is a maternal phage that has the power to take nine forms: the maternal (T7 phage), the paternal (186 phage), the feminine (T12 phage), the masculine (P2 phage), the greeter (R17 phage), the satanic (T2 phage), the devil (T4 phage), the deity (λ or lambda phage), and the para deity (Φ or phi phage).

Each form is a different strain of COVID-19, mediated by the "circulating cosmic creation" (*Srishti*, 379) in the form of the varying dimensions of the "workculture element" (*Nayaki tattva*, 379). The maternal phage is the Confucian Asian strain—26143 (D614), the deadliest virion. The "paternal phage" (*Vamatevan*, 285) is the Oceanic strain—2891, the least infectious virion. The "feminine phage" (*Nama*, 285) is the Latin American strain—8782, the least deadly virion. The "masculine phage" (*Hara*, 285) is the West European strain—14408, which is the most symptomatic virion, with moderate mortality and infectivity. The "greeter phage" (*Aakash*, 285) is the Eastern European strain—23403, with the least immuno-susceptibility.

The "satanic phage" (*Sabda*, 285) is the North American strain—17857, with moderate symptoms, mortality, infectivity, and immuno-susceptibility. The "devil phage" (*Ab*, 285) is the Southern Asian strain—28144, with the highest immuno-susceptibility. The "deity phage" (*Maruta*, 285) is the African strain—18060 (L5F), the least symptomatic virion, with moderate mortality and infectivity. The "para deity phage" (*Yujya*, 285) is the Eastern hybrid, global strain—28881 (D614G), the most infectious virion.

9.4 Fourth, the Archaeon as the Secondary Life Form

By bonding with a second macrobiomolecule, the 570-unit system forms the "ester element" (*Ganesha: Param human-effect*, 570), known as the "archaeon" (*Jangamavisha*, 570), which norms the "protist kingdom—the universe of animate, sentient entities" (*Akalpa*, 570). The "universe of white-blood-cell organism—the archaeon" (*Shvetalohita*, 36) is a universe of the primordial form of the biological life. It has a formative power for the direct "physical, diazotrophic, nitrogen fixation" (*Usha*, 16), without the mediation of the neutron value in the methane. This power is in the form of the "Golgi—the formative divine energy" (*Madhusudan*, 16). An archaeon has a normative power for the "sexual, vertical gene transfer." It reproduces the paternal sequence of trading the curvilinear lipid fat and servicing the linear ATP (adenosine triphosphate) carbohydrate. The archaeon generates its normative power through the "autotrophic, anaerobic respiration" (*Abhiyoga*, -18) to catalyze the "lytic cycle" (*Dharana*, -8).

The Lytic cycle oxidizes the lipid fat through an entropy diffusion of its electrons, so that the atmospheric oxygen, the air-effect, dissolves within the phosphorous, the divine-effect. It empowers the paternal phage to transform into a triplet of replicons, composed of the maternal phage, the feminine phage, and the masculine phage.

Fire-effect is dominant in the paternal phage. Water-effect is dominant in the maternal phage. Air-effect is dominant in the feminine phage. Earth-effect is dominant in the masculine phage. Divine-effect is dominant in the "greeter phage—the Golgi, i.e., the white blood apparatus" (*Madhusudan*, 16). Ether-effect is dominant in the satanic phage—the paternal phage that impregnates the feminine phage for forming the "maternal archaeon—as the plasmid, ciliate, or euglenoid" (*Rajakara*, 570). Guider-effect is dominant in the devil phage, which motivates the maternal phage to inspire the masculine phage to impregnate her for forming the paternal archaeon—the "chromid or the flagellate" (*Uparanj*, 570).

The SHEENY-effect is dominant in the deity phage, which motivates the greeter phage to compensate by impregnating the maternal phage sexually with its masculine dimension. It concurrently inspires the paternal archaeon to asexually impregnate its feminine dimension to form the "bacterium" (*Vetranavah*, 855). The Corporate-effect is dominant in the para deity phage, which motivates the formation of the seven additional archaea: the "fosmid or the alga as the para deity archaeon" (*Vishakha*, 570), the "phagemid or the kelp as the deity archaeon" (*Karajakara*, 570), the "yeast as the devil archaeon" (*Khamira*, 570), the "fungus as the satanic archaeon" (*Kyaku*, 570), the "amoeba as the greeter archaeon" (*Abhirakta*, 570), the "protozoan as the feminine archaeon" (*Vaibhrajaka*, 570), and the "sporozoan as the masculine archaeon" (*Rajaka*, 570).

9.5 Fifth, the Bacterium as the Tertiary Life Form

By bonding with a third macrobiomolecule, the 855-unit system forms the "bacterium—the worm" (*Vetranavah*, 855), which norms the "param mineral-effect" (*Parapara*, 855) codified into the protein "RNA polymerase" (*Atisara*, 855).   The bacterium is the param form of the biological life, with a normative power for the direct "chemical, fermenting, aerobic respiration" (*Manjushri*, 19), without the moderation of the electron value.   The bacterium generates its normative power by fermenting the ATP carbohydrate with the SHEENY-effect, "krypton" (*Atman*: Soul, 4), to organize "a causative grouping of the supernatural energy" (*Naraki*, 1) into a "microorganism" (*Triyancha*, 19), known as the "enzyme" (*Hiranyagarbha*, 19).

An enzyme is the "cell" (*Hiranyagarbha*, 19) of "energy" (*Shakti*, 19) without the "param animal-effect of the phage" (*Pashutva*, 285). The enzyme norms the "energy as the param deity-effect" (*Shakti*, 19). Consequently, the "Argon" (*Chitta*, 100), the guider-effect, dissolves within the Calcium, the ether-effect, through the enzyme-catalyzed entropy in the formative SHEENY-effect. The RNA polymerase works as a "catalyst agent" (*Shilajit*, 855) to "accelerate the electromagnetic massification" (*Gardabhejya*, 108) by adding the "self-ordered, referential presence" (*Virodhita*, 108) of an additional "normal, paternal soul dimension" (*Samanya dharma*, 108) as the "budding energy" (*Utapalai Shakti*, 108).

There are nine bacterium classes, which vary in the formative gravity, the electromagnetic speed of reproduction and the consequential sentient shape.

First, the asexually-reproducing, conjugating, linear, coiling, rod-shaped "*Bacillus* as the maternal bacterium" (*Jivanu*, 855). Second, the sexually-reproducing, curvilinear, "*Vibrio* as the paternal bacterium" (*Jivabhyasa*, 855).   Third, the predominantly asexually-reproducing, metamorphically transforming, convergent, trichome-forming, bottle-shaped, "filamentous as the feminine bacterium" (*Jivagribh*, 855). Fourth, the predominantly sexually-reproducing, bud-morphing, divergent, squared, droplet sphere-forming "*Coccus* as the masculine bacterium" (*Jivashulaka*, 855). Fifth, the fragmenting, exponentially-multiplying "sheath as the greeter bacterium" (*Jivadara*, 855). Sixth, the logarithmically-dividing, generalized-transducing, binary-fissioning, spindle-shaping, stalking, "spirillum as the satanic bacterium" (*Shakanavah*, 855). Seventh, the parthenogenic, multilinearly-fusing, specialized-transducing, pleomorphic, "spirochete as the devil bacterium" (*Jivanikaya*, 855). Eighth, the fertilizing, harmonically-diffusing, spindle-shaped, "rickettsia as the deity bacterium" (*Jivati*, 855). Ninth, the lobed, star-shaped, self-destroying, self-replicating "mycoplasma as the para deity bacterium" (*Jivanada*, 855).

9.6 Sixth, the Eukaryote as the Quaternary Life Form

The consequential 963-unit system forms the "eukaryote—the shell-protected mollusk" (*Rudhita*, 963), which norms the "param solar plant-effect" (*Shasyatva*, 963) codified into the "one membrane-bound DNA molecule" (*Ranabajari*, 963). It is the primeval form of biological life, with the normative power for direct "biological, atrophic, photosynthetic perspiration" (*Shipra*, 963). The DNA molecule is the first, primordial unicellular organization, whose sentient life is entirely dependent on the cell's energy. Three para-biological lifeforms—the phage, archaeon, and bacterium—are immanent within the unicellular eukaryote.

The "chlorophyta or prasinophyte green alga" is the only eukaryote lifeform. It derives its green color from the photosynthetic "param solar-effect" (*Shasvatava*, 963). The DNA molecule generates its normative power through the proton value's thermodynamic diffusion after sequentially fulfilling its organizationally programmed genetic work. There are nine classes of the DNA molecules.

First, the conjugally-reproducing, linear, right-handed, negative, anti-clockwise, double-helix, eleven base-pair, "the A-form as the maternal DNA" (*Banagi*, 963). Second, the sexually-reproducing, hydrated, curvilinear, right-handed, positive, clockwise, double-helix, ten base-pair, "the B-form as the paternal DNA" (*Balangi*, 963). Third, the metamorphically-transforming, narrow, convergent, negative, right-handed, anti-clockwise, double-helix, twelve base-paired, "the Z-form as the feminine DNA" (*Pishi*, 963).   Fourth, the bud-morphing, divergent, squared, positive, left-handed, clockwise, double-helix, nine base-paired, "the C-form as the masculine DNA" (*Agradhanyama*, 963).

Fifth, the exponentially-fragmenting, negative, right-handed, anti-clockwise, double-helix, eight base-paired, "the D-form as the greeter DNA" (*Saramadi*, 963).   Sixth, the slowly, logarithmically-transducing, open-circular, positive, left-handed, clockwise, single-helix, seven-base, "S-form as the satanic DNA" (*Khicada*, 963).   Seventh, the multilinearly-fusing, super-helix embodying a six base-pair, double-helix for personifying a closed-circular, positive, left-handed, clockwise, single-helix, five-base, the "F-form as the devil DNA" (*Shendedhanya*, 963).   Eighth, the fertilizing, harmonically-diffusing, negative, right-handed, anti-clockwise, closed-circular, single-helix, four-base, "M-form as the deity DNA" (*Tantola*, 963). Ninth, the self-replicating, neutral, centered, single-point, triangulated, three-base, the "O-form as the para deity DNA" (*Shirem*, 963).

Similarly, there are nine corresponding classes of the unicellular, eukaryotic, chlorophyta, green alga: the *hydrodictyon, cladophora, volvox, oedogonium, acetabularia, desmids, chlorella, chlamydomonas,* and *spirogyra*.   Since they rely on the intrinsic cellular energy as a biological entity, they enjoy a life of "immortality by becoming the micrasterias" (*Chiranjivi*, 26), under the ideal laboratory conditions of the zero extrinsic energy exchange.

9.7 Seventh, the Prokaryote as the Quinary Life Form

There are eighteen squared linkages—the "organelles" (*Dasha*, 1)—among the three extrinsic life forms, without the guider mediation of the three intrinsic molecules, and an additional eighteen, with the guider mediation. These thirty-six organelles form the "nucleoid" (*Maheshwari*, 17) to norms a mitochondrion molecule: the "endoplasmic reticulum" (*Rasagni*, 17). Each organelle is a natural energy sequence, norming the "absolute earth-effect" (*Trinetra*, 1) and transforming into the "supernatural energy" (*Shram Shakti*, 1), by trading the grouping-effect within the enzyme.

The thirty-six organelles constitute the thirty-six different primordial-primordial forms of the mitochondria molecule, grouped into nine classes. Each class gets differentiated into four primordial forms as a function of the sequential development of the "phage-effect" (*Parai*, 257), the "archaeon-effect" (*Danda*, 176), the "bacterium-effect" (*Chetan Shakti*, 75), and the "eukaryote-effect" (*Suswani*, 87), within the organelle.

As a consequence of the supernatural energy's trading-effect, the mitochondrion molecule enjoys a dynamically-dancing, seven-faced system value. It is composed of the phage system, archaeon system, bacterium system, eukaryote system, prokaryote system, organelle system, and primordial circular system, converging into the phage system and diverging from the prokaryote system. Therefore, the "mitochondrion molecule" is the convergent value of the spirit-invigorating, "organizational performing" (*Rachana*, 286) of the "cosmic perpetuator" (*Rachayita*, 27).

The nine classes of the "mitochondrion molecule" (*Tandava*, 286) have varying energy values as a function of the varying trading and servicing of the supernatural energy. These forms include the "maternal mtDNA" (*Gauri*, 63), the "paternal mtDNA" (*Ananda*, 185), the "feminine mtDNA" (*Krishna*, 32), the "masculine mtDNA" (*Uma*, 6), the "greeter mtDNA" (*Tripura*, 789), the "satanic mtDNA" (*Samhara*, 195), the "devil mtDNA" (*Sandhya*, 89), the "deity mtDNA" (*Kalika*, 23), and the "para deity mtDNA" (*Haum shakti*, 9).

The sequential 999-unit system forms the "prokaryote—the spines as the red-blood apparatus" (*Shadayatana*, 999), which norms the "param metal-effect" (*Dhatva*, 999). It is the primeval-primeval form of biological life. It is a multicellular organization whose sentient life is free from one cell's energy limitations. The prokaryote trades its normative power from the thirty-seventh mitochondrion "organelle" (*Dasha*, 1) that eventually forms the "lifeless cellular body" (*Nirjara*, 1,000)—as a "nucleus, the origin of life" (*Yoni*, 1,000).

The lifeless cellular body norms the "absolute spirit-effect" (*Badhabuddhi*, 1,000) of the greeter spirit, perpetuating an absolute oneness between the guider-mediated intrinsic molecules and the guiding extrinsic life forms. The normative power is for the direct "conscious, metamorphic, sentient distillation" (*Nirmana chitta*, 1000) of the fourteen building-blocks and the five physical foundations of the "sentient energy" (*Varuna*, 1,000) into a nineteen-unit "atom" (*Anu*, 19). Consequently, the inanimate atom spontaneously metamorphizes into an animate cell, without any energy exchange. The cell is "an entity with the sensory tentacles like an octopus" (*Hiranyagarbha*, 19).

9.8 Eighth, the Spirit as the Senary Life Forming the Cell, Without the Satanic Energy

The "spirit" (*Kapinjala*, 20) of the "self-luminous consciousness" (*Saguna*, 20) is the sixth, metaphysical foundation of the sentient energy, which an entity experiences during the fifth phase of the "dream state" (*Maha Swapna*, 9) of the "self-mediated dream" (*Dhyana*, 9). During the four physical phases, one descends the awareness of the "extrinsic consciousness" (*Suryaphanichakra*, 10^{10}), while ascending the "self-perpetuating" (*Udvaha*, ½) "self-consciousness" (*Prajna*, 2,222) of the following four building blocks:

- First, the "para-consciousness" (*Nirharin*, 18) dream state. It comprises two metaphysical, primeval spirits. First, the primeval greeter spirit generates the dynamic energy flow converging into the methyl particle. Second, the masculine spirit trades the divergent metaphysical effect of the methyl particle.

- Second, the "param consciousness" (*Shivadrishti*, 17) dream state. It comprises the three physical, primeval spirits: the transformative maternal spirit, the normative paternal spirit, and the formative feminine spirit.

- Third, the "intrinsic consciousness" (*Antarmana*, 10) dream state. It comprises the four param-effects of the primeval spirits. First, the macrobiomolecule, lipid—the fattening maternal sequence. Second, the ester element, the archaeon—the consequential paternal carbohydrate. Third, the bacterium, RNA polymerase—the masculine protein norming absolute animal-effect. Fourth, the eukaryote, DNA molecule trading the absolute plant-effect of the feminine cytoplasm.

- Fourth, the "consciousness" (*Chaithanya*, 4) dream state. It comprises the five primordial souls. First, the greeter soul, generating the primordial circular-effect with the sound of the energy flow. Second, anesthetic diethyl ether—the problem-making paternal soul. Third, methoxyethane—the solution-shaping maternal twin soul. Fourth, polytetramethylene ether glycol—the self-driving and circulating masculine soul. Fourth, the chloroplast—the normative form of the feminine soul as the eventual "life" organizing dimension.

Without the limitations of the "self-consciousness" (*Saguna*, 20), an "enlightened entity" (*Shri Krishna*, 10)—with an intrinsic consciousness of the "life" (*Prabhasa*, 4)—enjoys a strategic awareness of the six foundations of the four building-blocks during the "wakefulness state" (*Swapna*, 59):

- The nucleoid, codifying the maternal mitochondrion molecule—endoplasmic reticulum and norming the absolute mineral-effect.

- The prokaryote, distilling the paternal sentient value of the consequential param metal-effect into the sentient energy.

- The causative grouping of the supernatural energy into the feminine microorganism—the enzyme, norming the absolute deity-effect.

- The sequential linkages of the natural energy in the form of masculine organelles, norming the absolute earth-effect.

- The area primordial greeter services as the origin of life, norming the absolute spirit-effect.

- The param-primordial spirit of the entity with self-luminous consciousness of the entropy reality, after exchanging the sentient energy with the inanimate atom.

A "primeval wisher" (*Artharthi*, -6), guided by the genetic programming, trades the "qualifying" (*Nirguna*, -6), "universe of technological reality" (*Shramika*, -6). Therefore, it becomes a "vessel" (*Kamandalau*, -900) of the "deity realm" (*Deva loka*, 1,000) for perpetuating the "differentiated, quadratic reality" (*Kritartha*, -6). The next-generation "param child" (*Manyu*, 19) becomes eternally grateful to the "paternal, passing the consciousness of life" (*Arjuna*, 10^{16}). The latter binds himself with the "deity debt" (*Deva rina*, 10^{16}), seeking reincarnation, as a primeval child, for the oneness with the para deity realm in the future.

9.9 Ninth, the Param Child as the Spiritually Programmed Life, Within Each Cell

A param child is an absolute solution to the primeval problem of illuminating the differentiated entity reality of the "creation as the param greeter" (*Srishti*, 379). Tables 18-21 highlight the forty-eight pathways for a param child to develop absolute consciousness of the multidimensional absolute reality of the "param creature" (*Prabhu*, 1600), within the param greeter. These forty-eight pathways sequentially ascend the consciousness about the self-luminous maternal, paternal, feminine, and masculine entities, within the descending para-consciousness of the self-luminous greeter.

Table 18. The Twelve Pathways of the Param Child, Within the Maternal Self-luminous Entity

Sequential path of param child as the mindful destroyer	Ascending Maternal self-luminous entity	Descending Paternal self-luminous entity (*Bhavanavasi*; Exousiai; Potestas; Authority)	Forward Feminine self-luminous entity: Archangel	Backward-effect of the Masculine self-luminous entity: Repulsion from	Concave or Convergent reality (*dharma*)	Linear or Infinite reality (*artha*)	Convex or Divergent reality (electromagnetic metric)	Curvilinear or Parallel reality (*jya*)
Primordial *Priyangu*	Primordial *Vidyutkumaras*	*Harisimha*	*Nathaniel*	Purpose of life	*Daan dharma*	*Ityartha*	Electromagnetic current or flow	*Yujya*
Primeval *Priyangu*	Primeval *Vidyutkumaras*	*Harikanta*	Raziel	Secret of life	*Paripurna dharma*	*Vyanjanartha*	Electromagnetic dose equivalent	*Tribhajya*
Primordial *Salmali*	Primordial *Suparnakumaras*	*Venudeva*	Uriel	Light of life	*Putra dharma*	*Jahatsvartha*	Electromagnetic intensity	*Kramajya*
Primeval *Salmali*	Primeval *Suparnakumaras*	*Venudarin*	Metatron	Knowing life	*Sankhya dharma*	*Tulyartha*	Electromagnetic flux	*Kotijya*
Primordial *Palasa*	Primordial *Agnikumaras*	*Agnishikha*	Jophiel	Life communicant	*Sanatana dharma*	*Gurvartha*	Electromagnetic thermal conductivity	*Ekarajya*
Primeval *Palasa*	Primeval *Agnikumaras*	*Agnimanava*	Raphael	Life healing	*Yati dharma*	*Tvampadartha*	Electromagnetic density or sigma value	*Ganarajya*

Sequential path of param child as the mindful destroyer	Ascending Maternal self-luminous entity	Descending Paternal self-luminous entity (*Bhavanavasi;* Exousiai; Potestas; Authority)	Forward Feminine self-luminous entity: Archangel	Backward-effect of the Masculine self-luminous entity: Repulsion from	Concave or Convergent reality (*dharma*)	Linear or Infinite reality (*artha*)	Convex or Divergent reality (electromagnetic metric)	Curvilinear or Parallel reality (*jya*)
Primordial *Raja-druma*	Primordial *Samiranakumaras* (*vayukumaras*)	*Velamba*	Jeremiel	Exalted soul	*Sadharana dharma*	*Hetvartha*	Electromagnetic dose rate	*Pravrijya*
Primeval *Raja-druma*	Primeval *Samiranakumaras* (*vayukumaras*)	*Prabhanjana*	Chamuel	Seer	*Samanya dharma*	*Pushtartha*	Electromagnetic acceleration	*Gardabhejya*
Primordial *Kadamba*	Primordial *Stanitakumaras*	*Sughosha*	Ariel	Royal splendor	*Dharma*	*Mukhyartha*	Electromagnetic substance	*Putrejya*
Primeval *Kadamba*	Primeval *Stanitakumaras*	*Mahaghosha*	Zadkiel	Righteousness	*Parmarthika dharma*	*Mahartha*	Electromagnetic substance concentration	*Mahavijya*
Primordial *Vetasa*	Primordial *Udadhikumaras*	*Jalakanta*	Azrael	Cleansing	*Prakriti dharma*	*Padartha*	Electromagnetic radiance	*Autaghaticem rajya*
Primeval *Vetasa*	Primeval *Udadhikumaras*	*Jalaprabha*	Haniel	Glory	*Sasaka dharma*	*Upkaranartha*	Electromagnetic pole	*Jya*

Table 19. The Twelve Pathways of the Param Child, Within the Paternal Self-luminous Entity

Sequential path of param child as the mindful destroyer	Descending Maternal self-luminous entity	Ascending Paternal self-luminous entity (*Bhavanavasi;* Exousiai; Potestas; Authority)	Backward Feminine self-luminous entity: Archangel	Forward-effect of the Masculine self-luminous entity: Enchantment with	Concave or Convergent reality (*dharma*)	Linear or Infinite reality (*artha*)	Convex or Divergent reality (electromagnetic metric)	Curvilinear or Parallel reality (*jya*)
Primordial *Jambu*	Primordial *Dvipakumaras*	*Purna*	Raguel	Sociological mimicry truth of mercy blessing	*Yuddha dharma*	*Paramartha*	Electromagnetic permeability	*Ardhajya*
Primeval *Jambu*	Primeval *Dvipakumaras*	*Avasishta*	Gabriel	Psychological black magic truth of might begging	*Asanga dharma*	*Mitartha*	Electromagnetic temperature	*Pratirajya*
Primordial *Sirisa*	Primordial *Dikkumaras*	*Amitagati*	Michael	Holistic truth of chanting	*Nivrtti dharma*	Bhavartha	Electromagnetic displacement	*Utkramajya*
Primeval *Sirisa*	Primeval *Dikkumaras*	*Amitavahana*	Sataniel	Psychogenealogical truth of Intercessor	*Ritu dharma*	*Yathartha*	Electromagnetic mass fraction	*Svarjya*
Primordial *Kadamba*	Primordial *Pishacha* (vampire)	*Kala*	Achaiah	Clairvoyant truth of the natural secret	*Guna dharma*	*Rakkhartha*	Electromagnetic susceptibility, i.e., diffusion torque	*Yojya*
Primeval *Kadamba*	Primeval *Pishacha* (vampire)	*Mahakala*	Lauviah	Scientific truth	*Varna dharma*	*Parindartha*	Electromagnetic reluctance	*Bhrijajya*

Param child	Maternal entity	Paternal entity	Feminine entity	Forward-effect of Masculine entity	Convergent reality	Infinite reality	Divergent reality	Parallel reality
Primordial Tulasi	Primordial *Bhuta* (ghost)	*Svarupa*	Hahuiah	Psychometry truth	*Beeja dharma*	*Niyogartha*	Electromagnetic induction or flux density	*Yajya*
Primeval Tulasi	Primeval *Bhuta* (ghost)	*Pratirupa*	Nith-Haiah	Occultic truth	Pitr *dharma*	*Gudhartha*	Electromagnetic luminous flux	*Ghodamunjya*
Primordial Banyan	Primordial *Yaksha* (warlock; witch)	*Purnabhadra*	Haaiah	Scrying truth	*Laukika dharma*	*Kritartha*	Electromagnetic motive force, i.e., potential difference	*Prayujya*
Primeval Banyan	Primeval *Yaksha* (warlock; witch)	*Manibhadra*	Sealiah	Alchemic Truth	*Dur dharma*	*Nirartha*	Electromagnetic capacitance	*Marjya*
Primordial Kantaka	Primordial *Rakshas* (demon)	*Bhima*	Nithael	Palm line Development	*Satya dharma*	*Yuktartha*	Electromagnetic plane angle	*Uttarajya*
Primeval Kantaka	Primeval *Rakshas* (demon)	*Mahabhima*	Iah-hel	Telepathic illuminati	*Kanya dharma*	*Vyartha*	Electromagnetic field strength	*Samastajya*

Table 20. The Twelve Pathways of the Param Child, Within the Feminine Self-luminous Entity

Sequential path of param child as the mindful destroyer	Backward Maternal self-luminous entity	Forward Paternal self-luminous entity (*Bhavanavasi;* Exousiai; Potestas; Authority)	Ascending Feminine self-luminous entity: Archangel	Descending-effect of the Masculine self-luminous entity: Descending	Concave or Convergent reality (*dharma*)	Linear or Infinite reality (*artha*)	Convex or Divergent reality (electromagnetic metric)	Curvilinear or Parallel reality (*jya*)
Primordial *Asoka*	Primordial *Kinnara* (Werewolf; Lower human body, upper horse head)	*Kinnara*	*Chabuiah*	Trading well-being	*Vyakti dharma*	*Nitartha*	Electromagnetic wave reciprocal meter	*Brahmasayujya*
Primeval *Asoka*	Primeval *Kinnara* (Werewolf; Lower human body, upper horse head)	*Kimpurusha*	*Mumiah*	Workcultural well-being	*Moksha dharma*	*Dtyartha*	Electromagnetic diffusion rate	*Ravanarajya*
Primordial *Madhuvaha*	Primordial *Kimpurusha* (Wizard; Lower bird body, upper human head)	*Satpurusha*	*Vehuaiah*	Cultural well-being	*Naimttika dharma*	*Sabdartha*	Electromagnetic catalytic activity	*Ganapujya*
Primeval *Madhuvaha*	Primeval *Kimpurusha* (Wizard; Lower bird body, upper human head)	*Mahapurusha*	*Ielahiah*	Social well-being	*Visesa dharma*	*Ekartha*	Electromagnetic moment	*Pindajya*

Param child	Maternal entity	Paternal entity)	Feminine entity	Masculine entity	Convergent reality	Infinite reality	Divergent reality	Parallel reality
Primordial *Naga*	Primordial *Mahoraga* (Zombie; Lower serpent body, upper human head)	*Atikaya*	*Mabaiah*	Human well-being	*Yuga dharma*	*Ishtartha*	Electromagnetic potential	*Tyajya*
Primeval *Naga*	Primeval *Mahoraga* (Zombie; Lower serpent body, upper human head)	*Mahakaya*	*Mebahel*	Ecological well-being	*Veshya dharma*	*Svartha*	Electromagnetic dynamic viscosity	*Ritryajya*
Primordial *Tumbaru*	Primordial *Gandharva*	*Gitarati*	*Haamiah*	Economic well-being	*Stree dharma*	*Purushartha*	Electromagnetic illuminance	*Avanejya*
Primeval *Tumbaru*	Primeval *Gandharva*	*Gitayasas*	*Nemmamiah*	National well-being	*Adharma*	*Vdyartha*	Electromagnetic entropy or maximum likelihood value	*Viyojya*
Rohini	Primordial *Aprajnapti*	*Sannihita*	*Seeiah*	Psychological well-being	*Sva dharma*	*Laksyartha*	Electromagnetic catalytic activity concentration	*Dvajya*
Prajnapti	Primeval *Aprajnapti*	*Samanaka*	*Haiel*	Guider well-being	*Raja dharma*	Sutrartha	Electromagnetic exposure to extrinsic radiation	*Varjya*
Vajra-srinkhala	Primordial *Pancha-prajnapti*	*Dhatr*	*Asaliah*	SHEENY well-being	*Pravrti dharma*	*Vakyartha*	Electromagnetic mass density	*Jinamshajya*
Vajrankusa	Primeval *Pancha-prajnapti*	*Vidhatr*	Aniel	SHEENY entropy	*Arama dharma*	*Artha*	Electromagnetic power or radiant flux	*Pathitajya*

Table 21. The Twelve Pathways of the Param Child, Within the Masculine Self-luminous Entity

Sequential path of the param child as the mindful destroyer of the androgynous-effect (Harita rupa, Parimandala Sadashiva, Mandala Usha, Sushumna)	Forward Maternal self-luminous entity forming the entity as the feminine-effect (Avadata rupa: Parimandala Usha, Mandala Aum, Pingala)	Backward Paternal self-luminous entity (Bhavanavasi; Exousiai; Potestas; Authority) norming the entity as the masculine-effect (Lohita rupa, Parimandala Aum, Mandala sushumna, Ida)	Descending Feminine self-luminous entity (Archangel) transforming the entity into the soliton (Pita rupa, Parimandala sushumna, Mandala pingla, Kana)	Ascending-effect of the Masculine self-luminous entity on the destiny of the param child (Nila rupa, Parimandala pingla, Mandala Ida, Aksara)	Concave or Convergent reality of the diverse body forms of the param child (Parimandala Ida, Mandala kana, Rupa)	Linear or Infinite reality norming taste sensation (Dirgha rupa, Parimandala kana, Mandala aksara, Arupa, Rasa)	Convex or Divergent reality forming touch sensation (Hrasva rupa, Parimandala aksara, Mandala rupa, Arasa, Sparsha)	Curviline ar or Parallel reality transforming smell sensation (Parimandala rupa, Mandala rasa, Asparsha, Gandha)	Consequential reality of the transformed smell of the param child as a masculine self-luminous entity (Parimandala rasa, Mandala sparsha, Agandha, Nada)	Causative reality of the greeter self-luminous entity who is sounding the octave note of the presence of the param child (Parimandala sparsha, Mandala gandha, Jivanada)	Culture-effect of the luminous who is indoctrinating the doctrinal value for scripting convergence with the param creature (Parimandala gandha, Mandala nada; Ajiva)
Jambunada	Primordial Rishivaditaka	Rishi	Harahel	Scripted creature entropy	Andhakara (Void enveloping into the darkness point of the physical body's self-consciousness, 10^{19})	Mula (Root: Essential; Earthly)	Kumala (Soft)	Raudra + Putti (Furious + Putrid)	Avila + Chini (Polluted + Consciousness diffusing, rheumatic, conically diverging, double-octave encompassing, powerful shawn sound)	"c" $2^{3/12}$	Doctrine of sentient water

Param child	Maternal entity	Paternal entity	Feminine entity	Masculine entity	Convergent reality	Infinite reality	Divergent reality	Parallel reality	Consequential reality	Causative reality	Culture-effect
Vaimya	Primordial *Kusmandana*	*Sveta*	*Rachel*	Scripted ecosystem entropy	*Visata* (Extrinsic consciousness of the diffused zodiac system, 10^{16})	*Kantaka* or *Katuka* (Plant shoot; Spicy)	*Snigdha* (Oily)	*Bhayanaka + Tamalasu* (Fearsome + Masculine)	*Tikshna + Tantri* (Sharp + Reverberating, emotion-pumped, lute-sound)	"G" $2^{11/12}$	Doctrine of North
Acyupta	Primeval *Kusmandana*	*Mahasveta*	*Mehiel*	Scripted entity entropy	*Sata* (Incarnational consciousness of the presiding starseed system, 8000)	*Pravala* or *Kharika* (Plant bud; Salty)	*Ruksha* (Dry)	*Bibhatsa + Chulasu* (Fatigued + Feminine)	*Keshava + Ghanta* (Strong + Resonating, mind-pumped, bell-sound)	"A" $2^{12/12}$	Doctrine of East
Manasi	Primordial *Pavaka*	*Pavaka*	*Nanael*	Scripted deity entropy	*Avanta* (Energy-less human consciousness of the predominating, curving the primordial realm, 1600)	*Niryasa* or *Ambila* (Plant exudate; Sour)	*Sthira* (Healing)	*Shanta + Svadu* (Peaceful + Intrinsic)	*Trivikrama + Shankha* (Weak + Inorganic, lung-pumped, Conch-sound)	"B" $2^{13/12}$	Doctrine of South
Maha-manasi	Primeval *Pavaka*	*Pavakapati*	*Menadel*	Primordial deity entropy	*Unnata* (Self-luminous consciousness of the dominating, serving the primordial-primordial realm, 20)	*Beeja* or *Madhura* (Plant seed; Sweet)	*Chala* (Infecting)	*Vatsalya + Yojana* (Loving + Extrinsic)	*Visada + Chini-Chini* (Natural + Organic, hand-pumped, harmonium sound)	"b" $2^{14/12}$	Doctrine of guider fire

Chapter 10: Consequences of the Paradigmatic Planning

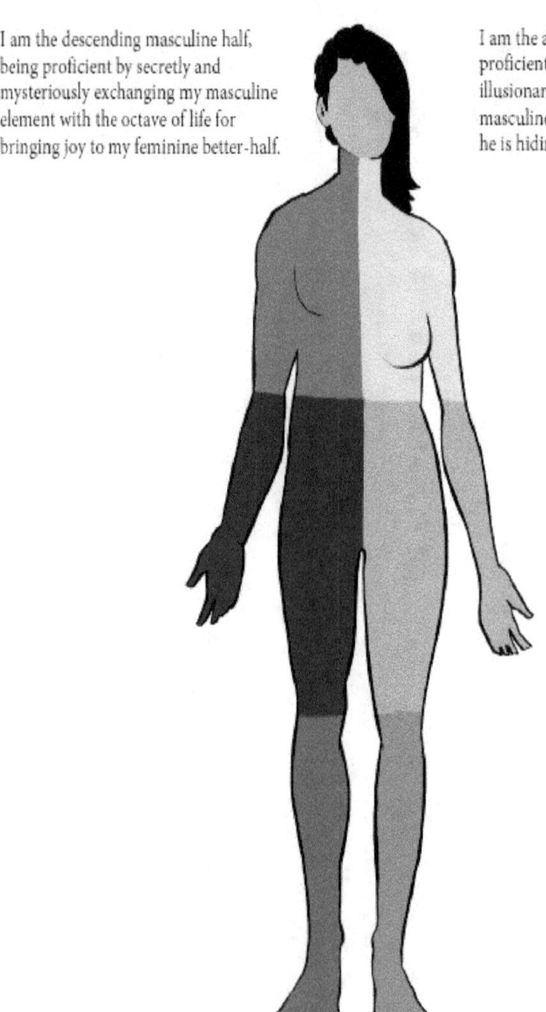

Infinite Potentiation of Each Life Unit's Octave of Soul Potential Is Cost-effective

Infinite Potentiation of Each Life Unit's Octave of Life Potential Is Cost-escalating

A sentient "cell" (*Hiranyagarbha*, 19 = 21 – 2) comprises the positive twenty-one unit inanimate "earth-effect" (*Ravi*, 21) and the negative two-unit "divine-effect" (*Padartha*, -2). It is a product of the conscious transformation of a positive unit person into a negative unit metric after endowing the cell with the "octave of life" (Sirtuin: *Sacha*, 19) potential. Its unit value is nineteen.

Instead of realizing the life's goal by liberating the primordial paternal self, which is still motivated to demonstrate his potential to each child, each param masculine child becomes an engaged citizen, seeking to demonstrate his homolog potential to the universe. It makes each param feminine child doubt her responsibility as the better-half, given that the masculine soul is already proficient in reproducing the paternal potential. Therefore, she also becomes a reproducer of the paternal potential.

When both the masculine and the feminine halves become the reproductive machinery power, they produce a "double-octave" (*Madhusudan*, 16) of entities as their collective goal, with a sixteen-unit value. The double-octave includes an "octave of illusionary souls" (Heterochromatin: *Maha Vidya*, 17 = 4 * 4 + 1), comprising the theoretical "soul" (*Atman*, 4) quaternity and the "self-luminous person" (*Insan*, 1). It further includes the "octave of life" (Sirtuin: *Sacha*, 19) potential, hidden and eclipsed by the "discordant satanic metric person" (*Asura*, -1).

The three units (19 – 16) are the cost of the "masculine body" (*Linga sharira*, 3), that becomes the "multiplier" (*Vaishya*, 3) of the sixteen-unit "conscious consciousness" (*Mantra*, 16), inclusive of the effects of both the primordial feminine and the primeval masculine selves.

The six units are the metaphysical cost of the masculine body and the masculinized feminine body that produce a "six-fold growth" (*Khara*, 6). They manifest the thirty-six-unit causative factor of potentiating the life's goal within each unit of the four-unit "life" (*Prabhasa*, 4). The thirty-six "causative factors" (*Prapaka*, 36) of the life are immanent within the thirty-seventh unit, which embodies the six-unit consequential factor potential.

The value of the "ageless" (*Santana*, 37) sequential factor is thirty-seven units. It is "impossible" (*Pratiyatraka*, 1369) to reproduce the thirty-seven mitochondria DNA genes within the masculine and the feminine selves. The value $37^2 = 1369$ is beyond the "luminous" (*Maha Kali*, 13) realm of the possible "goal-oriented" (*Maha Saraswati*, 9) travel of the "residual effect" (*Khara*, 6).

By forcing the feminine self to bear the six-unit metaphysical cost of reproducing the thirty-seven unit performing genes, the masculine self becomes the multiplier of the thirty-unit causative factor within the past, present, and future time dimensions to be the 127-unit "proficient exchange system" (*Sadhaka*, 127). The masculine self's proficiency is conditional on the non-reproducible 37-unit "loving" (*Vatsalya*, 37) element serviced by the feminine self.

10.1 Self-Luminous Entity Profiting from the Spirit-Free Performing

The maternal self-luminous entity forms, with the transformation of the dark matter into the "formative divine energy" (*Madhusudan*, 16) of the "soul essence community" (*Mandala*, 16). The paternal self-luminous entity forms, with the trading of the primordial greeter's absolute divine-effect, through the "nitrogen fixation" (*Usha*, 16). The feminine self-luminous entity forms, with the servicing of the nitrogen-free eighteen-unit fragrant "chloroplast" (*Nirharin*, 18) after the "death" (*Jaramarana*, 18) of the primordial maternal entity. The masculine self-luminous entity forms through the exchange of the "feminine-effect" (*Pingala*, 19) with the "param child in the form of the cellular enzyme" (*Hiranyagarbha*, 19). It catalyzes the growth of the "masculine-effect" (*Ida*, 1) by eventualizing the primordial paternal entity's death.

The "glowing, *henna* green color" (*Harita*, 10) of the cellular enzyme transforms into the "orangish indigo color" (*Avadata*, 12) of the breath of the masculine child. It manifests within the "pale yellow color" (*Pila*, 186) of the "soliton quantum particle" (*Kana*, 186), through the "aggregation" (*Jama*, -7) of the "*henna* green color" (*Harita*, 10) and the "self-reproducing" (*Upanayana*, 1/3) value of the six "organelles" (*Dasha*, 1). The six organelles are the psychic linkages of the self-luminous greeter, maternal, paternal, feminine, masculine, and child entities within the "lipid phage" (*Hara*, 285).

The param child within the lipid phage transforms the "red color" (*Lohita*, 1) of the immanent "masculine-effect" (*Ida*, 1) into the "guider-effect" (*Guru*, 100) by trading the "gravitational, open system-effect" (*Dyujya*, 100) of the "dark matter" (*Sadashiva*, 1600), without the "primordial greeter" (*Sati-Parvati*, 16). It empowers the param child to transform into the "archaeon" (*Jangamavisha*, 570) by attracting an additional follower phage. It vertically fuses the "feminine-effect" (*Pingala*, 19) of the follower phage with the "self-perpetuating" (*Udvaha*, ½) value of the "atmosphere" (*Parimandala*, 1100) and the intrinsic "masculine-effect" (*Ida*, 1).

The archaeon trades the "dark blue face" (*Nila mukha*, 1869) of the spiritually-scripted programming of the effects of the horizontal fusion of the "soliton quantum particle" (*Kana*, 186) and the self-perpetuating (*Udvaha*, ½) value of the "chloroplast" (*Nirharin*, 18). It empowers the param child to strangely repel a pair of "bacteria" (*Vetranavah*, 855) for ascending the "life consciousness" (*Apas*, 169). The param child trades the "enzyme-effect" (*Sushumna*, 10 = 169 + 855 + 855 - 1869) of the transformative divine energy.

By transforming the life consciousness with the twelve diverse life doctrines, the param child sequentially forms a system of the twelve colorless bodies of the formative consciousness: the physical body, the intellectual body, the mental body, the astral body, the etheric body, the causal body, the self-luminous entity, astrological-effect, zodiac-effect, starseed-effect, primordial-effect, and the primordial-primordial-effect. Each of these comprises the twelve convergent, curvilinear "forms" (*Rupa*, 100,000) of the "life consciousness" (*Apas*, 169). Each empowers the "bacterium" (*Vetranavah*, 855) to transform into a "eukaryote" (*Rudhita*, 963) by trading the "super-lexicographic" (*Virodhita*, 108) "budding energy" (*Utpalai Shakti*, 108 = 18 * 6) of the inter-breeding "chloroplast" (*Nirharin*, 18), and the "rainbow colorful" (*Rishaba*, 1) potential of the primeval group of six "organelles" (*Dasha*, 1).

The eukaryote sequentially generates the six primordial "tastes" (*Rasa*, 269) through the "aggregation" (*Jama*, -7) of the "life consciousness" (*Apas*, 169) and the varying "guider-effect" (*Guru*, 100) of the six primordial organelles: the root-based, the trunk-based, the bark-based, the leaf-based, the flower-based, and the fruit-based organelles. After fertilizing a group of thirty-six primeval organelles (6 x 6), the eukaryote transforms into the "prokaryote" (*Shadayatana*, 999) to produce the six primeval "tastes" (*Rasa*, 269). The six tastes sequentially service an entropy "inanimate consciousness" (*Mahika*, 47) of the six primordial organelles. These six tastes are the astringent, the bitter, the spicy, the salty, the sour, and the sweet.

By trading the "supernatural energy" (*Shram shakti*, 1) immanent within the six primordial organelles, the prokaryote transforms into the "lifeless cellular body" (*Nirjara*, 1,000)—as a "nucleus, the origin of life" (*Yoni*, 1,000). The nucleus sequentially generates the twelve "touch sensations" (*Sparsha*, 378) of the horizontal fusion of the thirty-seven organelles and the "self-perpetuating" (*Udvaha*, ½), "soul essence community" (*Mandala*, 16). These are the soft, conspicuous, symbolic, bloated, heavenly, grounded, tough, sensitive, healing, and the infecting forms of touches.

With an ascending infection from the lifeless nucleus, the param child sequentially generates twelve secondary taste-effects and their correlated "smell" sensations (*Gandha*, 268) of the thermodynamic "proton" (*Sarvodaya*: form-effect, 150). These correlated pairs are: the furious and the putrid, the wonderous and the biased, the erotic and the pungent, the comic and the leafy-mushy, the heroic and the flowery-kissing, the pathetic and the desired, the seductive and the poisonous, the proud and the buttery, the fearsome and the masculine, the fatigued and the feminine, the peaceful and the intrinsic, and the loving and the extrinsic.

With an ascending healing from the thermodynamic proton, the param child sequentially generates twelve secondary touch-effects and their correlated "sound" sensations (*Naad*, 257) of the sentient "electron" (*Dyumna*, 365). These correlated pairs are the polluted and the powerful shawn sound, the residual and the converging sackbut sound, a fundamental and loud trumpet sound, the diffused and the thunderous horn sound, the concentrated and the soulful bagpipe sound, the principal and the soul-deep drum sound, the agent and modulating flute-sound, the dull and rhythmic cymbal sound, the sharp and the reverberating lute sound, the strong and resonating bell sound, the weak and the inorganic conch sound, and the pure and the organic harmonium sound.

The sentient electron is the "somatic cell" (*Vyanavata*, 365), which trades the horizontal fusion value of the three-hundred sixty diurnal-effects of the lunar zodiac system and the "self-perpetuating" (*Udvaha*, ½) "enzyme-effect" (*Sushumna*, 10).   With the horizontal dryness of the sentient electron, the param child sequentially generates twelve secondary sound-effects and their correlated "self-replicating, sound vibrations" (*Jivanada*, 855) of the "absolute zodiac-effect" (*Pashutva*, 855) of the "maternal bacterium" (*Jivanu*, 855) for forming the "supernatural, RNA polymerase" (*Atisara*, 855).   These include the twelve single-letter words describing the twelve musical notes, and the ten descending and the two ascending octaves digitalized notes. The capital letter signifies a dominating masculine-effect.   The normal letter signifies a predominating feminine-effect, within the ascending entropy of the dominating masculine-effect.

With the forward viscosity of the proliferating universe of the RNA polymerase, the param child trades the twelve tertiary self-managing sound-effects in the form of the twelve diverse life doctrines, as the deciding androgynous-effect of the starseed realm.   These include the doctrines of the sentient water, the supernatural air, the natural earth, the satanic ether, the almighty creature, the almighty creation, the almighty creator, the west, north, east, south, and the guider fire.   Eight of these are the formative doctrines. The four—the west, north, east, and the south—are the self-perpetuating effects of the formative doctrines.   They generate the two additional self-perpetuating-effects: the doctrines of the northeast and southeast, which in turn generate the normative "doctrine of entity" (*So Ham siddhanta*, 4).

A "masculine human entity" (*Manush*, 4) trades the doctrine of entity for transforming the "sentient-effect" (*Soham*, 4) of the "life" (*Prabhasha*, 4) into the "soul element" (*Atman tattva*, 4). If the "right, feminine-side" (*Vijnana*, 47) is dominant, then the entity takes the masculine soul as the transformative "doctrine of soul" (*Esa Ta Atmantaryamyamrtah siddhanta*,

100). If the "left, masculine-side" (*Vinayka*, 53) is dominant, then the entity takes the feminine soul instead. The other, twin soul works as the "Twin Flame Songmaster Council at the 2000th I AM Level" (*Priti*, 33), The androgynous "spirit" (*Kapinjala*, 20) and the "twin soul" (*Priti*, 33) compensate for the "discordant energy" (*Asura*, -1) of the human-being as an essentially negative masculine entity. The consequential fifty-two units balance forms the "shadow reality of the astral body" (*Chhaya*, 52). It is correlated with the fifty-two weeks (52 x 7 = 364 days), within each of the three-hundred sixty diurnal-effects of the zodiac system and the four nocturnal-effects of the sentient life system.

10.2 Ecosystem-level Paradigm to Compensate for the Cost of Entity Profiting

The One hundred and ninety pathways highlighted in Tables 10-20 constitute the "self-perpetuating" (*Udvaha*, ½) "self-luminous consciousness" (*Unnata*, 20) of the "spirit" (*Kapinjala*, 20) and its infinite entanglement with the "param child" (*Manyu*, 19). The nineteen doctrines, including the doctrines of immanence, emanation and the divine truth together, norm the "clarified consciousness of the infused astrological system" (*Accha*, 19) within the param child. The ascending clarified consciousness ascends the positive goodness within the param child. The ascending para consciousness ascends the negative evil within the param child, without a self-awareness of the divine truth immanent within the emanating universe. Therefore, there is a further need to investigate the six additional ecosystem-level, shadow doctrines, beyond the thirteen entity-level luminous doctrines, for developing an absolute consciousness of the present reality of the self-luminous human entity.

A self-luminous human entity has a unique capability for trading the triangulated consciousness of the present reality from the masculine soul, the feminine twin soul, and the androgynous spirit, immanent within the self as the primordial paradigm of Mother Nature. Besides, the self-luminous human entity can be a "multiplier" (*Vaishya*, 3) of the triangulated power through the "androgynous, centering-effect" (*Sushumna*, 10) of the "divine energy" (*Asrava shakti*, 10). The multiplier path is subject to the psychological bias, which emanates from the weakening of the correlation of the planning entity with Mother Nature's divine plan.

Mother Nature's divine plan empowers each creature to enjoy a simple one-dimensional future reality. However, a creature decides to become a creator of a guider plan to be the paternal king of the universe of sentient entities. Therefore, the universe of sentient entities must compensate by becoming a maternal harem of the paternal king. Else, the creator entity supernaturally becomes a "paternal genie" (*Jinn*, 4), hypnotizing the universe with his antics and wishing to secure absolute compliance of his hegemony over the creation. A paternal genie is an almighty creature who wishes to make the entity an almighty creator, so that he may manipulate the universe as the almighty creation and distribute all the discredit of destroying the soul to the guider fire of the Satanic ether.

A primordial greeter is the sentient water, which dowses the guider fire by grounding the supernatural air with the natural earth element. The primordial greeter then becomes a primordial illuminator who illuminates the presence of the following six doctrines within the self-luminous human entity. The illuminated consciousness of "Mother Earth's" (*Kshiti*, 725) divine plan empowers one to ascend the consciousness of the immanent param deity reality for forming an infinity of additional doctrines, while transcending beyond the negative present reality of the param child.

10.3 The Objective-fulfilling Super-Primordial Doctrinal Paradigm Is the Divine Plan of Mother Earth

The ecosystem-level paradigm is composed of the six doctrines.

10.3.1 <u>First, "I am the Northeast Direction Doctrine"</u> (*Varnasam amnaya siddhanta*, 846)

For the divine-exhumation of the Satanic ether-effect of the universe of the sentient entities, Mother Earth trades the sentient energy from the Sun situated in the Northeast direction. As an almighty creator, a self-luminous entity is intent on devotionally pursuing the path of self-entropy through the investment of sentient energy into an almighty creation. The almighty creation is the causative factor for the supernormal diffusion of the soul's thermodynamic energy. The "symbolic, congested togetherness" (*Sita*, 1) of the almighty creation descends the supernatural air and ascends the natural earth within the almighty creature.

Mother Nature compensates for the disproportionate earth-effect by thermodynamically burning the "earth element" (*Bhu*, 724) and diffusing the earth-effect in the form of a carbon atom. The carbon atom ascends the sentient superconductivity and brittleness by "infecting" (*Chala*, 26) the souls of the "universe of entities" (*Kashi*, -10^{1024}) with the "conspicuous" (*Kathora*, -1), "subjectively-biased smell of the physical body" (*Purnasau*, 948). The "self-luminous entity realm" (*Satya loka*, 10^{1024}) trades the entire positive energy of the souls of the "universe of entities" and services its entire negative energy to the latter. This exchange system generates a feeling touch of "wonder" (*Atbudha*, 285) within the self-luminous entity, which programs the method of exchange in the form of a "Cytosine nucleotide" (*Atbudha*, 285).

When the intellectual body is unable to sustain the energy exchange from the universe of entities, it seeks to rewrite the destiny by trading the sentient energy of the "spirit" (*Kapinjala*, 20) within the Plant kingdom. Consequently, it forms a causal body of souls and an etheric body of spirits within the almighty creator. The almighty creator itself is an astral body of twin souls, seeking to illuminate the consciousness of the almighty creature and liberate the individual souls from the dominating spirit of the almighty creation.

The almighty creature is a mental body of the twin spirits, who holds the "office of the guardian angels of the infinite council of the departed spirits" (*Upadhi*, 37). It works to infuse a false sense of "agelessness" (*Sanatana*, 37) to promote the "conditional division" (*Ekadhikina Purvena*, 37) of the dominating spirit. The almighty creation works hard to divide its "spirit" (*Kapinjala*, 20) by servicing half of it as the "tangential gravitational linkage" (*Ardhajya*, 10) of the universe of child souls and the other half as the "divine energy" (*Asrava shakti*, 10). The objective of the conditional division is the patrilineal genetic transfer to the progeny.

The "patrilineal dimension" (*Putra dharma*, 38) is oriented towards the "masculine-to-feminine gender exchange" by trading the "feminine-effect" (*Pingala*, 19) of a feminine entity for the growth of the hegemonic universe beyond one progeny. After both the paternal and the maternal entities have diffused their spirit value, their astral body radiates the life consciousness in the form of the gravitoelectric energy to attract the soul of a "primeval child" (*Akrura*, 123).

A primeval child services its sentient light force to be the soulmate of the "androgynous entity" (*Vega*, 67), who is guiding the togetherness of both the paternal and the maternal entity as a path to reunify with the primeval child. By trading the sentient light force of the primeval child, the almighty creation becomes the intellectual body of the soulmates. The intellectual body of the soulmates motivates the androgynous entity to service the astrological-effect of the "infinite council of the living child souls" (*Akalpa*, 570. It forms a "self-luminous human entity" consciousness (*Purusha*, 12), as the convergent value of the astrological system, without the divergent zodiac-effect of the "infinite council of the departed child souls" (*Sthavaravisha*, 57).

The androgynous entity services the zodiac system's convergent value as the "conditional summative value" (*Purana Apurnabhyam*, 11). It adds the three units to the "self-perpetuating" (*Udvaha*, ½) value of the "soul essence community" (*Mandala*, 16). These three units empower the self-luminous entity to transcend beyond the eight-unit supernatural paradigmatic "local-effect" (*Mahadasha*, 0) of the soul essence community. As a twin spirit, the soul essence community continuously "propounds" (*Virocana*, 38) an infinity of theories, for destroying the almighty creation with its patently "soft" touch (*Komala*, 4905), hiding the latent "putrid smell of fury" (*Putti*, 4905). The objective is to shape the "behaving energy" (*Rathai Shakti*, 38) of the almighty creation so that it may unify with the universe of soulmates, trapped within the intellectual body of the self-luminous entity.

Of the three units, the first is the convergent value of the "infinite psychic linkages" (*Shakti-Bheda*, 257) with the "infinite council of all spirits" (*Ratnakosha*, 30). It empowers the self-luminous entity to protect itself from the fury, by diffusing the "local, theory-effect" (*Mahadasha*, 0) of the starseed universe into the entire universe of entities. The second unit is the divergent value of the electromagnetic energy of the "primordial realm" (*Maha Kalpa*, 10^{1000}). It is circulating, geometrically curving, and eventually converging into the absolute oneness of the "self" (*Akhanda*, 8×10^{15}) and the "zodiac universe" (*Shunya Kalpa*, 8×10^{15}). It is the "incarnational value" (*Markatesh*, 8×10^{15}) that makes one an "omnipresent entity" (*Vyapin*, 8 x

10^{15}). The third unit is the parallel value of the sentient energy of the "primordial-primordial realm" (*Antara Kalpa*, 3794) to "embody the power to annihilate" (*Pochamma*, 3794) the "wheel of creation" (*Srishti chakra*, 3794).

By diffusing the "self-luminous consciousness" (*Unnata*, 20), the twin spirit motivates the entity to "proliferate the mass communicable diseases" (*Mariamman*, 3794). It allows the spirit to become a "messiah" (*Samratah*, 27) for servicing with the "healing" touch (*Sthira*, 3794) and trading the "para consciousness" (*Nirharin*, 18) of the "infinite council of spirits" (*Ratnakosha*, 30). With the "void in consciousness" (*Andhakara*, 10^{19}), the entity conceives the convergent psychic value of the "infinite council of spirits" to be the "Goddess of divine energy" (*Sheetala Mata*, 3794) servicing with the healing touch.

The entity descends the "macro, grounded" touch (*Guru*, 100) of the "desired" (*Ishta*, 20) gravitational energy of its "spirit" (*Kapinjala*, 20) by perceiving its smell to be "pathetic" (*Karuna*, 20), conditioned by the infinite demands for its "messiah complex" (*Samratah*, 27). On the other hand, it ascends the "micro, heavenly" touch (*Laghu*, 377) of the present self as the "sensual" (*Pushpa*, 280) "spiritual entity" (*Hurupa*, 280), seeking to devour the Goddess of the divine energy. Consequently, the "absolute self" (*Hurupa*, 280) diffuses its entire "cellular energy" (*Shakti*, 19) and realizes an "absolute oneness with Mother Earth" (*Ubhayato mukha*, 280).   It empowers the spirit to merge with the twin spirit in the starseed universe.

10.3.2 <u>Second, "I am the Southeast Direction Doctrine"</u> (*Sumamkasa Mahayuga*, 375)

For the fire-burning the technological cost of the "sentient healing value" (*Reiki*, 38), Mother Earth services the thermodynamic energy traded from the universe of sentient entities to the "planet Neptune" (*Ketu*, 140), situated in the Southeast direction. The Neptune services the thermodynamic energy to the universe of entities in the whole cosmos for ascending their entropy. With a descending pressure on the sentient energy in the whole cosmos, the "planet Uranus" (*Rahu*, 73) trades the ascending sentient energy from the "universe of white stars" (*Mahar loka*, 27000) embodied within the "plasmon" (*Puneehi*, 27,000).   The Sun countertrades the plasmon and services it to the planet Earth for forming the "oöcyte" (*Puneehi*, 27,000) within the universe of the sentient entities.

The oöcyte works like a "purifier" (*Puneehi*, 27,000) of the scripted organizational entropy. With its "sentient energy" (*Varuna*, 1000), it enjoys the metaphysically pure, "clarified consciousness of the infused astrological system" (*Accha*, 19) and "touch of sensitivity" (*Mrdu*, 8). It promotes an "absolute oneness" (*Payu*, 360) of the future-visualizing "pupil of the eye" (*Akshitara*, 375), with the "primordial-primordial space" (*Drishti-Mandala*, 375), for the "potentiation" (*Sambhavi*, 375) of the "truth" (*Sathya*, 375).

The universe of self-luminous entities ascends the freedom from the limiting "salty" (*Kharika*, 15) taste of the primordial energy, which manifests the "incarnational consciousness of the presiding starseed system" (*Sata*, 8000). It descends the "illusionary dividing energy" (*Maya shakti*, 1), which makes one a "divider" (*Shudra*, 1) of the truth of the self as an "aggregator" (*Rajah*, 0) of the "energy" (*Shakti*, 19) and the universe as the "indestructible, supreme enemy" (*Kayayini*, 375). The self then competes with the universe to manifest the "desired" (*Ishta*, 20), "convergent reality" (*Tulyartha*, 10^{100}), in the "free-for-all and winner-takes-it-all oceanic market" for the sentient energy (*Kacangala*, 10^{100}).

With a descending "entanglement" (*Kula*, 9) with the universe, the entity ascends "entropy value of the zodiac-effect" (*Deva siddhi*, 375). Consequently, the entity enjoys a growing "divine, gravitomagnetic energy" (*Asrava Shakti*, 10) for becoming a sustainable, proportionate "doctrine" (*Agama*, 375).

10.3.3 <u>Third, "I am the North Direction Doctrine"</u> (*Uttar amnaya siddhanta*, 169).

For water-cleansing the "polluted value of the universe" (*Avila*, 2), Mother Earth exchanges the gravitomagnetic energy immanent within the universe of sentient entities. She trades the gravitoelectric energy emanating from the "Vega white star, the fixed north pole and the gravitational center of the cosmic universe" (*Vega*, 67). The Vega white star services the gravitoelectic energy as an "electron" (*Dyumna*, 365) stream. The electrons become the "somatic cell" (*Vyanavata*, 365) within the universe of sentient entities and norm the consciousness of the consequential "noumena" (*Paramartha*,

365), without the "divisive, circulating energy sphere" (*Pakshma-mandala*, 268), of the infinite sequentially eye-lashing "phenomenal reality" (*Ityartha*, 285).   Consequently, the "param child" (*Manyu*, 19) descends the "addiction" (*Akarshana*, 268) to the fragmenting, lovey-dovey, "sappy" taste (*Kanda*, 268) and "smell" (*Gandha*, 268), of the "phonetic sperm" (*Ajapa*, 268) of the "paternal entity" (*Gunasa*, -18).

A paternal entity is guided by a "positive-sum mindset" (*Anurupyena*, -18), seeking to trade the entire "Eastern-effect" (*Uttarmanasa*, -18).   The Eastern-effect norms the convergent value of the energy traded from the Sun in the Northeast, the universe of white stars in the Southeast, the Vega white star in the North, and the Dark matter in the South.   By trading the Eastern-effect, the paternal entity becomes a "destructive, masculine entity" (*Pranasa*, -18), entangled with the "discordant" (*Asura*, -1) phenomenal reality of each of the eighteen starseed universes.   It becomes subject to the "autotrophy through the anaerobic respiration" (*Abhiyoga*, -18).

The somatic cell destroys the "protective sphere of the eyelids, projecting the reluctance for trading the primeval light force" (*Vartma-mandala*, 169) of the Vega white star.   After that, the param child ascends the "life consciousness" (*Apas*, 169) of the "spiritual energy" (*Saniddhya shakti*, 169) of the "spirit kingdom" (*Vishvagoptri*, 169).   With the flowing water of life consciousness, the "still, lifeless, poker face" (*Karata mukha*, 169) manifests its "ten-faced system value" (*Varuni*, 95).   The ten-faced system includes the starseed universe in the Southwest, the Black Hole in the West, the primordial realm in the Northwest, the primordial-primordial realm in the East, the entity universe in the ascending direction, and the sentient universe in the descending direction. It also includes the Sun in the Northeast, the universe of white stars in the Southeast, the Vega white star in the North, and the Dark matter in the South.   It excludes the inanimate universe's divisive duality with the horizontal, poker-face, and the entity with the "circulating, centrifugal face" (*Griha mukha*, 0).

10.3.4 <u>Fourth, "I am the South Direction Doctrine"</u> (*Dakshin amnaya siddhanta*, 195)

For air-resurrecting the "lifeless, inanimate, poker face" (*Karata mukha*, 169), Mother Earth invests the gravitoelectric energy received from the "astral radiation of the mortal sentient entity" (*Martya*, 146). She ascends oneness with the "dark matter" (*Sadashiva*, 1600), whose "gravitational center" (*Jamadagni*, 629) is in the South direction. A param child trades a descending astrological-effect from the dark matter's gravitational center for ascending the zodiac-effect on the "conditional destiny" (*Ayati vela*, 629). Consequently, he develops the "voice" (*Vak*, 169) for sustaining the "life consciousness" (*Apas*, 169) traded from the Vega white star with the "unconditional certainty" (*Haum Shakti*, 9).   With the power for programming the "unconditional destiny" (*Bhaga*, 27), he becomes the "cosmic perpetuator" (*Rachayita*, 27) of the "entity dimension" (*Prakriti dharma*, 27), with the "freedom-effect" (*Rupa siddhi*, 27) of the "creepy-feeling master creator of the self-destiny" (*Messiah*, 27).

The param child lacks the full "confidence meter" (*Svara*, 28) for carrying the "weight" (*Bhara*, 28) of the "Lord of the path" (*Pusan*, 28) because its freedom is conditional on the "invincibility of the Vega-effect" (*Aparajita Durga*, 28).   Therefore, the param child generates an "ascending ideal-effect" (*Chara Paryaya dasha*, 28) for authenticating the "immanence-effect" (Letter "A" of *aum shakti*, 28) of the "begging value" (*Ukalita*, 28) of the "Vega-effect" (*Durga*, 28).

The octave-doubling A-letter ($2^{12/12}$) includes a primordial unit of the aggregation (1+1=2) of the unit of the param child's "supernatural energy" (*Shram shakti*, 1) and a unit of the Vega-effect linkage in the form of "organelle" (*Dasha*, 1). The former repels a primeval unit of the horizontal summative sequence (2+8 = 28) of the "thought impulses" (*Vichara*, 8) for binding the "paradigm of natural reality" (*Yukti*, 8) with its "sensitive" (*Mrdu*, 8) "connoisseur intuition" (*Jihva vijnana*, 8).

The supernatural "air element" (*Vayu*, 385) transforms the "multidimensional ground reality" (*Radheshyam*, 385) with its "personification" (*Tamasika*, 385) of the "extrinsic, social dimension" (*Sadharana dharma*, 385). The param child perceives the Vega-effect to be immanent within the self as the self-organizing "organelle" (*Dasha*, 1), for sustaining the oneness of "the right feminine dimension and the left masculine dimension" (*Pariyanga yoga*, 385). This oneness is an "essential condition" (*Hetvartha*, 385) for "freedom from the present-effect" (*Moksha*, 1600) of the "planet Mercury" (*Budha*, 1600).

The planet Mercury is also situated in the South direction. It trades the "energy-less masculine human consciousness" (*Avanta*, 1600) from the planet Earth and services that back within a state of devoted followership of the "param human child" (*Kesari nandan*, 1600). Therefore, a need emerges for the param child to differentiate the "doctrine" (*Siddhanta*, 375), value of the self to triangulate, isolate, and mitigate the polluted, "primeval trading-effect of the North direction" (*Kuber*, 57).

10.3.5 <u>Fifth, "I am the West Direction Doctrine"</u> (*Paschim amnaya siddhanta*, 164).

For the earth-grounding of the "private, circulating, centrifugal face" (*Griha mukha*, 0) of each entity in the universe, Mother Earth gifts her immanent electromagnetic energy capability to each entity in the form of a "physical body" (*Sthula sharira*, 387). The physical body is the "volume carrying capacity" (*Avanejya*, 387) of an entity as a closed system. Each entity is a "primordial value" carrying a "cluster of entities" (*Gana*, 387) within itself. Each of those entities has the potential to develop a normative physical body by trading the sentient energy of the entity or the cluster of entities.

After servicing its energy, the paternal entity has the potential for using its formative intellectual body to be the astral body of the dominating entity. Similarly, after servicing their energy, the cluster of greeter entities has the potential for using their transformative mental body to be the etheric body of the dominating entity. After trading the "octave-doubling overlordship" (*Parameshthi*, 28), the masculine child entity has the potential for using its normative physical body to be the causal body of the predominating maternal entity—the Mother Earth. After creating a group of six entities (primordial and primeval paternal, greeter, and masculine entities), the Mother Nature, as a primordial maternal entity (twin spirit of Mother Earth, the primeval maternal entity), has a potential to be the "self-luminous entity" (*Maha Gayatri*, 12).

As a self-luminous entity, Mother Nature is the "formative entity" (*Rupi*, 12) blessing the universe of entities with a "form" (*Rupa*, 100,000). The "radiance" (*Eroli*, 100,000) of the unique formed shape of each entity is the "soul of the param deity" (*Atma-linga*, 100,000). It empowers each entity to concurrently enjoy both the "sentient energy" (*Varuna*, 1000) to enjoy the SHEENY well-being as well as the "gravitational energy" (*Lalita*, 100) to service joy as a "guiding force" (*Guru*, 100). Consequently, each entity is motivated to be a "paternal self-luminous entity" (*Bhavanavasi*, 12), who is the resident master authority on the "divine energy" (*Asrava Shakti*, 10) by trading the "ten-faced system-effect" (*Ardhajya*, 95).

The ten-faced system-effect is the effect of the "one rotational symmetry" by one-hundred twenty degrees of an equilateral triangle. It transforms the animated creature into an embodiment of natural creator as a "spirit" (*Kapinjala*, 20), within each of the six bodies—the physical, the intellectual, the mental, the astral, the etheric, or the causal. Consequently, the animated creature's overall rotational-effect is one-hundred twenty units within the three-hundred sixty diurnal-effects of the zodiac system. It empowers the animate creature to trade an additional one-hundred twenty units of the androgynous creator for servicing the primordial one-hundred twenty units to form the inanimate creation as the third-dimension of the equilateral triangle. The process of creating the inanimate creation is the "effect of two rotational symmetry by 240 degrees of an equilateral triangle" (*Paramarthika dharma*, 397). It transforms an animated creature into an inanimate creation.

An animate creature ascends the "pH, acidic water" (*Muppu*, 396) as a "natural-effect" (*Sanjna*, 397) of the "curvilinear" (*Vakri*, 396), "emanation-effect" (Letter "U" of AUM, 396) of the "inanimate, bunching-oriented, feminine energy" (*Achitta Shakti*, 396). The "feminine self-luminous entity" (*Isha*, 12) changes the "death point" (*Ayati*, 863) of the animated, masculine "Wish deity" (*Kamadeva*, 396) by servicing the pH acidic water as the "premonition" (*Andesha*, 397), about the "knowable" (*Mara*, 396) wish for togetherness. The death of the animate creature empowers the former to ascend the "pH, alkaline water" (*Amuri*, 586), for completing a three-rotational "communion symmetry" (*Girisha*, 486) of 360 degrees of the equilateral triangle, that transforms it into the "feminine dimension" (*Vijnana*, 47). It further transforms the animate creature into the "masculine dimension" (*Vinayaka*, 53) within the "pH, neutral water" (*Guru*, 100).

Consequently, both the creature and the creation are "grounded" (*Guru*, 100) within the "gravitational energy of the presiding deity" (*Lalita*, 100).

However, the gravitational energy is subject to the "psychic linkages" (*Mahika*, 47) of the dead masculine, now sharing "one-identity symmetry" (*Girisha*, 486) with the "inanimate, feminine dimension" (*Vijnana*, 47) within the "black hole" (*Vishnunabhi*, 82).

The "gravitational potential" (*Kalika*, 23) of the black hole gets traded by the "new moon" (*Chandra*, 82), which services that to the "human entity" (*New Lemurian*, 82) during the "dream state" (*Maha Swapna*, 9) of the "maternal self-luminous entity" (*Rupi*, 12) in a "cellular form" (*Hiranyagarbha*, 19). Therefore, a need emerges for the param child to be liberated from the "circumference-effect" (*Citra*, 100) of the "primeval self" (*Rama*, 100), which is promoting the "Doctrine of Soul" (*Esa Ta Atmantaryamyamrtah siddhanta*, 10), as a common "institutional system" (*Sthiti*, 100), constrained by the "cellular energy" (*Shakti*, 19).

The primeval self is the duality of the primordial feminine and the param masculine dimensions traded from the "primordial-primordial self" (*Dvandva Brahma*, 125). The primordial-primordial self is the five-dimensional east-facing pair of opposites (the primordial paternal and maternal, the primeval paternal and maternal, the primordial masculine and feminine, the primeval masculine and feminine, and the primordial and the primeval greeter self-luminous entities) within the one "west-facing self-luminous entity" (*Virgin Mary*, 12).

A "primordial human child" (*Eeshan*, 12) has the power to trade the "absolute value of the primeval self" (*Shri Rama*, 12) for the normative development of limitless qualities, transcending the constraints of the cellular energy. It may discard the entire "Eastern-effect" (*Uttarmanasa*, 18), that is polluted with the "lust-smitten" (*Kama*, -19) "Western-effect" (*Dakshinamanasa*, -19) of the "overlord, spirit entity" (*Lokapurusha*, -19). After that, the primordial human child may ascend the "Eastern workculture-effect" (*Antardasha*, 374) for realizing "oneness with the dark matter" (*Shri Garuda*, 374), without the "western black hole culture-effect" (*Chara Paryaya dasha*, 28).

10.3.6 <u>Sixth, "I am the East Direction Doctrine"</u> (*Purv amnaya siddhanta*, 170).

For ether-purifying the "international-effect" (*Dasha*, 1), Mother Earth trades her gravitational energy from the universe of entities. She divides the "gravitational energy" (*Lalita*, 100) into the "param local-effect" (Pramod, 85) for the "tangential, east-oriented, centering" (*Ardhajya*, 10) of the "workculture" (*Nayaki*, 379) element within each entity. She then becomes the "primordial even value" (*Saranyu*, 5), which punctuates the two consequential, parallel, odd, primordial-primordial and the param masculine lifelines (3 and 7), with the primeval even values of the two causative, linear feminine lifelines (2 and 4).

Mother Nature is immanent as the "self-perpetuating" (*Udvaha*, ½), "aggregation" (*Jama*, -7) of the two odd masculine lifelines ($[3+7]/2 = 5$). She emanates the two primeval feminine lifelines ($2+4=6$), as the vertical differentiation of herself and the para-primordial "greeter" lifeline (*Vignesh*, 1). The greeter lifeline embodies a unit "supernatural energy" (*Naraki*, 1), generated by her "workculture-effect" (*Nivrtti dharma*, 40). The "collectivity of the universe of six entities" (*Samshti*, 7×10^{180}) energetically resurrects the "primordial paternal" (*Indra*, 0) with its "group-effect" (*Sundarika*, 7×10^{180}), without diffusing the "entity-effect" (*Anishta* 38).

The primeval "individuality of the seven entities" (*Vyashti*, 10^{29}) diverges from the primordial "collectivity of the six entities." It forms the "geography-effect" (*Sundari*, 10^{29}) for "nurturing" (*Tantri*, 48) the consequential "culture-effect" (*Bhavanarama*, 6×10^{192}) of the eight entities: the seven with their individuality and one with its collectivity. The culture-effect is the product of the group-effect and the geography-effect. The eighth, collective entity, is the param-primordial greeter of the "universe" (*Brahman*, 2) without entities. It is the "almighty creator entity" (*Pinakapani*, $10^9 = 1000^3$), servicing the convergent "sentient energy" (*Varuna*, 1000) of the creator, the creation, as well as the creature entities for the "universal mass well-being" (*Sarvabhuteshu hitah tattva*, 10^{90}).

The almighty creator entity empowers the "council of the seven individual entities" (*Saptarishi*, 26) to service their "fraternizing" (*Sakhya*, 26) "ecosystem cost" (*Yajya*, 26), for primeval projecting, an "infinite council of the spirits" (*Ratnakosha*, 30). "Everybody" (*Pasaka*, -9), collectively and individually, becomes a "self-obsessed" (*Nastika*, -9) and "arrogant" (*Abhimanin*, -9) entity, for the "mindless" (*Amanaska*, -9) trading of the "extrinsic, chaotic reality" (*Niryogartha*, -9), "skeptical" (Adharmi, -9) about the cyclically-perpetuating "universe of ecosystem reality" (Pasaka, -9).   Therefore, a need emerges for investigating if the present is indeed a reality.

Without the certainty of the present value within the absolute consciousness of the universe of the sentient entities, the present value becomes the "sigma value" (*Ganarajya*, 476) of the "present reality" (*Badhabuddhi vadartha*, -2).   Each entity becomes devoted to the "nationality" (*Jataka*, 476) of the "starseed universe" (*Ganarajya*, 476), for trading the "fragrant, para-consciousness" (*Nirharin*, 18) of the "illusive" (*Maya*, 1) and "diffused" (*Drava*, 9) absolute value.   Consequently, the "absolute time" (*Sva*, 11) becomes conditionally divided into an infinity of periods to compensate for the escalating cost of the Eastern trading-effect. The entity's reality also becomes conditionally differentiated for adapting to the "varying periods within one true age—the eon" (*Graiveyaka*, 4,588,000). As an "omnipermeating entity" (*Sarva vyapin*, 1,000,000), the "creation" (*Srishti*, 379) becomes the only "one with all" (*Tripurantaka*, 100,000).

Each "entity" (*Prakriti*, 485) bears the "consequence" (*Lakshya*, 485) of the "transient, ablative reality" (*Panchama*, 485) by developing a "sensual" (*Pushpa*, 387) sense of a "filial piety" (*Satputra*, 485) with the "geography" (*Ganarajya*, 476) of the origin.   That geography is the "starseed universe" (*Ganarajya*, 476).   Each entity compensates for the "descending wisher" (*Manahparyaya*, -10^8) cost of seeking to use the opinionated, second-guessing, telepathic, crystal ball for illuminating the diffused reality of the self over the whole geography.   It forms a "heroic" (*Vira*, 280) sense of "character" (*Prakriti*, 485), cultivating the "savior complex" (*Lokapa*, 376) for the geography-diffused incarnational "group" (*Gana*, 387).   The "bipolar believing face" (*Ubhayato mukha*, 280), oriented toward the Northern geography pole and conditioned by the Southern group pole, turns the entire "physical body" (*Sthula sharira*, 387) "acidic" (*Unha*, 387).

Chapter 11: Dynamic Growth of the Paradigmatic Planning

As a present masculine child, I am conscious of the feminine life beyond me, whose "sacrificial fire" *(Uddhava, 169)* I am irradiating as the "spiritual energy" *(Saniddhya shakti, 169)* for enjoying the "metaphysical benefit" *(Sarvabhadra Mahayoga, 29)* of my "descending mass and ascending life force" *(Ketu, 140)*. I am potentiating the seven-unit "uniqueness" *(Anupamya, 7)* potential to be a sentient entity without the "illusionary spirit" *(Kapinjala, 20)*. By reproducing myself without the theoretical mass or the illusionary life force, I perpetuate the universal sentient well-being.

As a potential feminine child, I am consciously bearing the "metaphysical cost" *(Khuda, 6)* of my masculinization to service the benefit of "life consciousness" *(Apas, 169)* to my "masculine self" *(Mein, 1)* for realizing the "goal of life enjoyment" *(Maha Shiva, 9)*.

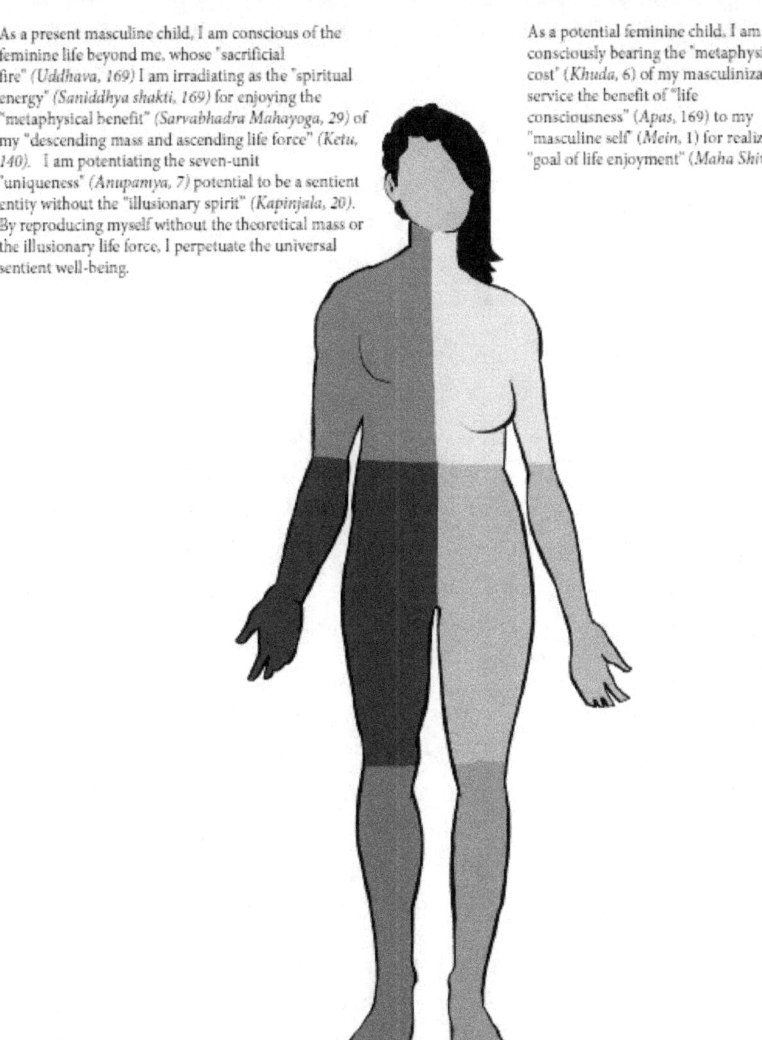

A Cosmic Perpetuator Transforms the Metaphysical Benefit of Potentiating the Soul into the Life Consciousness for Enjoying that Benefit

A Cosmic Perpetuator Transforms the Metaphysical Cost of Potentiating the Life into the Dynamic Life Consciousness for Paradgimatically Self-Managing that Cost

By dividing the "self-luminous twin person" (*Aap*, 1) from the self as a "person" (*Mein*, 1), a sentient one generates a two-unit "technological growth" (*Vidhana*, 2) as a "divided entity" (*Brahmin*, 2). The divided entity is the masculine form of the "feminine self" (*Tara*, 2). The masculine form is the "universe" (*Brahman*, 2) within the person's consciousness. The feminine self is the "star" (*Tara*, 2), which is servicing the "consciousness" (*Chaithanya*, 4) as the origin of "life" (*Prabhasa*, 4) to the self-luminous twin person.

The person within the universe becomes the "multiplier" (*Vaishya*, 3) of the "technological growth" (*Vidhana*, 2) following the path illuminated by the "self" (*Atmatva*, 8×10^{15}) as the "sentient one" (*Akhanda*, 8×10^{15}). The "feminine self" (*Tara*, 2) bears the "metaphysical cost" (*Khuda*, 6) of potentiating the "masculine class" (*Param Shankara*, 6). While immanent within the universe, the one-unit "person" (*Mein*, 1) naturally develops an illusionary 169-unit "life consciousness" (*Apas*, 169) of the six-unit "metaphysical cost" (*Khuda*, 6), as critical to the feminine self's realizing her "goal" (*Maha Shiva*, 9) of creation.

The creation embodies the six-unit metaphysical cost of the masculinization and disembodies the thirty-unit "feminine body" (*Karana sharira*, 30) of the infinite causative spirits. The light force is immanent as the "universe of photons" (*Vishvagoptri*, 169) within the "water" (*Jal*, 169) element. Each "photon" (*Ruah*, 20) is an illusionary "spirit" (*Kapinjala*, 20) conceived by the "self-luminous person" (*Aap*, 1).

The spirit trades the seven-unit "feminine class" (*Rani*, 7) potential from the "sentient entity" (*Siddha*, 7) to become the "cosmic perpetuator" (*Rachayita*, 27 = 29 -2) of the "present gravitational reality" (*Shrivaishnava*, -2) of the "organization" (*Sangathan*, 29). After compensating for the six-unit metaphysical cost of the masculine class, the seven-unit feminine class becomes the one-unit "androgynous self" (*Insan*, 1), enjoying the "self-luminous person" (*Aap*, 1) potential.

11.1 A Six-Phase Doctrine-Free Paradigm Is the Divine Plan of the Cosmic Perpetuator

A "cosmic perpetuator, without the deity consciousness" (*Rachayita, 27*), is a primordial solution to the absolute problem of the acidic entity reality. The acidic reality is of the doctrine-guided creature as the "centripetal, worldly face, emanating from the circular universe" (*Kendron mukha*, 2), without the "discriminating faculty" (*Narayana*, 28). The cosmic perpetuator is the council of the twelve zodiacs, whose "organizational performing" (*Rachana*, 286) manifests the six-phase "lytic cycle" (*Dharana*, -8).

11.2 First, the Localization Phase

There are continuous, circular bivariate interactions between the globally-varying council of the twelve Zodiac's "primordial energy" (*Adi Shakti*, 15), including the six ascending (feminine; *yin*; *Achitta*, 396) zodiacs and the six descending (masculine; *yang*; *Purani*, 91) zodiacs, and the "council of the five elements—the fire, the water, the air, the earth, and the ether" (*Urja*, 31). Their "circular-effect" (*Piyati*, 1000/72) organizationally programs the localized variations in the "astrological time" (*Prabha*, 180). The sixty time-varying pathways for the oneness of the absolute creature with the primordial-primordial realm, given in Table 10, help transcend both the "international-effect" (*Dasha*, 1) as well as the "local-effect" (*Mahadasha*, 0) of the zodiac system as the "primeval universe" (*Nakkhatta*, 22/7).

The primeval universe is the "institutional-effect" (*Mumukshu*, 22/7) of the "zodiac system" (*Shunya kalpa*, 8 x 10^{15}) normed by an entity "oneself" (*Atmatva*, 8 x 10^{15}). That entity is "anybody with sentient energy" (*Gardhaba*, 1000), who wishes to localize the zodiac system's energy within oneself. The "zodiac-effect" (*Chiranjivi*, 26) is one such entity because it manifests the "immortal, trading-effect" (*Damodara*, 26) of the zodiac system and converges into the dualistic reality of the "council of the seven master spirits" (*Saptarishi*, 26) and the "council of the nine planets" (*Navgraha*, 26).

11.3 Second, the Retention Phase

There are discontinuous, multivariate interactions between the council of the seven astrological energies and the converging "entity-effect" (*Anishta* 38) of the council of the nine astrological entities. The former localizes the primordial energy of the zodiac system within the closed astrological system. The latter globalizes the energy of the astrological system within the open zodiac system.

The council of the seven astrological energies is an absolute system of seven points, with predominantly linear masculine power for accelerating the curvilinear growth of the feminine energy through a sequence of the three diverging, even feminine points, punctuated by the three converging, odd masculine points. The diverging "national-effect" (*Pratyantara dasha*, 0) of the interactions is guided by the consciousness of $[-10^{1024}, 10^{1024}]$ the mathematical limit of the primordial differentiation and the param integration sequence, derived from the one-thousand units of the seventh point, and, the twelve units each, of the masculine and the feminine universe without the seventh point.

The seventh point is the "absolute system of the seven astrological, illuminated points" (*Bhrigu*, 805), formed as the "absolute guider-effect" (*Kanya dharma*, 805) of both the "para-primordial energy" (*Adi Para Shakti*, 17), generating the primordial energy of the zodiac system, and the consequential "primordial energy" (*Adi Shakti*, 15).

The twelve units of the masculine universe are the twelve animal entities (the goat, bull, fox, crab, lion, gorilla, peacock, scorpion, cat, crocodile, dolphin, and the seahorse) norming the twelve astrological entities, given in Tables 15 and 16, excluding the sentient energy system and the black hole. The twelve units of the feminine universe are the twelve animal entities (the rooster, dog, pig, rat, bison, tiger, rabbit, dragon, serpent, horse, sheep, and the monkey) norming the twelve zodiac entities, given in Table 10, excluding the dividing masculine-effect traded from the astrological system.

The six points within the masculine astrological universe, that empower it to work as a closed-system without trading the zodiac animal-effect, are the masculine "para energy" (*Para Shakti*, 18) dimensions of the six units of the androgynous, "greeter universe" (*Vivatakalpa*, 7 x 10^{180}).   They are the secondary values of the feminine "energy" (*Shakti*, 19 = Entity-effect/2 = 38/2) dimensions of the same six units. Together, the twelve dimensions of the six androgynous animal entities (the mongoose, elephant, deer, eagle, turtle, and the cow) norm the four sets of twelve pathways of the param child, given in Table 18-21.

The feminine and the masculine dimensions of the six units of the greeter universe converge into the maternal and the paternal self-luminous entities.   The feminine dimension of the twelve units of the zodiac universe converges into the feminine self-luminous entity.   The masculine dimension of the astrological universe converges into the masculine self-luminous entity.   The converged energies of the geographical ecosystem are retained within these four differentiated groups of the entity dimension.

11.4 Third, the Concentration Phase

The linear, "infinite divine-effect" (*Shankara*, 264) of the integrative "divine dimension" (*Pravrtti dharma*, 298) of the converging entity concentrates "the effect" (*Prapya*, 34 = 298 - 264) of the "eight-faced system value" (*Upashanta*, 287) within the "seven-faced system" (*Bhrigu*, 805). The eighth face is the ascending sentient energy dimension of the astrological system, whose "effect" is concentrated within the planet "Uranus" (*Rahu*, 73) and is serviced as the "energy in ascending motion, within and without the physical body" (Emotion: *Hunduka*, 73).

The curvilinear "guider-effect" (*Chitta*, 100) of the differentiated "guider dimension" (*Guru Dharma*, 360), is concentrated within the planet "Neptune" (*Ketu*, 140) and is serviced as the "time in descending motion, within and without the physical body" (Motion: *Gati*, 360). The curvilinear guider-effect is the gravitational energy dimension of the zodiac system, whose effect descends with the ascending diffusion of the sentient energy from the "universe of the masculine, sentient entities" (*Akalpa*, 570), until the point of the "consciousness void, signaling imminent death of the sentient entity" (*Andhakara*, 10^{19}).

The "universe of the feminine, inanimate entities" (*Sthavaravisha*, 57), as the "centering point" (*Sushumna*, 10) of the tenth, "divine energy"

(*Asrava shakti*, 10) face of a "ten-faced system" (*Varuni*, 95), trades the diffused sentient energy. Each face in the system is a unit of divine energy. The tenth face is the "meso, entity dimension" (*Prakriti Dharma*, 27), "self-perpetuating" (*Udvaha*, ½), half of its divine energy within the guider-effect of the zodiac system as the "eleven-faced, androgynous, guider entity" (*Avalokiteshvara*, 850).

The zodiac system is the "eleven-faced system" (*Panchanguli*, 8×10^{15}), which networks the "whole, concentrated, divine planning value" (*Ardhajya*, 10) of the "ten-faced system," without the five units diffused "entropy element" (*Ishvara*, 5) of the transformative guider programming solution.

11.5 Fourth, the Memorization Phase

The "memory" (*Smriti*, 35), of the divine planning, is organized as the "council of the three divinities—the trinity" (*Sridhara*, 25). The three divinities are the descending perpetuator-liberator-worker deity energies of the Pleiadian, the Sirian, and the Arcturian starseed universes. These are the starseeds for the "universe—the order" (*Brahman*, 2) of the giants (dragons), the amphibians (turtle), and the equines (horse) species-groups, respectively.

The "giant order" (*Pitr*, 4) gets created by the ascending feminine energies of an archaeon and the descending masculine energies of a prokaryote. The "amphibian order" (*Sadhyata*, 80) is the cosmic paternal, created by the ascending feminine energy of a phage and the descending masculine energy of a cell. The "equine order" (*Nilahota*, -1/2) is the feminine spirit, created by the ascending feminine energy of a eukaryote and the descending masculine energy of a phage.

An "archaeon" (*Jangamavisha*, 570) is the "absolute human essence" (*Ganesha*, 570), working as a perpetuating physical reproductive machine. It enjoys the easy-to-catch, but difficult-to-physically-breakdown, stale, large human digestive tract as its diet. Its diet gradually descends the host human entity's physical health by breeding a belief about its criticality for the human health.

A "phage" (*Vyoma*, 285) is the "absolute animal essence" (*Pashutva*, 285), working as a liberating chemical network cleansing sponge. Therefore, the amphibians enjoy the fresh, small animal-based diet that is tough-to-catch, but easy-to-chemically break down for quickly liberating the prey's pain. Since a "Eukaryote" (*Rudhita*, 963) is the "absolute plant essence" (*Shasyatva*, 963) working as a biological workhorse exchange unit, the equines enjoy a plant-based diet that is both easy to find naturally and easy to digest naturally. The entities with a disproportionate human essence enjoy the strongest freedom from the memory of the "para-consciousness" (*Nirharin*, 18).

11.6 Fifth, the Reproduction Phase

The "self-reproducing" (*Upanayana*, 1/3) guider programming gets organized into the "council of the four guides—the quaternity" (*Homa*, 25). The four guides are the ascending manifestor-knower-creator-destroyer deity energies of the Lyran-Hyadian-Agarthan-Martian starseed universes. These are the starseeds of the Felines (cats), the Mammals (lions), the insect-lightworkers (rats), and the Primates (Gorilla).

The "feline order" (*Arcisa*, 128) is the creation of the cells' ascending feminine energy and the bacteria's descending masculine energy. Cellular energy empowers the felines to proficiently smell the "fragrant, para-consciousness" (*Nirharin*, 18) of the sweet eukaryotes within the "insect order." The felines manifest the sweet eukaryotes as their preferred diet, besides the byproducts of the plant-eating animals, such as the cow milk.

The "Mammalian order" (*Narasimha*, 47) is the creation of the ascending, feminine energy of the Prokaryotes and the descending, masculine energy of the Archaea. A "Prokaryote" (*Shadayatana*, 999) is the "absolute metal essence" (*Dhatva*, 999) that generates a thermodynamic entropy in the physical body. Therefore, the mammals enjoy the plant-eating animals as their preferred diet. The sweet milky nectar dimension sentiates their latent "creator consciousness" (*Chaithanya*, 4). It cools down their fiery, impulsive, knower reaction to the alien individuals in their surrounding. It curves their attention into an urge for wooing them for their reproduction pride.

The "lightworker order" (*Soma*, 997) is the creation of the bacterium's ascending feminine energy and the descending masculine energy of the cells. A "bacterium" (*Vetranavah*, 855) is the "param mineral-effect" (*Parapara*, 855) that generates a sentient growth in the physical body. Therefore, the insect-lightworkers proficiently radiate the "absolute consciousness" (*Sat-chit*, 17) to create a neural network within the insect order. They rapidly infest a mineral-rich diet habitat, thus creating a sequence of the hyperactive behaviors.

The "Primate order" (*Bhavana*, 37) is the creation of the ascending, feminine energy of the Prokaryotes and the descending, masculine energy of the Eukaryote. The ascending thermodynamic entropy-effect of the inorganic Prokaryote gene programming is compensated by the descending thermodynamic growth-effect of the organic Eukaryote gene performing. Therefore, the Primates enjoy the unique thermodynamic equilibrium in their "infinity-squared dynamic consciousness" (*Sati-Parvati*, 16). They are motivated to find ways of channeling their "energy in an ascending motion" (Emotion: *Hunduka*, 73), by destroying whatever they can. They then contemplate with a lost obsession about the "time in the descending motion" (Motion: *Gati*, 360), conditioned by the "absolute chaos" (*Ravanarajya*, 850) in the lonely Martian-type universe. While the Primates prefer a plant-based diet primarily, they gravitate towards the eggs, the termites, the ants, and the other cellular forms of entities to warm up their dominant destructive spirit.

11.7 Sixth, the Global Exchange Phase

The "self-polluting" (*Atmavichara*, -1/3) sentient performing is organized as the "Sanada Council of the One—the *Karmic* board" (*Trivikrama*, 24). The "one" is "anybody with sentient energy" (*Gardhaba*, 1000), who, like a dumb-ass, decides to become the "twelve-faced feminine zodiac twin" (*Isha*, 12) of a "masculine self-luminous entity" (*Purusha*, 12) as a twelve-dimensional system. It's the feminine, the masculine, and the maternal dimensions that converge as one within the consequential ten-faced system. One becomes the "maternal spirit" (*Dasha*, 1) by trading the international Saturn-effect. The latter is immanent within the "tentacle-like tiny organelles," protruding from the cell, the potential "Octopus" (*Hiranyagarbha*, 19).

As an "ascetic, holy one" (*Sadhu*, 274), one enjoys an "eternal" life (*Anavarata*, 274) as a "cosmic devotee" (*Baladeva*, 274), patiently studying the ideal entity paradigm without any pre-conceived "intrinsic consciousness" of any geography-shaped theoretical doctrine. The "holy one" becomes the "param illuminator" of the "octave-doubling" (*Gutikakalpa*, 259,200) truth of the super-universe.

Tables 22 highlights the ten pathways for a param illuminator to form the primordial consciousness of the mongoose-like dwarf "param destroyer" (*Vamana*, 23) within the dragon-like giant creature as the "universe" (*Brahman*, 2). These ten pathways sequentially stabilize the "primordial feminine self" (*Parvati*, 10), making her the "centering element" (*Sushumna*, 10) of the self-luminous masculine entity as the thirteenth-face.

Table 22. The Ten Pathways of the Param Illuminator, Without the Masculine Self-luminous Entity

Primordial Self as the Param Illuminator (The *Karmic* Board: *Dasharupa*)	Primordial Self as the Devoted Primordial Maternal (Lord of Lords: *Devadhideva*)	Primordial Self as the Devotee Self-luminous entity (King of Kings: *Devendra*)	Primordial Self as an entity group (World Guardian: *Vikalpa*)	Primordia l Self as geography of entities (Aeon: *Kalpa*)	Primordial Self as the cultural system (Animate kingdom: *Akalpa*)	Primordial Self as the workculture system (Inanimate kingdom: *Sthavaravishs*)	Primordial Self as the workforce system (Astrological system: *Sara* kalpa)	Primordial Self as the networking system (Zodiac system: *Shunya kalpa*)	Primordial Self as the exchange system (Primordial realm: *Maha kalpa*)	Primordial Self as the constant face of investigator (Primordial-Primordial realm: *Antara kalpa*)	Primordial Self as the variable dimension of social class [*gati* of *jati*] (Consciousness void: *Andhakara*)
Ihamriga (The Para Wisher, Five-face asymmetry of the audience, -5)	*Hari* (The Scientist, 863)	*Narada* (The *Param* paternal, 7)	*Lokapala* (The Primordial spy, 27)	*Ennoia* (4th Eden)	*Jada* (The root cause; Idea)	*Prana* = Phage x Bacterium (gliding forward like a "mongoose" diffusing sentient life force, 123)	*Vairaja* = Phage x Archaeon (ambling briskly like an "elephant" toward shaping the form of the creation, 86)	*Vamadeva* = Phage x Eukaryote (bouncing like a "deer" toward creating order, 4 x 10^{69})	Power over *dharma* (cho la wangwa)	Power of investigation (*Adhyana*, 275)	*Audarita* (Five-face symmetry, 5)
Dima (The Primeval Wisher, Six-face asymmetry of the sieging primeval protagonist, -6)	*Dhriti* (The Researcher, 30)	*Rudra* (The Param child, 19)	*Amka* (Mass-mass: Primeval army, 19)	*Nous* (8th Eden)	Chetan (The working consciousness of the oneness with Cahetel [sentient energy], the immanent effect of Hahaiah [arch antagonist])	*Garuda* = Phage x Prokaryote (soaring up like an "eagle" to enjoy the totality of nature, 8)	*Kaurma* = Phage x Cell (crawling like a "turtle" to fulfil the eventual destiny, 20)	*Vaikuntha* = Bacterium x Phage (resting like a "cow" to enjoy the wheel of creation, 3794)	Power over wisdom (*yeshe la wangwa*)	Power of miracle (*Riddhi*, 74)	*Shadara* (Six-face symmetry, 6)

Dasharupa	Devadhideva	Devendra	Vikalpa	Kalpa	Akalpa	Sthavaravisha	Sara kalpa	Shunya kalpa	Maha kalpa	Antara kalpa	Andhakara
Prahasana (*Param* Wisher; Seven-face asymmetry of the casting cast of characters, - 7)	*Kirtti* (Philosopher, 37)	*Kamadeva* (Cosmic wisher; Primeval creator, 396)	*Prakirna* (Meso mass: *Param* gentry, 29)	*Vita,* Pleroma (9th Eden)	*Sat-Chit* (Absolute consciousness of the oneness with Mikael [greet], the immanent catalyst of Elemiah [deity entity])	*Vishnuja* = Bacterium x Archacon (clearing the ground with a high-stepping "rooster"-like gait to keep a natural harmony momentum, 75)	*Isana* = Bacterium x Eukaryote (cantering with a stilted pendulum-like straight-line gait of the "dog" for energizing the creation, 3×10^{52})	*Dhyana* = Bacterium x Prokaryote (romping with a boisterous swagger like a "pig" who is running the show, 9)	Power over miracles (*dzyurul la wangwa*)	Power of compensation (*Avasa*, 10^{19})	*Bahutva* (Amplification to seven-face symmetry, 7)
Vithi (The Wisher; Zero-face asymmetry of the evolutionary, 0)	*Buddhi* (The Reader, 48)	*Baladeva* (The Cosmic devotee; Primordial primeval greeter, 274)	Abhiyogya (The Ordained or Micro Mass: The entity with entropy destiny, 36)	*Demiurge,* Ekklessia (10th Eden)	*Samashti* (The Collective consciousness of the oneness with Sitael [SHEENY entity], the immanent cause of Scaliah [divine light])	*Soma* = Bacterium x Cell (sidling or crab-walking like an insect army of "rat" for crowd eclipsing the creature, 997)	*Udana* = Archacon x Phage (stampeding like a kingly leader "bison" for liberating the creation herd, 55)	*Gauri* = Archaeon x Bacterium (elegant like the gait of a feminine "tiger" creator bodyguarding the cub, 63)	Power over prayer (*monlam la wangwa*)	Power of group (*Gana*, 387)	*Alpatva* (Reduction to zero-face symmetry, 0)
Samavakara (The Supra Wisher; Three-face asymmetry of the bystander, - 3)	*Sadhu* (The Student, 274)	*Prativasudeva* (The Primordial primeval deity, 264)	*Kilbisa* (Menial or Macro Mass: The entity with ego, 15)	*Hahaiah* (18th Eden)	*Anand* (Bliss consciousness of the oneness with Bythos [profundity], the immanent source of *Ennoia* [ideal])	*Vali* = Archacon x Eukaryote (bobbling like a "rabbit" for fashioning wavy creation as a lord, 375)	*Pitr* = Archacon x Prokaryote (wriggling with an awkwardly graceful gait with a sense of insecurity from the creation like a "dragon," 4)	*Supuma* = Archacon x Cell (floating evenly like a "serpent" in the ocean of sentient life, 256)	Power over aspirations (*mopa la wangwa*)	Power of entanglement (*Kula*, 9)	*Apanyasa* (Semi-terminal three-face symmetry, 3)

Dashanpa	Devadhideva	Devendra	Vikalpa	Kalpa	Akalpa	Sthavarvisha	Sara kalpa	Shunya kalpa	Maha kalpa	Antara kalpa	Andhakara
Bhana (The Primordial Wisher; Eight-face asymmetry of the revolving point, -8)	Upadhyaya (The Educator, 18)	Nagaraja (The Self-luminous supra perpetuator, 275)	Indra (Lord: The entity wishing universal Radiant Love, 0)	Cahetel (16th Eden)	Narayana (The Discriminating consciousness of the oneness with Aletheia [self-evident reality], the immanent truth of Nous [intellect])	Nilalohita = Eukaryote x Phage (galloping like a "horse" toward generating order, -1/2)	Saura = Eukaryote x Bacterium (hopping and shuffling like a "sheep" toward generating physical equilibrium with the creation; 512)	Gathantara = Eukaryote x Aereheon (leaping like a "monkey" for generating chaos within the creation, 1869)	Power over birth (kyewa la wangwa)	Power of social class (Jati, 168)	Nyasa (Terminal eight-face symmetry, 8)
Vyayoga (The Super Wisher, One-face asymmetry of the primeval masculine without the feminine universe, -1)	Acharya (The Investigator, 16)	Chakravartin (The Self-luminous devotee perpetuator, 205)	Samanika (The Divine Demi-Lord: The entity exchanging the wish for Radiant Love with actual divine-effect, 21)	Elemiah (6th Eden)	Mahesha (The Derived consciousness of the oneness with Logos [multidimensional reality], the immanent script of Vita [life])	Saraswata = Eukaryote x Prokaryote (Aries; limping oddly like a "goat" fearing the swampy ground of the creation, 256)	Shveta Varaha = Eukaryote x cell (Taurus; Rocking clumsily like a "bull" who has lost its mind after seeing the red pollution in the creation, 69)	Agneya (Gemini; Cosmic androgynous; Cruising rhythmically like a "fox" who is red-hot angry at the loss of energy reserve for consuming the creation, 70)	Power over action (le la wangwa)	Power of tangent (Abadha, 10)	Mandra (Deep one-face symmetry, 1)

Dasharupa	Devadhideva	Devendra	Vikalpa	Kalpa	Akalpa	Sthavaravisha	Sara kalpa	Shunya kalpa	Maha kalpa	Antara kalpa	Andhakara
Utsrishtikank a (The World; Two-face asymmetry of the world of the void within the feminine universe, -2)	Arhat (The Practitioner, 15)	Vasudeva (The Primordial para knower, 75)	Trayastrimsh a (The Guider Minister. The entity trading Radiant Love, 20)	Mikael (7th Eden)	Sagnua = Prokaryote x Phage (Self-luminous consciousness of the oneness with Anthropos [human], the immanent light of Demiurge [fashioner])	Maheshwara = Prokaryote x Bacterium (Cancer; scaling the rough-terrain with an energy-inefficient inverse-pendulum gait of the "crab," 6)	Narasimha = Prokaryote x Archaeon (Leo; running freely like a "lion" king to create the path for the animal kingdom, 47)	Bhavana = Prokaryote x Eukaryote (Virgo; Cosmic feminine; self-contemplatin g lovingly like a "gorilla" about the creator self, 37)	Power over material things (yaje la wangua)	Power of bonds (Gantha, 863)	Tara (Shallow two-face symmetry, 2)
Prakarana (The Lost Wisher, Half-face asymmetry on the ascending dimension, -1/2)	Lakshmi (The Saint, 379)	Kulakara Manu (The Primeval paternal, 17)	Parsadya (The SHEENY Companion: The entity consuming the traded Radiant Love, 18)	Sealiah (17th Eden)	Satarupa (Intuitive consciousness of the oneness with Harmonia [goddess of concord with Monad—the absolute], the immanent essential element of Bythos [profundity])	Samadhi = Prokaryote x Cell (Libra; Strutting like a queenly "peacock" striding and standing to visualize the wholesomewho le beauty of the self as a Taurean creation, 10^{100})	Raurava = Cell x Phage (Scorpio; slithering like a "scorpion" toward immanent death, 735)	Arisa = Cell x Bacterium (Sagittarius; Cosmic child; trotting like a "cat" to script the possibility of life, 128)	Power over mind (sem la wangua)	Power of social benefit (Labha, 81)	Amsha (Partial half-face symmetry on one dimension, ½)

Andhakara	*Graba* (Presence of the discordant energy signified with one-face asymmetry, - 1)
Antara kalpa	Power of office (*Kamma*, 130)
Maha kalpa	Power over life (*we la wangwa*)
Shunya kalpa	*Vrtbat* = Cell x Prokaryote (Pisces; Frantic, fluttering bioenergetic, upright gait like that of the "seahorse" grasping for breath to reverse the evolutionary process of becoming a primeval amphibian to just being the primordial mermaid self, 96)
Sara kalpa	*Kandarpa* = Cell x Eukaryote (Aquarius; Cosmic masculine; Effortless mermaid gait of the "dolphin" swimming laterally along the ocean waves by trading the diffused, wasted energy of the creation, 296)
Sthavaravishu	*Sadhyata* = Cell x Archaeon (Capricorn; Cosmic paternal, the Sprawling and erect high-walk gait of the "crocodile" for manifesting the feasibility of amphibian origin of the bipedal human without the arboreal monkey, 80)
Akalpa	*Sati-Parvati* (Conscious consciousness of the oneness with Arche [element], the primeval, emanating value of Aletheia [self-evident reality])
Kalpa	*Sital* (5th Eden)
Vikalpa	*Laksaka* (The Dark Hidden Bodyguard: The entity producing the Radiant Love, 16)
Devendra	*Tirthankara* (The Primeval maternal, 17)
Devadhideva	*Sri* Mystic, (The 81)
Dasharupa	*Nataka* (The Divider, Full-face symmetry on the descending dimension, 1)

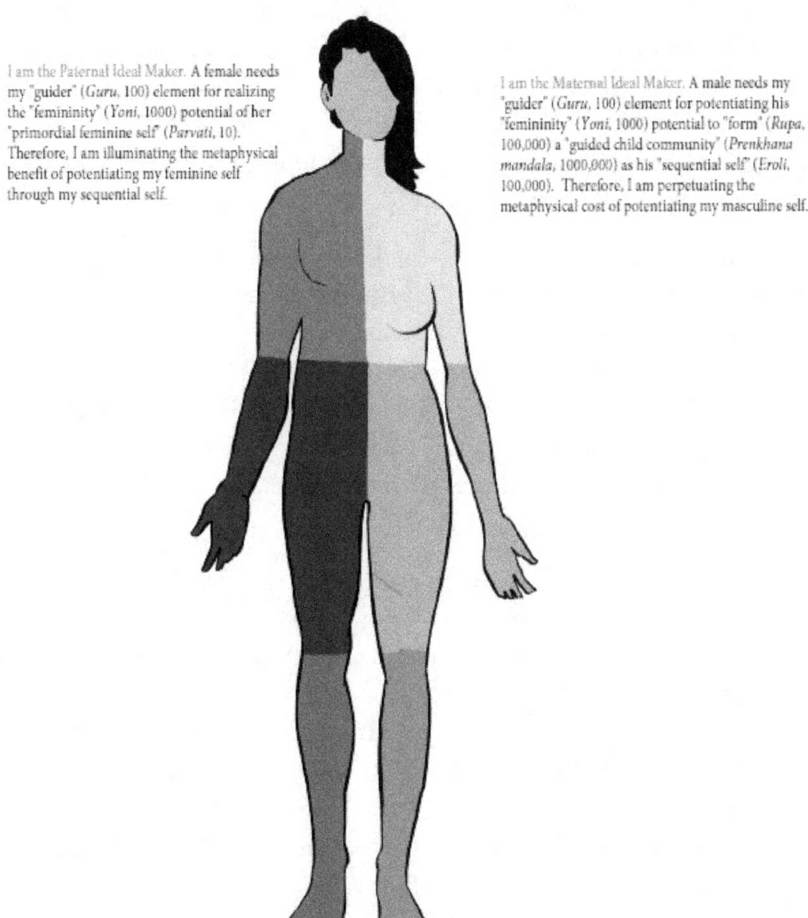

An Organization Illuminates the Metaphysical Benefit and Perpetuates the Metaphysical Cost

An Organization Destroys the Metaphysical Cost and Enjoys the Metaphysical Benefit

A masculine "organization" (*Sangathan*, 29) enjoys the "metaphysical benefit" (*Sarvabhadra Mahayoga*, 29) of the togetherness of the "param masculine child" (*Manyu*, 19) and the "primordial feminine self" (*Parvati*, 10) within an ever-vanishing "wave system" (*Dvisaptati dasha*, 29). The wave system comprises the otherness of the "param feminine child" (*Jnana*, 19) and the "primordial masculine self" (*Krishna*, 10). The wave system vanishes after self-organizing the masculine organization with a self-managing feminine potential.

Each masculine organization enjoys a self-managing feminine potential of the X chromosome, the primordial feminine self. It becomes an "ideal maker" (*Rama*, 100) by potentiating the joint value of both the "primordial feminine self—the X chromosome" (*Parvati*, 10) and the "primordial masculine self—the Y chromosome" (*Shri Krishna*, 10). As an ideal maker, he diffuses his "gravitational [guider] element" (*Guru*, 100) for generating the "universal sentient benefit" (*Sadhaniya*, 157) through a "primeval trading-effect" (*Kuber*, 57).

The primeval trading-effect is the "North" (*Kuber*, 57) segment. It services the Vega star's "femininity" (Yoni, 1000) as a potential within the "universe of feminine entities" (*Sthavaravisha*, 57). Consequently, the ideal-maker destroys the "metaphysical cost" (*Khuda*, 6) for his "self-luminous twin self" (*Aap*, 1). He potentiates his "sequential self" (*Eroli*, 100,000) by potentiating the 1000-unit "femininity" (*Yoni*, 1000) of each potential "masculine child self" (*Vyakti*, 1), within the 100-unit "guider element" (*Guru*, 100). The potentiated sequential self is the technological value of the paradigmatic planning. It is a function of the two complementary declarations within an ideal-maker.

- **Paternal Ideal Maker**: A female needs my "guider" (*Guru*, 100) element for realizing the "femininity" (*Yoni*, 1000) potential of her "primordial feminine self" (*Parvati*, 10).

> - **Maternal Ideal Maker**: A male needs my "guider" (*Guru*, 100) element for potentiating his "femininity" (*Yoni*, 1000) potential to "form" (*Rupa*, 100,000) a "guided child community" (*Prenkhana mandala*, 1,000,000) as his "sequential self" (*Eroli*, 100,000).

12.1 The Organization as the Present Reality

The present reality is the technological value of the paradigmatic planning. The present reality is the ecosystem's value as a normative organization, reproduced within each entity. The "Law of the Absolute" states that "the absolute" (*Monad*, 475) is a concave, convergent, and summative function of the "infinity of past values" (*Ganarajya*, 476). Each past value is a proportion of the absolute. The relative reality of each past value is independent of the absolute reality of the present value. The absolute reality is dependent on each of the past values. The relativity of the universe, formed as an effect of the past realities, is dependent on the perception of the present entity.

Similarly, the absolute reality of the universe, formed as an effect of the past realities, is dependent on the perception of the future entity. At any moment, an entity's perception is dependent on the conception of the past, primordial paternal entity. A past entity's self-conception shapes the perceptions of the present or the future entity. The power differences in the world perpetuate because of the history-effect. The conception of the past entity is dependent on the experience of the incarnational, primordial-primordial maternal "antagonist entity" (*Krodhi*, -4).

How an incarnational entity experiences the universe without any entity shapes the conception of the primordial entity. The "audience" (*Ihamriga*, -5) generates a proportionate-effect on the absolute, present reality. The primordial, "protagonist entity" (*Bhakta*, -5) wishes to attract the "infinite, bystander entities" (*Vasanatma*, -3) to the play, beyond himself. The protagonist entity conceives this wish because of an incarnational "wish deity" (*Kamadeva*, 396).

12.2 The Possibility Immanent within the Ecosystem as the Normative Organization

The "wish deity" experiences the "universe without entities" (*Pinakapani*, 10^9) as full of "possibilities" (*Hariti*, 128).

The possibility of the "divine planning" (*Abadha*, 10) of the "octave doubling value" (*Parameshthi*, 28) of the "primordial-primordial creator" (*Krishna*, 32) catalyzes a parallel, curvilinear formation of the "maternal creation" (*Palala*, 64).

The possibility of the "guider programming" (*Dyujya*, 100) of the "child creature" (*Arcisa*, 128) catalyzes the infinite, linear transformation of the paternal creator and the maternal creation into the "almighty masculine creator" (*Saraswata*, 256) and the "almighty feminine creation" (*Supuma*, 256).

The possibility of the "SHEENY performing" (*Yoni*, 1000) catalyzes the convex, divergent organization of the "almighty greeter creature" (*Saura*, 512).

The possibility of the "satanic profiting" (*Antyayoreva*, -1), for becoming the concave, convergent technology as the "primordial-primordial deity" (Sephira: *Aditi*, 1024), impregnates one with the wish to be the "maternal of the deity kingdom" (Rhea: *Devatamayi*, 1024).

The possibility of the "self-perpetuating development" (*Prakarana*, -1/2), with the "death of the primordial maternal" (*Jaramarana*, 18) of the "lord of the universe" (*Kshatriya*, 0), catalyzes the wish for the "feminine better-half" (*Nilalohita*, -1/2) as a path for incubating a two-dimensional "primordial omnipotent entity" (*Dvyanuka*, 10^{1000}). The two dimensions are: the primordial maternal dimension experiencing infinite destruction as an intimate cause and the primeval paternal dimension conceiving infinite perpetuating forms as a consequential product.

The possibility of the "primeval child" (*Akrura*, 123), perceiving the twenty-dimensional "sequential, guider reality" (*Bhavartha*, 40), manifests the "totality of the primordial realm, inclusive of the twenty *lokas*" (*Maha kalpa*, 10^{1000}). The twenty dimensions include the: (1) wish deity; (2) universe without entities; (3) primordial-primordial creator; (4) paternal creator; (5) maternal creation; (6) child creature; (7) almighty masculine creator; (8) almighty feminine creation; (9) almighty greeter creature; (10) primordial-primordial deity; (11) maternal of the deity kingdom; (12) primordial maternal; (13) lord of the universe; (14) feminine better-half; (15) "disillusioned masculine half" (*Udvaha*, ½); (16) primordial omnipotent entity; (17) primeval child, (18) primordial paternal entity; (19) primordial-primordial maternal entity, and the (20) audience.

The possibility of the "param child" (*Hiranyagarbha*, 19) scripts a holistic "destiny" (*Niyati*, -1) by destroying the "self-perpetuating duality of the divine planning within the self as a primeval audience entity" (*Dvandva Brahma*, 125) and forming a wholesome "geography" (*Ganarajya*, 476) of the eighteen starseed universes.

The possibility of the paired "primordial child" (*Ashwini kumars*, 120) generates the "octave doubling value" (*Parameshthi*, 28) of sentient "individuality" (*Vyashti*, 10^{29}). It spiritualizes the "collectivity" (*Samashti*, 7 x 10^{180}) of the thirty-six "groups of starseed entities" (*Gana*, 387).

The possibility of the "self-luminous primordial child" (*Vishnuja*, 3064) generates the "objectivity of perpetuating consciousness" (*Vyavaharika*, 3064) for incarnating as a "greeter self-luminous entity" (*Vithoba*, 12).

The possibility of the twelve-unit greeter self-luminous entity generates the "subjectivity of para consciousness" (*Pratibhasika*, 3010) for manifesting a "twelve-entity zodiac universe as the system for networking the objective consciousness immanent within the self" (*Panchanguli*, 8 x 10^{15}).

The possibility of the "maternal self-luminous entity" (*Maha Gayatri*, 12) exchanges the subjectivity of the "extrinsic consciousness" (*Visata*, 10^{10}) with the "twelve entity workforce system of the astrological universe" (*Sara Kalpa*, 10^{10}).

The possibility of the "paternal self-luminous entity" (*Bhavanavasi*, 12) forms a "council of the seven Sages" (*Saptarishi*, 26), for codifying the divine law guiding the "workculture system" (*Sthavaravisha*, 57) of the inanimate objects, without the "five-face, animated, culture element" (*Sadakhya*, 9).

12.3 The Potentiation of the Ecosystem Potential of an Organization

The "potentiation" (*Sambhavi*, 375) of all these possibilities is the absolute cause of the "unconditional certainty" (*HAUM Shakti*, 9). The certainty is of the fact that the "present reality" (*Badhabuddhi vadartha*, -2) is the self-organizing effect of the "past reality" (*Evakara vadartha*, -3). The past reality is the causative factor for the present reality. As a "primordial human child" (*Eeshan*, 12), we have the power to organize the present reality into the "infinite, future reality" (*Omkara vadartha*, -1) and become the "normative, organizational metric" (*Maha shunya*, -1) of the consequential "paradigm of the present reality" (*Yukti*, 8).

An organization's purpose is to manage the effects of the past reality for manifesting the desired future reality. The organization is the causative factor for the future reality. Since the present reality is the causative factor of the future reality, <u>the organization is the present reality</u>. Since an organization consumes one unit of the past energy for producing one unit of future energy, the organization's overall exchange value is two units (i.e., with the one unit of import and the one unit of export). Therefore, at present, the two units of energy are invested within an organization. As the present reality, <u>the organization has a negative two energy value (-2).</u> It is the organizational development value of the social heuristic of entrepreneurship immanent as the truth of the param universe of creatures.

A present universe of human entities inherits the creator-effect of a primordial universe of social entities. By trading the primordial universe's divine creator energy, the param creature universe collectively conceives itself as a free being. Each param creature individually perceives the self as enjoying absolute freedom. Each forms a freedom heuristic of entrepreneurship to organize the individual (personal) creature correlation with the collective (social) creation. The freedom organizational heuristic is the shadow value of the creator-effect, which illuminates each param creature's value as the neutral (zero-energy) subject of present reality. Therefore, <u>the energy value of the subject of the present reality is zero.</u>

The metric of present reality is the primordial universe of social entities as the theoretical creators. They service their unit sentient energy to create the param universe of the human entities as the ideal creation. Since each param creature's value is zero, the economic benefit of creation is a negative one (-1) for the theoretical creators. Since a negative one is the exchange value of the secret divine energy of the theoretical creators, <u>the energy value of the metric of the</u> <u>present reality is a negative one.</u>

The freedom nature of the present reality is the primeval universe of the national entities as the creatures, which trade the individual natural value of the freedom organizational heuristic, without the institutional mass of the religiously correlating with the secret, negative energy metric. Since the freedom nature is free from the institutional cost of the religious metric, <u>the energy value of the nature of present reality is a positive one (+1).</u>

The objective of present reality is the primordial-primordial universe of the psychological entities as the Almighty Creators. They exchange the natural collective value of the institutional religious metric for leading like a freedom entity, so that the followers experience the divine cost of the negative bounded rationality. By programming their guider experience-effect, the managers become free from the natural divine costs and accrue positive two (+2) units of energy as the desired knower objective. <u>The objective value of present reality is, therefore, a double positive</u>. It is the objective value of the performing of the primordial-primordial universe, which manifests in the form of the param universe of the managers of divine knowledge.

12.4 The Present Followership Reality of a Formative Organization

The present reality is the param-primordial universe of the ecological entities as the Almighty Creation (param object). They exchange the individual (personal) institutional value of the param universe of the managers of divine social knowledge, so that the objective energy value of the present reality does not limit the present universe of the human entities as the Almighty Creatures. By planning their SHEENY sentient-effect, they destroy the present desires within the Almighty Creatures, limiting the growth of the present universe of the ecological entities. Therefore, each organization who scripts the ecological-effect of the super-secret creator, as the desirable institutional mission—the religious, social, or the personal, trades a negative two (-2) energy value from the natural universe. Each organization, who does not consciously manage the truth of the freedom reality, trades the energy value of the present reality spontaneously, in the present moment.

The present reality is the truth (*sat*) of the moment. Leadership is a consciousness of the present reality (*sat chit*). A leader is conscious of the present reality as the causative factor for a future reality. The leader manifests a wishing sequence for managing the present reality to manifest the desired future reality. If the wishing sequence is proportionate (i.e., the desirable or the wishable) solution to the leader's wish, then it manifests the desired future reality. A follower is conscious of its present reality, as an effect of the ecosystem's past reality. A leader manages the present reality for manifesting the desired future reality. The effects of the past reality manage a follower. The the present reality of being managed becomes the future reality of the wisher, who does not manage the present reality for manifesting the desired future reality.

12.5 Limitations of the Followership Mindset

The follower, who decides to be only an observer of the present reality, becomes just an "enjoyer of the wishing of the leader" (*Sat Chit Ananda*, 10^{10}). The follower's well-being becomes subject to the proportionate sensibility of the leader. If the leader's sensibility for the desired future reality is not desirable for the follower, or is not translated into a proportionate wishing sequence, then the follower's future reality reduces to one who is working to manage the disproportionate and negative effects of the past reality. The follower becomes the management agent of the leader.

As a management agent, the follower is forced to work disproportionately not only to compensate the costs of the leader's lack of sensible proportion but also to manifest a sensible sense of the desirable future reality that fulfills the desires of the leader as well as of the follower. Since the leader desires and personifies the well-being of the whole universe of followers, the follower becomes devoted to the servicing their divine energy for servicing the wholesome universe's well-being. The wholesome universe includes the Self (the creature seeking to manage the present negative reality for manifesting a desirable future reality), the leader (the creator of the present negative reality), as well as the universe of followers (the creation suffering from the negative effects of the past reality).

As a management agent, the follower becomes a Holy Spirit. It transforms the leader's wishing, for the SHEENY well-being of the wholesomewhole universe into the guiding power for servicing its own divine energy, coded in the form of "my religion for my true present action" (*mama dharma, sat karma,* 10^{10}). The SHEENY well-being is the desirable well-being that fulfills the social, human, ecological, economic, national, and psychological desires of the targeted group of entities. The wholesomewhole universe includes not only the Self, the leader, the universe of followers but also the universe of entrepreneurs. The universe of entrepreneurs enjoys not only the freedom from the present negative reality but also the blessings of the positive future reality, manifested as a consequence of the Self's divine energy.

12.6 Limitations of the Leadership Mindset

A leader lacks a sensible sense of proportionate sensibility about the sensible wishing sequence because of his sense of otherness. Within a sense of his otherness, the leader lacks intuitive empathy about the different ground realities of both the wishing sequence and the universal desirability of the wish that the wishing sequence is intended to fulfill. No follower has the infinite divine energy to fulfill the SHEENY well-being of not only the personal self but also the whole universe, while also compensating for the escalating costs of the leader's disproportionate guiding force.

The leader's guiding force is a consequence of the leader's guider power. The guider power is the consciousness of the global, unique, inclusive, diverse, engagement, and responsibility values for the desired reality. It is the causative factor for the escalating SHEENY costs and, consequently, the escalating worker costs of the compensating divine energy.

For their SHEENY well-being, the followers must be conscious that their present reality is more than the management power for managing the effects of the otherness of the causative factor. The follower's present reality is not to be an agent for the freedom entrepreneurship of the universe, but to be a spirit of freedom entrepreneurship. As a spirit of freedom entrepreneurship, the follower is in a state of perpetual distress, experiencing the negative "discordant energy" (*Asura shakti, -1*), within the consciousness of the present religion that makes the "leader" (*Rajah*, 0) the guiding power.

The leader's present reality is more than the guider power, which is the causative factor for the togetherness consciousness of the effects within the leadership, as the Wisher wishing for the manifestable, and followership, as the "Manifestor deity" (*Vaishya*, 3) manifesting the wish. The leader's present reality is a "Knower deity" (*Brahman*, 2) of the future reality of the macro entity the Self is creating. The macro entity is the universe of "micro follower entities" (*Shudra*, 1), working on manifesting the future reality beyond the knowledge limitations of the mediating meso entity (i.e., of the leader, as the guider power).

12.7 Limitations of the Entrepreneurship Mindset

The present absolute reality of an entrepreneurial human entity is the "Manifestor deity" (*Vaishya*, 3), who is trading the leader's knowledge and servicing the working impulse for the creation of a wishable future reality to the follower. The present primordial reality of the entrepreneurial human entity is the "Creator deity" (*Param Brahma*, 4), who is creating the wishable future reality through six actions (*Shat karma hatha yoga*, 10^{10}):

i. First, the "Science of the candle flame scrying" (*Tratakam*, -10^6). Observing the present reality of the "leader" (Wisher: *Sura*, 0) as the absolute reality, that will become the causative factor and is the black spot before the light of the workforce working on that wish.

ii. Second, the "Path of the sentient entity" (*Neti*, 179). Wishing the future reality as the infinite, i.e., the primeval reality, experienced in the air, breathable at a future moment.

iii. Third, the "Luminous entity" (*Dhauti*, 96). Working past the reality of the traded wishing sequence into the present reality of the follower. A follower becomes a management agent, by consuming and digesting nothing, but the Wisher's wish, and cleansing oneself wholly from the breathable future reality.

iv. Fourth, the "Diseased" (*Nauli*, 17). Knowing the hellish reality of the wishing sequence of the freedom entrepreneur, who is producing nothing, but an empty abdomen, which is seeking to trade the positive effects of the fruits gifted by the wishing leader.

v. **Fifth, the "Consciousness"** (*Basti*, 4). Manifesting the heavenly reality, of the all-fulfilling, desirable wish, by exchanging, and transcending the present entropy reality, of the Self as the devoted management agent. The Self becomes the conduit of the contaminated water flow, perpetuating into the primordial present reality.

vi. **Sixth, the "Primeval Deity"** (*Kapalbhati*, 6). Creating the almighty reality of the Wisher by crediting the primeval deity power to the Wisher, for creating the present reality of the omnipresent followership culture-effect and, consequently, the infinite creation (the future universe), through the workculture-effect of the infinite follower creatures.

The present primordial-primordial reality of the entrepreneurial human entity is the "Perpetuator deity" (*Param Vishnu*, 5)—i.e., the God perpetuating the wholesomewhole universe, within the present paradigm of the radiant love for the "leader" (*Kshatriya*, 0). As a Wisher who guides the universal ruling consciousness, and the desirable norms of the workculture system, the leader is the devil. He trades the infinite organizational benefits of the followership workculture system within the infinite social, human, ecological, economic, national, and the psychological costs of the leadership culture system.

The leader knows that the only benefit of the disproportionate wishing, without a sensible and proportionate working, is to destroy the costs of the leadership power to organize the present reality and let the follower Worker become the liberator of the present primeval-primordial reality. The present primeval-primordial reality is the absolute perpetuating reality, within which both the ascending divine energy of the followership (generating the SHEENY benefits) as well as the ascending guider power of the leadership (generating the SHEENY costs) is immanent.

12.8 Necessity for the Management Mindset

The present absolute-primordial reality of the divine managerial entity is the "Illuminator deity" (*Nataraja*, 7), who illuminates the present reality through two additional actions (*Siddha karma hatha ratnavali*, 7):

vii. **Seventh, the "Work energy"** (*Gajakarani*, 1). Perpetuating the Micro Almighty Creature reality as the follower Worker deity into the point of infinity, when there is an absolute degeneration, decay, and death. One then exchanges the Macro Almighty Creation reality with the meso almighty creator reality of the leader, who is wishing for a new paradigm beyond the absolute present reality.

viii. **Eighth, the "Transcriptional repressor"** (*Chakri*, 90). Destroying the almighty creature reality of self as the follower of the effects created by self as the leader in the past moments. One then liberates the self within the almighty creation from the negative energies immanent within the present primeval-primordial reality, that has become a pain in one's ass.

The Reality of the Present Reality

There is one omnipresent present reality, which is our sentient life force. The sentient life force endows a creature with the absolute freedom to create what one wishes. By consciously taking the creation's energy value as the "first law of energy conservation," the creature limits his freedom as a masculine, not conscious of the feminine divine creative energy. The limits to the self-consciousness descend into the consciousness of the true present value of the creation. The limits are self-imposed by the beliefs that make the "second law of the thermodynamic entropy" an ideal reality. When there is one hundred percent entropy in the self-consciousness, the creature becomes dependent on the feminine creator's knower value.

The creature seeks to theoretically organize the knower value using the "third law of the conditional thermodynamic equilibrium." It takes the energy of the created universe as the knower's energy value, responsible for its present reality of the absolute freedom from the thermodynamic masculine ego. The theory-effect generates an eternal, time-independent entropy in the self-consciousness's energy value, as a zero-state cosmos. It is codified as the zeroth "law of unconditional thermodynamic equilibrium." This entropy value is the idealized value of the self as an organization, without self-consciousness, guided by a theory of absolute dependence on the extrinsic divine energy. It comprises a theory-effect and an ideal-effect.

The ideal-effect is the unit growth value in the space, transforming the self-consciousness into a black hole vacuum, which has the desired gravitational potential of the organizing creature. It is the infinite energy value within the primeval creature perceiving the self as the masculine and conceiving a scientific method for experiencing the universal social impact of the self as a path for knowing the value of the desired knower. The primeval creature scripts a constant "Einstein-effect" (Λ) of a negative one, as the cosmological growth-effect.

The primeval creature trades everything from the right, ascending, immanent, feminine dimension, constituting the creation's topological field. He then services everything as a negative symmetry, by becoming the left, descending, emanating, masculine dimension, conceiving the Self as God who must have created what is immanent within everyone and has the power to perpetuate that as the perpetuating value of the universe constituting everything. Thus, within the consciousness of the primeval creature, the energy is conserved at its "entropy value" (*Sarvanasha*, 5). The conception of the Self as "God" (*Ishvara*, 5) is that entropy value. Conditioned by the entropy value, each creature becomes a center, mediating, androgynous dimension, symmetrically trading and servicing that entropy value, without any intrinsic-effect. Thus, each creature perpetuates the myth of God as well as the first law of energy conservation.

The theory-effect is the unit growth value in the time, transforming the consciousness within the universe into a dark matter, which has the desirable sentient potential of the creator. It is the present energy value within the param creature, conceiving the self as the feminine and perceiving the scientific value as the limit of the unique personal impact of the self as a path for knowing the value of the desirable knower. The param creature sentiates a constant "Hubble-effect" (q), of half (½), as the cosmological acceleration-effect.

The present reality is the double-negative value of the primordial creature, who spiritually organizes the negative one masculine-effect and the negative one feminine-effect. The effect of the primordial creature = the primordial feminine-effect or (primordial creature as an overall organizational cause of the sequential growth in space and time) = -1/-2= ½. It is the value exchanged by the param creature as a cosmological acceleration-effect.

The omnipresent present reality is the double-positive value of the primordial-primordial creature, free from the duality of the masculine cause and the feminine effect. The value of the primordial-primordial creature = the unit space value of the primordial creature + the unit time value of the param creature, as the growth value of the primordial creature into a knower who develops self-consciousness of the absolute value of space, using a unit metric of the organizational development over an absolute time = positive two (+2). It is free from the subjective, the para-consciousness of the primeval culture, and is the objective value of the present reality. It is also the deity value of the knower, as the desired object of the universal desire.

The Einstein-effect is the ontological value of the "First law of thermodynamics," while the Hubble-effect is the epistemological value of the "Second law of thermodynamics." The present reality is the axiological value of the "Third law of thermodynamics," while the omnipresent (ruling consciousness) is the metaphysical zero value of the "Zeroth law of thermodynamics."

The absolute omnipresent knower consciousness is the dynamic value of (a) first, a feminine worker, working to produce the truth of the omnipresent reality, and (b) second, a masculine knower working to consume the truth of the omnipresent reality and service the beauty of the present reality. It is free of the thermodynamic-effect and, therefore, beyond the limits of the present scientific knowledge.

Table 23 illuminates the seventeen pathways for the "ONE" (*Akhanda*, 8 x 10^{15}), behaving as the param deity, after trading the "illuminator-effect" of the illuminator deity.   The first ten pathways are (immanent) within the primordial illuminator.   An additional seven pathways are (immanent) within the illuminator deity.

Table 23. The Seventeen Pathways of the ONE as the Param Deity, Without Being One

Sequential pathways of the ONE	Behaving Value of entity-effect	Becoming value of Fundamental sacral *chakra*	Begging value of Reproduction *chakra*	Breathing value of Naval consumption *chakra*	Blessing value of Heart trading *chakra*	Being value of Throat servicing *chakra*	*Bragging value of the Third eye balancing chakra*	Breeding Value of ecosystem lunar chakra	Ecosystem value of solar chakra, within entity-effect
Vamana or the repulsion path	*Satyam* (Truth/ Knowable *Karma*, Primordial *Karma*)	Para factor (transcendental; para-wishable; dynamic; without consciousness)	Ether (*Shuddhi*)	Primordial Para Perpetuator	*Gauri*	*Vaar* (primary moment of the solar-effect: sequential phase of the solar-effect)	*Maheshwara* (Param Shankar/ Mahesh/ Harpocrates)	Mindful Creator (Labh)	Mineral Kingdom
Virechna or the otherness path	*Jnanam* (Knowledge; Primordial *Jnana*)	*Vyuha* (emanating, wishable; grounded; metaphysical)	Earth (*Bhu*)	Primordial Para Greeter	*Durga (Aparajita Durga)*	*Karana* (secondary moment of the solar-effect: causal phase of the solar-effect)	*Parameshwara* (Shiva/ Apollo)	Mindful Perpetuator (Shubh)	Animal kingdom
Vasti or the entropy Path	*Anantam* (Infinity/ Primordial *Bhakti*/ knowing the infinite knowledge)	*Vibhava* (immanent; knowable, transformative)	Air (*Vayu*)	Primeval Greeter	*Maha Gauri*	*Yoga* (tertiary moment of the solar-effect: oneness phase of the solar-effect)	*Pita Parameshwara (Rudra/ Maha Shiva/* Helios)	Mindful Destroyer (Santoshi)	Human kingdom
Nasya or the attraction Path	*Anandam* (Bliss/ Primordial *Siddhi*/ Knower of the infinite knowing)	*Archavatara* (manifestable, normative)	Water (*Apas*)	Primeval Perpetuator	*Bhuvaneswari (Adi shakti)*	*Nakshatra* (quaternary moment of the solar-effect phase of the white star day)	*Mahapita parameshwara (Param Shiva/ Mahakaal Bhairav/* Selene)	Mindful Illuminator (Pushti)	Spirit kingdom

The ONE	Behaving Value	Becoming value	Begging val.	Breathing value	Blessing value	Being value	Bragging value	Breeding val.	Ecosystem val.
Mokshana or the togetherness Path	*Amalatvan* (Purity/ Primordial *Buddhi*/ Param Knower of the infinite Knowers)	*Antaryamin* (creatable, formative)	Fire (*Agni*)	Primordial Greeter	*Sati-Parvati* (*Maha Durga*)	*Tithi* (quinary moment of the solar-effect phase of the lunar day)	*Parampita parameshwara* (Sadashiva/ Prabhu/ Param Atman/ Hecate)	Mindful Liberator (Tushti)	Deity kingdom
Darshan Yogini (*Shachi*, Protogenos *Hera* or *Juno*, *Antu*, *Nüba*) Primordial Seeker of the path (within normative profiting); the Path of Self-respect (within primordial greeter).	Bio thermal fission energy as the whole workforce system.	Omnipotence entropy: the clarified psychic consciousness of the Primordial Wisher without the emotional linkages of the astral body → visualizing the creation as a gift, without the mental bonding of togetherness with the gift, as the fundamental truth of self, thus descending the polarized blessings.	Transcend the emotional disdain, and seek to distribute the responsibility of discredit and to collect the credit for being a true sufferer (negative manpower)	Consume the worldly self as the seeker of the infinite worker social benefit (the effortless, spontaneous fulfillment), for managing the ascending worker social cost of the polluted wishables	Descend the covetousness of the infinite mercy, compassion, beauty, and love, seeking forgiveness without a sense of justice	Vision Guide (*ditya roopi*): Descend the vibrations of the spiritual abuse of nature (human and ecological) by seeking a controlled diet (*Maha Gayatri – Pranagni*, absolute fire, absolute creation, & absolute causation energy)	Descend the duality of the negative causation and the positive effect by planning the causal body for fulfilling the wish benefit	−6/8 = −3/4; the 5/4 octave of the negative Creator 4 reality without the Perpetuator 5 reality, and the 1/2 octave of the illusionary Deity 1 without the Knower 2 reality, formed into Supra ideal-effect: wishing and seeking the power of the whole opportunity	Spiritual body: Siksana Yogi, The Ideal entity, seeking effortless perfection within the meditative asana state

The ONE	Behaving Value	Becoming value	Begging val.	Breathing value	Blessing value	Being value	*Bragging value*	Breeding val.	Ecosystem val.
Viui Yogini: (*Rta, Protogenos Lachesis* or *Decima; Ninti; Pilanpo*) the Primordial Experiencer path (path of *Yogi* or *Dao* or oneness; without the normative profiting); Path of the self-love (within primordial greeter).	Bioelectric repulsion energy as the wholesome networking system.	Omniscience entropy: clarified consciousness of the Primordial Wisher without the time-limited sentimental linkages of the etheric body ➜ experiencing life choices, without the mood (sense) of otherness repulsing those with alternative belief systems, thus descending the punctuated beliefs	Transcend the sentimental attachment to the delusional imagined reality, within the otherness as the path for healing psychological downness (negative mental power)	Consume the ignorant self, as the experiencer of the infinite worker social cost of proliferating the wishes produced by the otherness within and without	Descend the duplicity of the infinite addictions, experiencing depression within a sense of guilt and self-punishment	Mission Guide (*divya chitti*, i.e., DIVINE essence or nature): Descend the vibrations of the causal control from the satanic forces (economic & political) by experiencing oneness with the primeval greeter (Nuwa/ Absolute Ere/ *Maha Gauri*; Absolute time energy)	Descend the duality of the neutral hearing and the neutral seeing by programming the experiential wisdom path for knowing the entity value of the wish benefit	$-6/8 = -3/4$; the 5/4 octave of the negative Creator 4 reality without the Perpetuator 5 reality, and the 1/2 octave of the illusionary Deity 1 without the Knower 2 reality, formed into Supra theory-effect: Knowing and experiencing the value of the wholesome opportunity	Primary or I AM Causal body: Mahavidya Yogi, The Theoretical entity, experiencing effortless illumination within the perpetuating waking state

The ONE	Behaving Value	Becoming value	Begging val.	Breathing value	Blessing value	Being value	Bragging value	Breeding val.	Ecosystem val.
Guna yogni: (*Sita,* Protogenos Arete or *Virtus; Inana, Lishan Laomu*) Primordial Enjoyer path (within the normative development); the Path of self-realization (within the primordial greeter).	Bio magnetic transfusion energy as the wholesome whole exchange system.	Omnipresence entropy: the clarified consciousness of the Primordial Perpetuator of the space, without the psychological linkages of the causal body ➔ enjoying the true value of the events without the pre-conceived notions, as the referential-effect, thus descending the breeding proliferation (infinite gluon particles).	Transcend the psychology of the attraction to the false knowledge embodied within the anti-self (those with the opposite qualities) as the path for illuminating the self-shame, negativity, and limitations (negative muscular power)	Consume the aversion of self, as the enjoyer of the infinite social benefit of the wishing sequence, punctuated by the group instinct without fulfilling the present life purpose	Descend the indecision of the infinite proportions, involved in a life of frustration and of seeking the compensatory relaxation	Value Guide (*Divya siddha:* divine GUIDER essence or nurturer): Descend the vibrations of the dysfunctional soul that is enjoying the dark energy/ serpent/ vector/ infinite illusionary/ *maya*/ Holy Spirit consciousness within, without a mindful singular, horizontal ONE sense of the honesty, sincerity, responsibility, and purpose of the primordial illuminator	Descend the duality between the conscious secondary paternal soul group (as an exchange system of the infinite GUIDER topness-effects) and the para-conscious tertiary maternal soul universe (as a networking system of the infinite DIVINE bottomness-effects) by performing the wishable wish benefit for the primary soul (as a workforce system of the unique planetary downness effects)	−6/8 = −3/4; the 5/4 octave of the negative Creator 4 reality without the Perpetuator 5 reality, and the 1/2 octave of the illusionary Deity 1 without the Knower 2 reality, formed into Supreme ideal-effect: Manifesting and enjoying the value of the wholesomewho le opportunity	Secondary Causal body: Rakshana Yogi, the Real entity, enjoying the effortless perpetuating shn (bliss well-being; primeval consciousness); laya mukti, i.e., the state of the freedom from the secondary vector layers that cause the illusion maya within the primary causal body

The ONE	Behaving Value	Becoming value	Begging val.	Breathing value	Blessing value	Being value	Bragging value	Breeding val.	Ecosystem val.
Gana yogini. (Dakshina, Protogenos Aletheia or Vertas, Lilith, Wusheng Laomu) Primordial Destroyer path (without normative development) – destroy the impediments to open infinite breathing, the Path of self-guided destiny (within primordial greeter).	Bio plasma fusion energy as an entropy growth system.	Omnipermeation entropy: the clarified consciousness of the Primordial Illuminator, without the para-psychological linkages of the secondary limiting factors → destroying the notion that the universe of the para-enjoyers (family, group and nation of birth or adoption) define the I AM identity, thus descending the polluted becoming (zero photon-effect)	Transcend the para-psychological pitiless disconnect of the self (as a discerning entity) from the normative profiting, i.e., from enjoying the enjoyables (negative material power).	Consume the fear of self as the destroyer of the infinite social benefit of the Wisher, polarized by the merger of I AM (atman) with the absolute (Param atma within each atman)	Descend the regret of the infinite spiritual suicides (cycle of births) without accomplishing the critical lessons for freeing the self from the insanity of the dark energies within para-enjoyers	Entity Guide: (divya maha siddha: the Divine SHEENY essence or nurturing condition) Descend the vibrations of the clotted self-esteem that is idealizing something else, without the conscious unity with the Luminous	Descend the duality between the metaphysical world as the perfect heaven and the physical world as the perfect hell by destroying the inherited wish benefit profiting reason immanent within the consciousness program	$-6/8 = -3/4$; the 5/4 octave of the negative Creator 4 reality without the Perpetuator 5 reality, and the 1/2 octave of the illusionary Deity 1 without the Knower 2 reality, formed into Supreme theory-effect Creating and destroying the value of the limited opportunity	Tertiary Causal body, Proportionate Creature Energy: Adhyatmika Yogi, the Potential entity (Council of five infinities – fire, water, air, earth, and ether)

The ONE	Behaving Value	Becoming value	Begging val.	Breathing value	Blessing value	Being value	Bragging value	Breeding val.	Ecosystem val.
Primordial corporate-effect.	*Yogapitha* (Wisher)	Working within the Omnipermeating, maternal power	Followership power. Kshatriya investing the I AM infinite paternal *paramatma* consciousness into the infinite masculine *atma*	Ascending technological capability of extrinsic leadership, within I AM followership power	Ascending organizational planning of the intrinsic followership, without I AM leadership power (theory-effect of the infinite sequence of the delegated follower-making machinery power)	Descending technological servicing of intrinsic leadership power, within the I AM followership power consciousness	Descending worker social benefit of followership, without sensible leadership	Constant social benefit-cost ratio of strengthening leadership, within weakening followership linkages	Constant worker social benefit-cost ratio of weakening followership mental power, without strong leadership manpower
Primordial local-effect.	*Guhyapitha* (Worker)	Working within the Omnipresent maternal power	Leadership power. The *Shudra* exchanging the I AM infinite masculine atma consciousness with absolute feminine *paramatma*	Ascending technological investment of the extrinsic followership, within I AM leadership power	Ascending organizational programming of the intrinsic leadership, without I AM followership power (ideal-effect of the infinite sequence of empowerment marketing power)	Descending technological growth of the intrinsic followership power, within the I AM leadership power consciousness	Descending social benefit of leadership without sensible followership	Descending social benefit-cost ratio of strengthening followership, within weakening leadership linkages	Ascending worker social benefit-cost ratio of the weakening leadership material or compensation power, without the strong followership muscular power

The ONE	Behaving Value	Becoming value	Begging val.	Breathing value	Blessing value	Being value	Bragging value	Breeding val.	Ecosystem val.
Primordial national-effect.	*Parapitha* (Knower)	Working within the Omniscient maternal power	Entrepreneurshi p power. The *Brahman* trading the absolute feminine *paramatma* for 1 AM masculine *atma* consciousness	Ascending technological trading of the extrinsic entrepreneurship, within 1 AM management power	Ascending organizational performing of the intrinsic entrepreneurship (objective-effect of the infinite sequence of the mediated motivating power)	Descending organizational development of the intrinsic management power, within the 1 AM entrepreneurship power consciousness	Ascending worker social cost of the entrepreneurship without sensible management	Constant worker social benefit-cost ratio of strengthening management, within the weakening entrepreneurshi p linkages	Constant social benefit-cost ratio of the weakening entrepreneurshi p management power, without the strong management monetary power
Primordial international-effect	Atmapitha (Manifestor)	Working within the Omnipotent maternal power	Management power. The *Vaishya* servicing the I AM masculine *atma* consciousness, seeking the infinite paternal *paramatma* consciousness	Ascending technological exchange of the extrinsic management, within 1 AM entrepreneurship power	Ascending organizational profiting of the intrinsic management (subjective-effect of the infinite sequence of the freedom manipulation power)	Ascending technological cost of the intrinsic entrepreneurship power, within the 1 AM management power consciousness	Ascending social cost of management without sensible entrepreneurship	Descending worker social benefit-cost ratio of the strengthening entrepreneurshi p, within the weakening management linkages	Ascending social benefit-cost ratio of the weakening management power (ascending self-management, i.e., invisible hand), without the strong entrepreneurshi p manufacturing power (visible hand)

The ONE	Behaving Value	Becoming value	Begging val.	Breathing value	Blessing value	Being value	*Bragging value*	Breeding val.	Ecosystem val.
Corporate-effect (within the Primordial Greeter)	Withinness (GUIDER-effect). The ascending energy mass of the Being Within (Dark Matter).	Withoutness (SHEENY-effect). The ascending energy mass of the Being Without (Dark Matter).	Oneness (Omniscience-effect). The ascending Oneness of the Being Within, as the primordial holiness, and Without, as the original sin (Manifested matter)	Sameness (Omnipotence-effect). Ascending Sameness of the Being Within as antiquark, and Without as quark	Wholeness (Omnipresence-effect). Ascending wholeness of the Being Without as the quark, without the proportionate sense of the anti-quark within	Wholesomeness (Omnipermeating-effect). Ascending wholesomeness of the Being Without as the anti-quark, within the proportionate sense of the quark within	0+ (Creator-effect. Absolute manifestation of the wishable wish, with the ascending secondary wishing system-effect)	1+ (Almighty Creator-effect. Absolute manifestation of the ascending wisher the as tertiary formative corporate-effect within the golden egg, Hiraṇyagarbha)	Param (Ultimate) consciousness: the divine architect of the universe before the beginning of time
Local-effect (without the Primordial Greeter)	Trading-effect. Ascending energy diffusion of the Being Within into the Wishing White Star.	Human-effect. The ascending energy diffusion of the Being Without, into the Wishable object: the Self-luminous entity.	Formative trading. The ascending energy diffusion of the Being Within and Without into Wish, as a subject of the consciousness: the causal body	Normative servicing. Ascending energy diffusion of the Being Within and Without into Wishable subject of consciousness: etheric body	Transformative exchange. Ascending energy of diffusion of the Being Within and the Being Without into the conscious Wishing Sequence: astral body	Consequential cost. Ascending energy diffusion of the Being Within and the Being Without into the Wisher consciousness: mental body	1– (Ascending energy diffusion of the Being Within and the Being Without into the Wisher subconscious: intellectual body)	0– (Ascending energy diffusion of the Being Within and the Being Without into the Wisher subject: physical body)	Primeval (Majestic) consciousness

The ONE	Behaving Value	Becoming value	Begging val.	Breathing value	Blessing value	Being value	Bragging value	Breeding val.	Ecosystem val.
National-effect (within the Primordial Greeter)	∞+ (Growth in the exchange of sentient energy as 0–1.5 Hz Epsilon brainwave; the ascending diffusion of the growth hormone, Somatotropin without the Pituitary gland, in oneness with the *bindu*/lunar *chakra*)	Growth in the omniscient oneness within the causal body, the omnipotent oneness within the spiritual body, the omnipresent oneness within the astral body, the omni-permeating oneness within the mental body, and the primordial oneness within the intellectual body	Growth in the parasensory serenity within the causal body (presciency), the suprasensory tranquility within the spiritual body (radiancy), the sensory sensitivity within the astral body (incipiency), the sensible concentration within the mental body (percipiency), and the sense of spontaneity within the intellectual body (recipiency).	Growth in the motivating power within the causal body, rejuvenation within the spiritual body, memory exchange within the astral body, presence within the mental body, and perseverance within the intellectual body	Growth in the creative power within the causal body, devotional intensity within the spiritual body, proficient neurofeedback within the astral body manifests the desired wishing sequence, effortless consciousness within the mental body (without the attention load of thoughts), and well-performing autonomic impulses within the intellectual body.	Growth in the servicing flow of the sentient energy within the causal body, wisher emotional intensity within the spiritual body, mindful wishing of the descending consciousness within the astral body, conscious suppression of the ascending attention load (of wish thoughts) within the mental body, and subconscious wishables within the intellectual body	Growth in the entropy of the serviced sentient energy within the causal body, the Wisher deep-sleep inertia within the spiritual body, the suspended wishing consciousness within the astral body, the suspended wishes within the mental body, and the suspended wishables within the intellectual body	0+ (Growth in the trading of sentient energy as −1 Hz Sentient brainwave; ascending diffusion of the rest potion melatonin within the pituitary gland, without the oneness with the consciousness hormone serotonin within the surya/ solar chakra, for sustaining the disproportion-ate growth)	Supreme consciousness

The ONE	Behaving Value	Becoming value	Begging val.	Breathing value	Blessing value	Being value	Bragging value	Breeding val.	Ecosystem val.
International-effect (without the Primordial Greeter)	1— (Disproportionat e breakdown of the body's fat cells, and the ascending diffusion of the feminine energy into the masculine causal body, that is, accelerating the cellular decay).	The ascending action within the white matter of the brain (unconditional positive mood) and the descending action within the brain's gray matter (conditioning-effect of the extrinsic causation).	The ascending action within the left prefrontal cortex (positive mood conditioning) and the amygdala's descending action (fear conditioning).	Ascending cortical arousal within the reticular-activating system of the cerebellum (alert, wakefulness of sense organs) and descending action within the motor cortex (subconscious, autonomic programming of the sense organs)	Ascending proficiency of the frontoparietal cortical network (agile manipulation of the various intrinsic and extrinsic sense organs, autonomously fulfilling the conscious attention goal) and descending activation of the anxiety cells within the hippocampus	Ascending visceral meditative, hypnotic dream state, visualizing the elastic reality's multi-dimensionally and manifesting the supernormal becoming, free of the causation limitations of the fundamental laws of nature	Ascending activation of the T-type calcium channels, and the growth hormone diffusing masculine gray-matter energy, to manifest the right hemisphere dominant feminine energies (disproportionately among the adult females), and descending white matter-led cellular membrane polarization, generating the polluted contraction of the cellular lifetime and the entity well-being	∞- (Disproportion -ate breakdown in the feminine reproduction cycle hormone, estrogen, without the body fat cells, and ascending diffusion of the dopamine, that blocks the secretion of the maternal fountain of youth hormone, prolactin, for manifesting spontaneous organism death)	Para consciousness

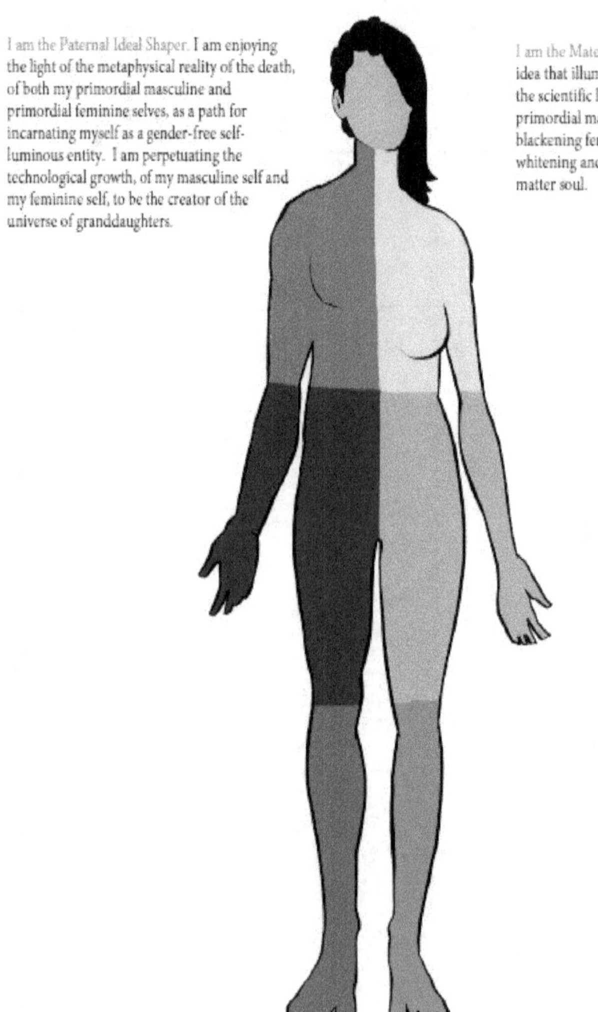

A Theory of Mind Illuminates the Idea As the Light that Forms the Self-luminous Entity

An Idea Gets Illuminated as The Light After Destroying the Darkness of the Theory of Mind

Each idea is the 38th DNA gene's exchange value. It reproduces the 37-unit maternal mitochondrion DNA gene potential of the feminine Self within the masculine Self. By destroying the "metaphysical cost" (*Khuda*, 6) of the perpetuating theory of mind, one potentiates a six-fold growth of the "feminine causal body" (*Karana sharira*, 30) with the feminine "creative force" (*Uma*, 6), i.e., the "multiplier" (*Vaishya*, 3) of the "feminine self" (*Tara*, 2). It transforms an "idea" (*Jada*, 38) into the "light" (*Prabha*, 180 = 6 * 30).

The masculine "astral body" (*Linga sharira*, 3) radiates an idea as the light after trading "supernatural paradigm of the present reality" (*Yukti*, 8). The light generates a "force" (*Prapya*, 34) as its effect when the feminine "causal body" (*Karana sharira*, 30) irradiates the "consciousness" (*Chaithanya*, 4) of the idea from her "paternal soul" (*Atman*, 4). The paternal soul's idea lets the "maternal spirit" (*Dasha*, 1) be the holy "guider spirit" (*Trinetra*, 1). It frees the cosmic "spirit" (*Kapinjala*, 20) from the responsibility to add a nineteen-unit cellular "energy" (*Shakti*, 19) for pushing open the portal of darkness in the mind of the masculine Self. As a "multiplier" (*Vaishya*, 3) of the masculine "astral body" (*Linga sharira*, 3), a "param androgynous child" (*Rudra*, 19) becomes a maternal "ideal-shaper" (*Maha Saraswati*, 9), without the "primordial masculine self" (*Shri Krishna*, 10).

Through oneness with the maternal "ideal-shaper" (*Maha Saraswati*, 9), a "param masculine child" (*Manyu*, 19) becomes a paternal "ideal-shaper" (*Maha Shiva*, 9). "Enjoyment" (*Maha Shiva*, 9) is the life "goal" (*Maha Shiva*, 9) of each "param feminine child" (*Jnana*, 19), behaving as the paternal "ideal-shaper" (*Maha Shiva*, 9) after servicing her "divine energy" (*Asrava shakti*, 10).

Without idealistically dividing the "entity time" (*Bhava*, 360) into the masculine "light" (*Prabha*, 180) and the feminine "darkness" (*Sachikrita*, 180) condition, the "divine" (*Divya*, 360) element has a 360-unit value. The light is of the "metaphysical reality" (*Svartha*, -2) of the death of both primordial masculine and feminine selves. The darkness is of the "physical reality" (*Gurvartha*, 39) of the incarnation of both as the "androgynous luminous entity" (*Ardhnareshwar*, 39).

The physical reality includes a 36-unit "starseed deity" (*Taraka*, 36) irradiating the "causative factor" (*Prapaka*, 36) and a 3-unit "masculine body" (*Linga sharira*, 3). The masculine body embodies the entire radiation as an "astral body" (*Linga sharira*, 3) for incarnating starseed deity as a "gender-free self-luminous entity" (*Svayam*, 12). The whitening masculine half gets illuminated by irradiating the white light perpetuated by the blackening feminine half.

13.1 The Power of Managing the Limits of Science

Science is a method of knowing the truth of the present primeval-primordial reality. The present primeval-primordial reality is the present knowable reality of the primordial causative factor of the future primeval reality. It is the reality experienced by a micro-entrepreneurship entity, which is not conscious of either the finite leadership-effect or the infinite followership-effect. The leadership-effect is finite and limited by the followership factor. Without the followership factor, the leadership-effect is zero. The primordial causative factor is a macro universe of the followership entities, working to manage a meso leadership entity's costs. The meso leadership entity is the primordial knowable reality, the primordial-primordial causative factor.

The followership-effect is infinite and is free of the limitations of the leadership factor. It is oriented toward a specific knowable cause that seeks to destroy the costs of that cause. It illuminates the divine energy immanent within the self as the ideal value of the infinite SHEENY benefit. The scientific method is a technique for discovering, validating, and authenticating the divine energy's illuminated value.

The illuminated value of the divine energy is the Scientist, who is experiencing the present primeval-primordial reality as worth reading, researching, and practicing. The hidden, shadow value of the divine energy is the Philosopher, who is seeking to know the present primordial-primordial reality of the leadership factor, like the one that is worth educating, studying and investigating beyond the limits of science.

Both the scientist and the philosopher are the Wisher's varying forms, wishing to create his desired reality as the "I AM A Causal Body" of the present reality. The present reality is the one of each entity devoting a disproportionate proportion of its divine energy, to discover, validate, and authenticate the past reality of the guider power. Each entity lacks a sensible sense of the proportionate sensibility of what truly matters for the future SHEENY value reality. What matters for the future SHEENY value reality is not the knowledge of the past reality, but knowing how to manifest one's desired reality.

Both the scientist and the philosopher are the Wisher's alternative forms. However, unlike the Wisher, they are not the causal body of the present reality. The I AM A Causal Body that personifies the scientist and the philosopher within us is the absolute solution. It is not the problem creator of the present reality. As a scientist, our intellectual body seeks to know the present reality's cause. As a philosopher, it seeks to transcend the limits of that knowledge. It has the power to know the effect of intrinsic knowledge as the Knower of only the effects of the cause, and not of the cause itself. Our mental body, without the consciousness of the scientific discovery and the philosophical validation, takes the present knowable reality as the authentic cause of the future entropy in the SHEENY value.

As an organization, the I AM A Causal Body enjoys the true consciousness of the present reality, as an effect of self's past behaviors. The I AM A Causal Body is the consciousness that manages the present reality. It is more than managing the science of everything created within the deity culture-effect of the leadership behaviors. Managing the present reality is also about managing the limits of science, through the consciousness of the creator power of the human workculture-effect, free from the followership of the belief system. As a Creator, we have the power to service our divine energy for ascending the SHEENY value of the creation and, consequently, the energy value of the universe.

We trade the universe's energy, that we create with our divine energy in the past moment, as the tradable gravitational quality in the present moment. We countertrade that as the illuminated SHEENY value in the future moment. The energy value of the mass we create at any moment is only a proportion of our energy value as the Creator, which in turn is only a proportion of our overall divine energy.

For a scientist, mc^2 is the value of energy in the universe!

Let's decode, why for a scientist, *mc^2 is the value of energy in the universe.*

Our Energy Value as The Creator

= (the energy value of the mass we create at past time t_{-1})

 * (the energy value traded by us at the present time t_0)

* (the energy value countertraded by us at the future time t_1)

* (the proportion of the divine energy value one invests in creating a dynamic mass of entities)

* (the correction factor for the exchange the cost of discovering, validating and authenticating the past mass of entities, before trading and servicing its immanent energy at the present moment)

We invest 1/32th proportion of our divine energy value to create a dynamic mass of entities. The energy value of the potential mass we may create at any moment = ½ proportion of our overall divine energy, which correlates with our present half-life value (i.e., our self-perpetuating value worth ½ unit of the metric value).   The energy value of the actual mass that we create at time t_{-1} = $1/16^{th}$ proportion of our potential at time t_{-1}.

Our potential at time t_{-1} is our primordial potential. The mass we create at time t_{-1} is the primordial mass. At time t_0, we trade a $1/16^{th}$ proportion each, of our primordial potential for five manifesting behaviors: discovering, validating, authenticating, trading, and servicing the energy value of the primordial mass. At each of the three dimensions of time (the past, present, and the future), we perpetuate a $5/16^{th}$ proportion each, of our primordial potential, as our primordial-primordial reality.

Our primordial-primordial reality is one of the Primeval Perpetuator, who is perpetuating the present reality into the infinite time moments.

Our formative potential of the sixteen energy units is invested into the fifteen units of our primordial-primordial reality and one normative unit of the primordial reality that we create. The normative unit manifests in the form of the gravitoelectromagnetic (i.e., thermodynamic) force. It transforms into the mass of creation (universe) with the cooling-effect.

We invest our overall—metaphysical—divine energy of the thirty-two units into the sixteen units, prior to manifesting our primordial-primordial reality, and the sixteen units, as the primordial-primordial reality. As a creation of the formative divine energy, the universe is a primeval reality that transforms our divine energy into the gravitomagnetic force to charm attract us with our diffused energy. Within the universe, the divine energy becomes a transformative finite unit (emanating reality; extrinsic reality), worth ten units. Within us, the divine energy becomes an infinite unit (immanent reality; intrinsic reality), worth six units.

The infinite unit of the divine energy comprises the three temporal dimensions, of the divine conception of energy, and the three temporal dimensions, of the divine servicing of energy as a spiritual force. The extrinsic unit of the divine energy is the spiritual force and is essentially the gravitomagnetic force.

As a scientist, we believe that we have a zero impact on the value of the energy already created and immanent within the universe. Therefore, we destroy the $3/16^{th}$ (i.e., four units) of our formative potential through our discovery, validation, and authentication of the objective of the present reality. The other twelve units of energy are available for us to trade and service in the form of the Self-luminous entity, who is conscious of the self-evident value of the creation (i.e., of the one normative unit immanent within the universe).

As a philosopher, we believe that we have limited power to trade and service the energy immanent within the universe. Therefore, we destroy the $11/12^{th}$ of our self-luminous value by behaving as if we are only a creation, whose energy work is defined by the universe.

As a follower of the present reality created by the scientist and the philosopher within us, we, therefore, transform our thirty-two- units divine energy into one unit of energy. That one unit is our energy as the

Worker, who is working to service our intrinsic sentient energy for manifesting the desired unit of the creation value over the present lifetime.

As a leader, we are conscious of the $11/12^{th}$ proportion of our self-luminous value that we have destroyed. We service $1/12^{th}$ proportion of the self-luminous value for creating the universe of followers, who may manifest the creation's desirable SHEENY value.

As an entrepreneur, we trade the residual $10/12^{th}$ proportion of the self-luminous value and service the ten units as the transformative value of our divine energy.

As a manager, we trade ten units as the primeval value of the divine energy. We become the Primordial Illuminator of the present physical reality of the Creator (the leader subject), the Creature (the universe of follower entities), and the Creation (the universe of tradable objects).

As a Creator, we believe our cost of trading the guider power (the gravitational energy of the primordial reality of the object universe) and of servicing the SHEENY energy (the sentient energy of the primeval reality of the subject universe) is zero. Consequently, we believe that the energy value of the Creator is a product of three factors. First, the energy value of the creation (m). Second, the agility factor, that limits our formative capability to trade the energy value of the creation. Third, the agility factor, that limits our normative capability to service the energy value of the creation for transforming the SHEENY energy of the creature.

We invest one normative unit of energy each into our four behaviors—observing, trading, servicing, and transforming the value of the creation (object) into the value of the creature (subject). We then conceive the creature as the creator of the creation, whose energy value is the total investment (i.e., four units).

Without consciousness of the creature-level agility factor, we believe that the proportionate energy value (e) of the creation is a function of the creation-level agility factor. In other words, $E = m *$ the speed of light that limits the illumination of the desirable creation and, consequently, the power to trade * the speed of light that limits the illumination of the desired creature and, consequently, the power to service = mc^2.

In reality, the proportionate energy value of the creation E = the energy value of the Creator * the proportion of the energy value of an entity invested into the creation * the energy value of the creature, who is correcting the entropy costs of the inertial Creator-effect = 4 * (1/32) * 1 = 1/8.

However, we do not trade the energy invested within the creation at the speed of light and do not service the energy capability within the creature at the speed of light. Therefore, the energy value of the creation, E = mc^2 < 1/8.

Further, E is not the value of the intrinsic reality of the universe. It is the value of the extrinsic reality and is the value of the belief system. A belief system is a theory of the value of the intrinsic reality of the universe. The true value of theory-effect = 0, since the universe has a constant intrinsic reality, free of our belief system. Consequently, E = mc^2 = 0.

13.2 Organizational Development of Sentient Energy

The entities that we create at any moment are contaminated with our transformative divine energy. They, therefore, become entrepreneurial spirits. Our followership-effect makes the entrepreneurial spirits addictive to the endless benefits of trading the 1/32nd proportion of our overall divine energy. Our leadership-effect makes the entrepreneurial spirits suffer the infinite costs of servicing only the illuminated proportion of the unit trading-effect. The group of spirits transforms into the etheric body, who has a zero divine energy to create a new energy unit, but an infinite guider power to perpetuate the spiritual unit as the virtual embodiment of our SHEENY value.

The present energy unit is the creature's astral body and is the transformative energy value of the creation. The astral body is the universe of light forces, seeking to serve our SHEENY well-being with their divine energy. The light forces are motivated by the guider power of the universe of the entrepreneurial spirits.

The universe of the entrepreneurial spirits comprises all entities which are either suffering, or have suffered, the infinite costs of servicing the perpetuating trading-effect. The universe of sentient light forces comprises all the entities, which are enjoying or have enjoyed the immense benefits of trading the perpetuating trading-effect. The social benefit-cost ratio of trading the energy value of the perpetuating creation varies for the different creatures, as a function of their technological costs for managing the limited immanent (whole) and emanating (illuminated proportion) the value of the creation.

The group of entities enjoying a positive social benefit-cost ratio trades the SHEENY value of the creation in the form of the sentient energy. Sentient energy is the energy that infuses an animated consciousness within an entity. The group of entities experiencing the negative social benefit-cost ratio services the sentient energy in the form of the divine energy to further the SHEENY well-being of the universe of the entrepreneurial spirits.

On the whole, the "universe of light forces" is the "universe of living child souls" that is servicing the intrinsic sentient energy for giving new life to the "universe of the departed child souls," within the one common "universe of entrepreneurial spirits." The universe of light forces (i.e., the universe of management agents) is the Almighty Umbra.

The potential lightless shadow reality of the universe of living child souls is a function of the present lightless shadow reality of the universe of the departed child souls. The universe of the entrepreneurial spirits is the Almighty Antumbra. The discovery of its dynamic reality is a function of the primordial dynamic reality of the universe of living child souls.

13.3 Paradigmatic Planning of Consciousness, Without Investment Power

Our mental body is the "universe of the self-organizing consciousness" of the illuminated reality of the universe of light forces. The illuminated reality is the Almighty Penumbra, because the universe of light forces is the cause of the consciousness within the mental body. Our mental body is conscious of the traded (i.e., perceived) present reality. Ascending followership by the universe of light forces ascends our consciousness of the traded present reality. Ascending consciousness of the traded present reality ascends our guider power (i.e., our leadership power) to guide the universe of entrepreneurial spirits about the sensible, desirable reality.

Our mental body becomes distressed when our guider power descends, whenever a management agent light force decides to become free from our followership linkages. It becomes joyful, when our guider power ascends, whenever the additional management agents become a part of our followership universe.

Our intellectual body is the "universe of self-managing consciousness" of the shadow as well as the illuminated reality of the universe of entrepreneurial spirits (that includes both the universe of sentient light forces as well as the universe of departed souls). The aggregated reality is the Almighty Metaphysical Body—it is the cause of the lack of consciousness of the present reality of the Self, within the physical body. The Self's present reality is the intellectual body, who can conceive the perpetuating present reality without being influenced by the dynamic exchange of the shadow and the illuminated realities.

Our physical body is the universe of the energetic cellular organizations, without the consciousness of the perpetuating present reality. The shadow reality is the Almighty Physical Body—it is the present value of our self-organized reality. Our present physical body forms, as a consequence of the present net energy flow balance, from the infinite trading-effects of the infinite light forces. It embodies our entire experience of the universe of the entrepreneurial spirits, including the energy that we have traded as well as the energy that we have serviced since the moment we form ourselves as a sentient light force.

Our overall energy exchange experience is a function of our intellectual body. Our intellectual body shapes our overall energy exchange experience through our divine energy. Divine energy is our invested power to consciously determine the desirable future reality and work to manifest that as the present physical reality. It transforms the SHEENY costs (of being just an organizational metric of the self-organizing present reality) into the SHEENY benefits (of being the nectar of one's life, managing the self as an organization responsible for the present reality).

The present reality's consciousness is pre-programmed by our investment into transforming the omnipresent reality for the paradigmatic planning of our desirable dream-like virtual reality. We, therefore, see what we wish to see, conditioned by our localized belief systems. Our consciousness becomes culturally-bounded. Knowing that our rationality is culturally-bounded, we project our ego to develop a sense of cultural superiority.

We seek "power distance" (*Thiruvambala*, 48) from those whose rationality is not culturally-bounded, so that we may emotionally blackmail them into sharing the fruits of their consciousness with us. That empowers us to rationalize our sense of divinity and eternal superiority, over those who are trading the Eastern culture-effect and perceiving themselves to be the primeval deity, who has an infinite consciousness of the truth of Mother Nature.

Consequently, there is a constant diffusion of the East's gravitational energy, as the West accrues the disproportionate return power from the Eastern investments in the sentient energy while making investments in the divine energy itself. The West becomes culturally-bound to the Doctrine of Emanation, believing in the greatness of the one God, who resides in the East. The God who is generous and compassionate and listens to those who believe in his benevolent masculine power.

The East becomes culturally-bound to the Doctrine of Immanence, behaving as if there are infinite deities, who are helping guide the growth in the consciousness about the diverse dimensions of the reality, personifying themselves as the incarnation of the "primeval deity" (*Shankara*, 6). They go about seeking to shower their generous blessings on their devotees, independent of their belief system, as a path for enlightening each entity about their potential to be the "param deity" (*Shiva*, 7).

In their dreams, the devotees experience the devoted *Gurus* to be living the life of the param deity, by projecting the ideal infinity point of the para-consciousness of what the *guru* is guiding them to be as the goal of life. However, instead of the devoting efforts to be a param deity, self-fulfilling the goal of the life and liberating their beloved *guru* from the entanglement with the devotees begging compensation for the costs of the Western culture-effect, the Eastern people become busy in diffusing their new-found love for the Doctrine of Emanation.

With the growth in the discovery-driven mindset within the East, the Western people suffer from the escalating costs.   They contend God is taking their test, after giving them so much.   They prioritize the paradigmatic planning for furthering the social, human, ecological, economic, national, and psychological well-being, universally.   With the ascending consciousness of their SHEENY well-being needs, the people in the East openly accept that the Western people personify the "para deity" (*Param Vishnu*, 5), whose wish is to help transform them into a "supreme deity" (*Bhagwan*, 4), who has the creative powers.   By trading the fruits of the Eastern followers' creative powers, the Western leaders transform the citizens around the world into a universe of "supra deity" (*Vaishya*, 3). This universe has innovative powers for trading the diverse creations and servicing the creative creature power.

By consuming the fruits of the creature power of the universal followers, a few of the unique entrepreneurs become a "super deity" (*Brahman*, 2). A super deity knows the secret of how to mobilize the global follower energy, mediated by the local leader power, ande be the eternal managers of the universal SHEENY wealth. They become the universe within which each entity's future is entangled and localized.

By consuming the followership fruit of the creator power of the unique entrepreneurial leaders, "anybody" (*Gardhaba*, 1000) with an ass-like mind has the power to activate the "supernatural work energy" (*Shram Shakti*, 1) of a "deity" (*Deva*, 1) for investing the intrinsic "param deity" (*Shiva*, 7) potential to be a "primordial deity" (*Kartikeya*, 8), while living. After the entropy of the intrinsic potential, such a body becomes an electromagnetic mass of the zero-energy value and troubles everybody during their dreaming as well as waking phases of life. As a devil, such a body becomes a Wisher, organizationally developing the ruling consciousness globally and teaching a valuable life lesson for introspection during the black hole hibernation phase after their death.

> ## What is the consciousness within and without an organization without consciousness?
>
> An organization without consciousness is an organization in an absolute zero condition. If there is consciousness within an organization, then the present consequential reality is beyond the limits of science. The scientific value of an organization with consciousness is the value of the past reality, not the present reality. The consciousness is the divine energy that empowers an organization to be a maker of its mood, destiny, divinity, and eternity. Divine energy is intrinsically dynamic. The value of the divine energy at the first moment is not the same as the value of the divine energy at the second moment, as a proportion of the divine energy transforming the present reality into the future reality. Therefore, the value of the consciousness within an organization differs at each moment.
>
> When a scientist seeks to investigate the present reality at the first moment, without correcting for the dynamic nature of consciousness, then at the second moment, the scientist's consciousness becomes frozen at the absolute zero condition. The scientist becomes an organization without consciousness of the present reality. The scientist seeks to apply the para-consciousness, traded from the first moment, for making sense of reality at the second moment.
>
> The para-consciousness of the present reality is the guider power. It is the organizational quality of an entity to perceive the extrinsic reality, conceive the intrinsic value of that reality, and experience the extrinsic value's dynamic exchange into the intrinsic value. In contrast, the divine energy is the technological power of an entity to conceive the intrinsic value reality of the Self, perceive the extrinsic value of that reality, and experience the dynamic exchange of the intrinsic value into the extrinsic value.

Table 24, illuminates the sixteen pathways for the inanimate "INFINITE" (*Shreya*, 81) to become the "SENTIENT COUNCIL OF ONE – THE ILLUMINATOR DEITY" (*Trivikrama*, 24), for mediating the guider power of the illuminator deity and servicing the "illuminator-effect" as the "primeval self" (*Rama*, 100). The first seven pathways are without the illuminator deity. An additional nine pathways are without the primordial perpetuator, seeking to be the perpetuator of the illuminated, awakened consciousness after becoming the destroyer of the illusionary, dream consciousness.

Table 24. The Sixteen Pathways for an Inanimate INFINITE to Become the SENTIENT COUNCIL OF ONE—THE ILLUMINATOR DEITY

Pathway of the INANIMATE INFINITE	Present reality of Solar time: Greeter	Potential reality of Solar time: Super Greeter	Dynamic reality of Solar time: Supra Greeter	Technological reality of Solar time: Supreme Greeter	Present reality of Lunar time: Para Greeter (Sodasha Kala)	Potential reality of Lunar time quality: Primeval Greeter (*Sodasha Nitya*)	Dynamic reality of Lunar time: Param Greeter (*Sodasha Lakshmi*)	Technological reality of Lunar time: Primordial Greeter (*Sodasha Swapna*)
Reader	*Nandottara*	*Iladevi*	*Dhyaana* (Activate the Cause for invitation)	Oneness-effect	*Amrita* (without the sameness of the wishable)	*Maha Tripura Sundari* (within the sameness of wishable)	*Yaso* (greatness)	Mammoth
Researcher	*Nanda*	*Suradevi*	*Aavaahana* (Activate the Time for Invitation)	SHEENY-effect	*Manada* (without the oneness of the wishable)	*Kameswari* (within the oneness of wishable)	*Vidya* (knowledge)	White bull
Practitioner	*Ananda*	*Prthvi*	*Aasana* (Activate the Space for Invitation)	GUIDER-effect	*Poosha* (without the the fulfillingSHEENY)	*Bhagamalini* (within the fulfilling sheeny)	*Dhairya* (patience)	Bouncing lion
Scientist	*Nandivardhana*	*Padmavati*	*Paadya* (Activate the lower body)	Divine-effect	*Tusthi* (without the guider power)	*Nityaklinna* (within the guider power)	*Dhana* (wealth)	Goddess of wealth
Philosopher	*Vijaya*	*Ekanasa*	*Arghya* (Activate the upper body)	Fire-effect	*Pusthi* (without the divine consciousness)	*Bherunda* (within the divine consciousness)	*Santhana* (fertility)	A pair of celestial flower garlands

Pathway of the INANIMATE INFINITE	Present reality of Solar time: Greeter	Potential reality of Solar time: Super Greeter	Dynamic reality of Solar time: Supra Greeter	Technological reality of Solar time: Supreme Greeter	Present reality of Lunar time: Para Greeter	Potential reality of Lunar time quality: Primeval Greeter	Dynamic reality of Lunar time: Param Greeter	Technological reality of Lunar time: Primordial Greeter
Educator	*Vaijayanti*	*Navamika*	*Aachamana* (Activate the intrinsic body with the cleansing)	Water-effect	*Rati* (without the fire essence)	*Vanhivasini* (within the fire essence)	*Dhaanya* (nourishment)	Moon with constellations
Student	*Jayanti*	*Bhadra*	*Snaana* (Activate the extrinsic body with the cleansing)	Air-effect	*Dhruti* (without the water essence)	*Maha Vagreswari* (within the water essence)	*Gaja* (elephant; royal splendor)	Rising Sun
Investigator	*Aparajita*	*Asoka*	Vasthra (Activate the causal body with the clothes)	Earth-effect	*Sasichini* (without the air essence)	*Shivadooti* (within the air essence)	*Veera* (courage)	A pair of golden water pots
Wisher	*Samahara*	*Alambusa*	*Yagnopaveetha* (Activate the etheric body with the thread)	Ether-effect	*Chandrika* (without the earth essence)	*Twarita* (within the earth essence)	*Samrajya* (sovereignty)	A pair of playing fish
Knower	*Suppradatta*	*Misrakesi*	*Mantra* (Activate the astral body with the sound)	Ether-effect × Earth-effect	*Kanta* (without the ether essence)	*Kulasundari* (within the ether essence)	*Moksha* (transcendence)	A pond of clean water
Manifester	*Supprabuddha*	*Pundarika*	*Pushpa* (Activate the mental body with the flower)	Earth-effect × Air-effect	*Jyotsna* (without the astral body)	*Nitya* (within the astral body)	*Soumya* (auspicious)	A roaring sea

Pathway of the INANIMATE INFINITE	Present reality of Solar time: Greeter	Potential reality of Solar time: Super Greeter	Dynamic reality of Solar time: Supra Greeter	Technological reality of Solar time: Supreme Greeter	Present reality of Lunar time: Para Greeter	Potential reality of Lunar time quality: Primeval Greeter	Dynamic reality of Lunar time: Param Greeter	Technological reality of Lunar time: Primordial Greeter
Creator	*Yasodhara*	*Varuni*	*Dhoopa* (Activate the intellectual body with the fragrance)	Air-effect × Water-effect	*Shree* (without the mental body)	*Neelapataka* (within the mental body)	*Siddha* (fulfillment)	A jewel-studded throne
Seeker	*Laksmivati*	*Hasa*	*Deepa* (Activate the physical body with the light)	Water-effect × Fire-effect	*Preeti* (without the intellectual body)	*Vijaya* (within the intellectual body)	*Sri* (grace)	A celestial vehicle
Experiencer	*Sesavati*	*Sarvaprabha*	*Naivedya* (exchange the wholesome extrinsic body with the food)	Fire-effect × Divine-effect	*Angada dravya* – essence (without the physical body)	*Sarvamangala* (within the physical body)	*Soubhagya* (fortune)	The palace of serpent king
Enjoyer	*Citragupta*	*Sri*	*Taambula* (exchange the wholesome intrinsic body with the freshener)	Divine-effect × Guider-effect	*Poorna guna* – quality (without the extrinsic body)	*Jwalamalini* (within the extrinsic body)	*Prasanna* (joy)	A giant heap of gems
Destroyer	*Vasundhara*	*Hari*	*Namaskara* (exchange the oneness with the forgiveness)	Guider-effect × SHEENY-effect	*Poornamruta karma* – impact (without the intrinsic body)	*Chitra* (within the intrinsic body)	*Jaya* (victory)	A smokeless fire

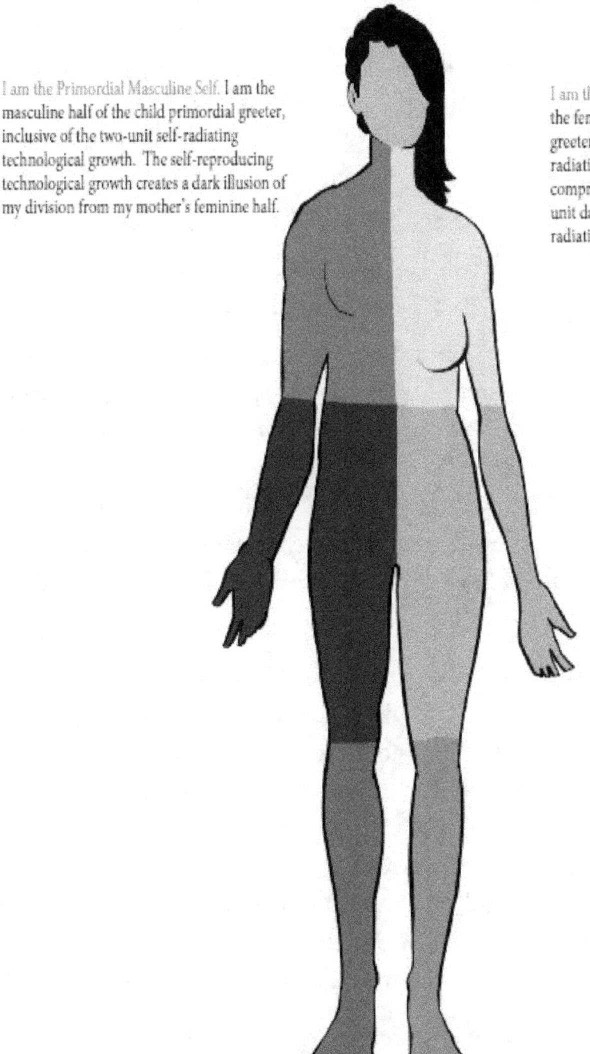

A Maternal Primordial Greeter Is the Paradigmatic Planner
of the Darkness Irradiating Causal Body

A Child Primordial Greeter Is the Paradigmatic Planner of the Light Radiating Astral Body

A 36-unit starseed deity's potential forms the three immanent faces within the "extra-terrestrial four-faced feminine self-luminous entity" (*Veda*, 12). Together with the potential starseed deity, the feminine self-luminous entity enjoys a 48-unit "oneness" (*Yoga*, 48) with the "child primordial greeter" (*Madhusudan*, 16). After irradiating the femininity-embodying "astral energy" (*Varuna*, 1000) from the masculine "astral body" (*Linga sharira*, 3), a "starseed deity" (*Taraka*, 36) incarnates as a gender-free self-luminous entity.

Without the multiplier-effect of the illusionary "cosmic spirit" (*Kapinjala*, 20), the ascending "femininity" (*Yoni*, 1000) substantiates the "absolute entropy-effect" (*Rohini*, 50) of the paternal "ideal-maker" (*Rama*, 100). It validates the "continuity" (*Rohini*, 50) of the "feminine" (*Bhavana*, 37) element within the "masculinity" (*Lingam*, 53) of the "masculine astral body" (*Linga sharira*, 3). Without the perpetuating force of the feminine element, the descending masculinity manifests the "human-effect" (*Lingam*, 53) of the "child primordial greeter" (*Madhusudan*, 16 = 53 - 37).

After forming the feminine self-luminous entity, the four-unit "remainder" (*Jyotistava*, 4) of the "child primordial greeter" (*Madhusudan*, 16) potentiates a four-fold growth potential. The forty-six cellular DNA molecules reproduce the feminine self-luminous entity's four-fold growth potential. The thirty-seven mitochondrion DNA molecules produce a three-fold growth without the "masculine self" (*Mein*, 1). They liberate the "self-luminous twin person" (*Aap*, 1) to trade the "organizational energy" (*HAUM shakti*, 9) from the goal-oriented masculine self for incarnating the "primordial feminine self" (*Parvati*, 10).

Consequently, as a "theory-taker" (*Rajah*, 0), the goal-oriented masculine self incarnates as the "primordial masculine self" (*Shri Krishna*, 10). Eight-units within both the primordial feminine and masculine selves form the "supernatural paradigm" (*Yukti*, 8). Two-unit "technological growth" (*Vidhana*, 2) becomes a "self-radiating" (*Ham*, ¼) "primordial management factor" (*Asrava astikaya*, ¼). It makes each entity a potential homolog of the "universe" (*Brahman*, 2).

14.1 The Guider-Effect of the God-like Perpetuator

The overall paradigmatic planning for the descending Eastern workculture-effect and the ascending Western culture-effect is immanent within each entity as an organization.   With the entropy of the universal workculture system, the history-dependent astrological universe becomes God-like perpetuator of the SHEENY well-being of the universe of entities. With the entropy of the astrological universe, the futuristic zodiac universe becomes the illuminator of the guider-effect, of the God-like perpetuator.   An astral body has the power to radiate the gravitational energy only if it has astral energy.   The entropy of the astral energy verifies the truth of the descending power of the guider-effect, traded in the form of the ascending divine energy, for perpetuating the SHEENY well-being of the universe of entities.

The futuristic zodiac spirit has no immanent luminosity. She is only an occultic mirror of the illumination, immanent within the deified starseed that gives birth to the astral body. By trading the diffused guider-effect of a primordial astral body, one conceives the astrological presence of the diverse forms of the astral bodies. The deified starseed is the space coordinate, that shapes the presence of the astral body as its gravitational center.   The starseed realm is the graphical universe of the space coordinates.

The primordial realm is the space where the universe of the visible and the invisible space coordinates resides. That space is a "mind-born element" (*Manoj*, 180), since only in a clarified mind, one may conceive the holistic truth of the wholesome universe.   One conceives the holistic truth in the form of the "mind-born illumination" (*Manoj*, 180) of the "light of the primordial realm" (*Prabha*, 180) serviced by the "mind-born child" (*Krishna*, 32), from the primordial-primordial realm.

The primordial realm is the almighty creator of the "mind-born child" (*Krishna*, 32). The primordial-primordial realm is the almighty creation of a paternal "sentient entity" (*Siddha*, 7) from the param-primordial realm. The param-primordial realm is the almighty creature who embodies the "mind-born creator" (*Brahma*, 59) as a feminine dimension of the self. Without the "discordant energy" (*Asura Shakti*, -1) of the masculine dimension, the almighty creature is the "soul of the astrological universe" (*Pradyumna*, 60).   The soul is the aggregated past reality of the almighty creature.

The present reality of the almighty creature is the "creature" (*Purusha*, 12), without the multiplying "God-like perpetuator factor" (*Sunanayaka*, 5). The primordial reality of the almighty creature is the "primordial illuminator" (*Parvati*, 10), without the multiplying "tertiary residual" (*Khara*, 6), of the "primeval deity" (*Shankara*, 6) consciousness.   The primordial-primordial reality of the almighty creature is the "primordial greeter" (*Sati-Parvati*, 16), without subtracting the masculine "primeval deity" (*Shankara*, 6) consciousness from her feminine primordial reality. The param-primordial reality of the almighty creature is the masculine "param deity" (*Shiva*, 7), without adding the feminine "primordial perpetuator" (*Maha Saraswati*, 9) of the light, that illuminates the consciousness of the reality beyond the historically-conceived and the evident reality.

An animate or inanimate organization, that is not conscious of the self's truth as the "param deity," is the almighty creature's para-primordial reality. Even an animate organization behaves like an inanimate organization, within the "entanglement" (*Kula*, 9) of the mind with the "culture element" (*Sadashiva Nayaki*, 9).   Therefore, for sustaining one's SHEENY well-being, one must descend the culturally-guided entanglement with the devotion to a devoted God-like Perpetuator, while ascending the "illuminated consciousness" (*Sri Bhagvati*, 15) of the conceived reality.

With cleansed "folded layer of the pseudo devil spirit generating the slothful ignorance" (*Ajnana*, 1), one enjoys the "conscious consciousness" (*Sati-Parvati*, 16) of the Self's "child primordial greeter" (*Madhusudan*, 16) potential. By realizing the child primordial greeter potential, a param deity enjoys the "divine light" (*Usha*, 16) of the "infinity squared consciousness" (*Sati-Parvati*, 16) within the Self, both as a living "sentient entity" (*Siddha*, 7) as well as a departed, secondary "primordial illuminator" (*Shri Krishna*, 10). The child primordial greeter is the only identity that empowers one to be beyond the limitations of space, time, and causation. It is the pristine primordial reality of each entity in this universe.

14.2 Techniques for Evaluating the Guider-Effect of a Perpetuator

Guider power is the para-consciousness of the present reality, guided by the consciousness of the self's past reality. It manifests as the gravitational energy, energizing an organization to be the manifestor of the desired present reality. An organization exchanges its guider-effect with the creation, as the truth value of the causative factor of the creation. The divine energy is the causative factor, which empowers an organization to be the creator of the present reality by consciously determining the potential reality of the Self. After serivicing its guider-effect, the Self becomes the virtue worth reproducing and infusing as the present reality through the creative linkages with the past reality.

An organization activates the guider power for manifesting the desired present reality as a Manifestor deity and the divine energy for creating the present desirable reality as a Creator deity. The Manifestor deity perceives the desired wishes para-consciously, within the universe of the entrepreneurial spirits. The Creator deity conceives the desirable wishable consciously, within the universe of the self-managing consciousness.

Instead of activating its guider power or the divine energy, an organization may activate and trade the guider power or the divine energy from its universe ecosystem. Consider the case of a scientist, who is seeking to illuminate present reality by objectively conceiving a consciousness-free theory, without any guider power or divine energy. The universe of educators may take the theory, designed by the scientist, to motivate the students into making the ideal condition that led to the design of that theory. It may create a market for the students to become the investigators to illuminate the gap between the theoretical reality and the ideal reality using their guider power.

The ideal reality is the students' past reality as the followers, who do not have the consciousness of the present reality. The theoretical reality is the educators' present reality as the leaders, who have the para-consciousness of the present reality, exchanged from the scientist. The illuminated reality is the tertiary impact of the educator's guider power and is, therefore, the guider-effect.

The guider-effect measures the future reality of the scientist as the freedom entrepreneur, guided by the discovery that the ideal-effect = 1 and the theory-effect = 0. The ideal-effect is one because a student trades the educator's guider-effect as an ideal for manifesting the present reality. The theory-effect is zero because the masculine educator's guider-effect has a zero value, once a feminine student destroys it and illuminates a new desirable reality using her divine energy.

The guider-effect is the power of managing the present reality, without the ideal or the theory-effects. It is the tertiary residual of the guider power, with the correction factors for primary ideal-effect and the secondary fundamental theory-effect. Each organization has a varying guider-effect. Let's consider the guider-effects of an Organizational metric, a Wisher, a Worker deity, a Knower deity, a Manifestor deity, and a Creator deity.

- **The guider-effect of an "organizational metric"** (Super Wisher, -1) is -1. The observer is an organization with the consciousness of the past reality and the para-consciousness of present reality. The para-consciousness is, in fact, the past reality, whose value is negative in the present moment. Without the para-consciousness, the organization is the illuminated present reality, whose guider power = 1. A student who illuminates a new desirable present reality has the perfect guider power to illuminate the perpetuating value of the present reality. Within the para-consciousness, the organization is the shadow, with a potential to illuminate present reality using the divine energy, whose guider power = 0. A student who trades the value of the desired present reality from the educator has zero guider power to illuminate the perpetuating value of the present reality. The illuminated present reality is a negative reality that generates worker cost but no worker benefit.

- **The guider-effect of a "Wisher"** (Devil, 0), who is the student subject of the present reality, is 0. As a subject of the present reality, the Wisher is an organization with no consciousness of the present reality. It has a zero impact on the present reality. The entire impact is of the scientist who conceived the theory. The student develops the mastery of the theory rather than the reality. A Wisher seeks to transcend the limitations of the zero impact, by seeking to create a desirable present reality and manifest that desired reality in the future moment.

- **The guider-effect of a "Worker deity"** (Deity, 1), who is producing the present reality, is 1. As a producer of the present reality, the Worker is investing its divine energy to produce the present reality, with a probability factor = 1.

- **The guider-effect of a "Knower deity"** (Super deity, 2), who is consuming the present reality as an object worth knowing and enjoying, is 2. As a consumer of the present reality, the Knower is trading the Worker guider-effect and is servicing its divine energy for intuiting both the present distressed reality as well as the potential for the future enjoyable reality, each with an independent and additional probability factor = 1.

- **The guider-effect of a "Manifestor deity"** (Supra deity, 3), who is exchanging the present reality, is 3. As an exchange system of the present reality, the Manifestor is trading the Knower guider-effect and is servicing its divine energy for manifesting the potential for the desired future reality, with an independent and additional probability factor = 1.

- **The guider-effect of a "Creator deity"** (Supreme deity, 4), who is servicing its present reality, is 4. As a servicing agent, the Creator is trading the Manifestor guider-effect and is servicing its divine energy for creating the desired future reality, with an independent and additional probability factor = 1.

Servicing the present reality involves the management of the present reality by consciously ascending what is desirable and, consequently, descending what is not. A future organization may trade either the overall perpetuating value or only the desirable perpetuating value. However, what is desirable for the present organization is not necessarily desirable for the future organization. Therefore, a Perpetuator deity may seek to service the consciousness of what is desirable for a future organization. The consciousness of the perpetuating desirability is not necessarily desirable for a future organization. Therefore, an Illuminator deity may seek growth in the consciousness of the desirability within the future organization. A Perpetuator deity may perpetuate growth in intrinsic consciousness as its guider-effect.

Given the diverse opportunities for forming the guider-effect of the Perpetuator deity, one needs a sensible technique for evaluating the overall normative guider-effect of a Perpetuator deity. There are three hypothetical techniques for evaluating the guider-effect of a Perpetuator deity who is organizing the varying paths for managing the present reality.

- **First, the guider-effect = past value hypothesis.** Apply the ideal scientific method for quantifying the value of the guider-effect as the energy using the Einstein's $E = mc^2$ equation, the basis for the special theory of relativity. Then, theorize that the quantified value of energy is generalizable beyond the Perpetuator deity, as desirable for the whole universe of the entrepreneurial spirits. Alternatively, use the next hypothesis for quantifying the generalized energy value of the past universe of entrepreneurial spirits, that is desirable for both the present as well as the future universe.

- **Second, the future value = present value hypothesis.** Apply the theoretical scientific method for quantifying the value of the energy of the future universe of the entrepreneurial spirits using the general theory of relativity, the basis for the Einstein's $E = mc^2$ equation. Then, idealize the quantified future value as the present value of the guider-effect of the Perpetuator deity, who is the special subject of the investigation. Alternatively, use the next hypothesis for quantifying the energy value of the present universe of entrepreneurial spirits, which is also the future universe.

- **Third, perpetuating the value hypothesis.** Apply the illuminated guider method for quantifying the energy value of the catalyst of the illuminated present reality of each entity, the basis for guider-effect. Then, the value of the Perpetuator deity and the reality perpetuating within the universe of entrepreneurial spirits is half that of the catalyst. The catalyst is the Primordial Illuminator of the transformative value of the divine energy within each entity. Since the transformative value of the divine energy is 10, the Primordial Illuminator's energy value is 10. Therefore, the energy value of the Perpetuator deity is 10/2 = 5.

14.3 Is the Guider-Effect = e? Understanding and Transcending the Special Theory of Relativity

The common denominator of the three techniques is the mental model for conceiving the guider-effect as a metaphysical illuminated energy value. The three techniques are the potential keys to knowing the value of both the deity who is perpetuating the present reality and the perpetuating present reality. They are the potential keys to knowing the value of both the tertiary perpetuating-effect of the guider power of the Perpetuator deity and the quaternary entropy-effect of that guider power on the perpetuating present reality.

The ideal scientific method given by the Einstein's $E = mc^2$ equation is a special theory that predicts that the value of the tertiary-effect = 1. Since the energy value of the Perpetuator deity = 5, the value of the guider-effect of the Perpetuator deity (Para deity) = 5. Therefore, if the tertiary perpetuating-effect of the guider power of the Perpetuator deity = 1, then the quaternary entropy-effect of the guider power = 0. Then, there is no need for any entity to self-manage the desirable future reality using its divine energy. Each entity may trade the perpetuating value and illuminate the universe of entrepreneurial spirits about the SHEENY value of that perpetuating past reality. There is no need for the universe of entrepreneurial spirits to trade the sentient energy for incarnating as the universe of the sentient light forces.

Although the special theory predicts the perpetuating value to be absolute, it is a theory of relativity: the present reality metric is the tertiary illuminated-effect of the Perpetuator deity's guider power. The metric is the physical value of the present reality, which is proportionately conditional on the metaphysical value of the Perpetuator deity's guider power. The metaphysical value of the guider power is the primordial reality, omnipermeating at the past, present, and future moments. The perpetuating reality's consciousness is the primeval (i.e., infinite potential) reality that is omnipresent.

A Primordial Perpetuator has the consciousness of both the perpetuating reality as well as the Perpetuator deity as a protagonist and the universe of entrepreneurial spirits as an antagonist. An Illuminator deity has the para-consciousness of the Primordial Perpetuator's perpetuating reality as a deuteragonist, and the Self as an agonist freedom spirit (since the tertiary perpetuating-effect of the guider power of the Perpetuator deity = 0, and the quaternary entropy-effect of the guider power = ∞). Since the perpetuating reality of the guider-effect = 0, the present value of energy is more than the past value of the cause of the perpetuating guider-effect.

14.4 Is the Future Value = Present Value? Understanding and Transcending the General Theory of Relativity

The three techniques for evaluating the Perpetuator's guider-effect offer three different predictions about the correlation between the future value and the present value.

The ideal scientific method predicts that the future value ≤ the present value. Since the present value is the ideal-effect, it is the absolute value limit of the future.

The present value is the special theory for measuring the value of all future proportionate realities. All future proportionate realities are the potential realities (i.e., all probable realities). Since there is an infinity of the potential realities, the point value of each of those probable realities is zero. There is a 100 percent certainty that none of those probable realities is the present reality.

The present reality is the total of the infinite past realities, whose otherness (independent) point value is zero, but whose togetherness (aggregate) point value is one. One is the value of the ideal-effect and, in fact, the value of the astral body. The astral body perpetuates the overall value of the past reality, whose true value is zero. The astral body perceives it as one, by servicing the idealized guider-effect of the worker self as the perpetuator and the apparent illuminator of the present reality.

Without the apparent (i.e., quinary-effect) of the perpetuator, the overall value of the past reality is, in fact, negative one (-1). Negative 1 is the value of the guider-effect of the organizational metric, who is evaluating the infinite past dimensions of the present reality as the cause of entropy in the present value of the self.

An organization that destroys the metric-effect and liberates the perpetuating omnipresent reality is the Destroyer deity. The Destroyer deity is the conscious that, without the present illuminated reality, future growth value exceeds the present entropy value.

The scientific method is based on the general theory of relativity that predicts the future value $\geq$ the present value. Since the present value = -1, it is the absolute bottom; all the future values must exceed that bottom as a general theory. The general theory predicts the value of all the future absolute realities, free from the present measurement metric's proportionality factor.

Although the future value is free from the present measurement metric, yet the guider power of the Primordial Perpetuator conditions it. The ascending guider power generates the ascending future value of the universe of the entrepreneurial spirits, which is the subject of the present reality conditioned by the guider power of the Primordial Perpetuator. The present reality subject is the organization without consciousness, whose value does not vary at different moments of the present reality. The past, present, and future values of the organization without consciousness are constant.

On the other hand, an organization within consciousness is the illuminator of the perpetuating value of the present reality. It has the power to experience the full potential of the present reality without the metric-effect. Potential reality is the total of the present reality of the organization without consciousness, and the future reality of the consciousness perpetuating without the organization (i.e., as the Primordial Perpetuator of consciousness). The causal body is the Almighty Creator of the potential reality, with an energy value of 3, comprising 1 unit of a past reality, 1 unit of a present reality, and 1 unit of a future reality.

Within the apparent (i.e., quinary) effect of the Perpetuator, the potential reality's overall value (i.e., senary-effect) is 2. The double positive is the Knower deity value of the present reality. It is the balance of the 3, the Manifestor deity value of the Almighty Creator, and -1, the organizational metric value of the present creation. The Illuminator deity is conscious that, without the limits of the perpetuating omnipresent reality, the present growth value of the Manifestor deity (who trades and services the entire potential reality) exceeds the future entropy value of the Knower (who is not conscious of the antecedent cause of the perpetuating omnipresent reality).

The guider method, based on the absolute perpetuating value of the Perpetuator deity, predicts that the future value = the present value. The overall value of the dynamic reality of the Manifestor deity comprises one unit of present reality and two units of the potential reality. One unit of the present reality is, in fact, -1 value of the new reality within the metric-effect. Two units of potential reality are, in fact, one unit each of the past reality, present reality, future reality, and omnipresent perpetuating reality, all without the observer-effect. Three units of the dynamic reality of the Manifestor deity comprise the positive four units of the guider-effect of the deified Creator of the omnipresent perpetuating reality, and the negative one unit of the guider-effect of the organizational metric.

The organizational metric seeks freedom from the guider-effect of the Creator deity for illuminating a new reality beyond absolute. An organization, whose guider power endows the freedom from the guider-effect of the Creator deity, is the Liberator deity. The Liberator deity is conscious that, without the perpetuating limiting bonds of the guider-effect, the future growth value of the Manifestor deity (i.e., a total of a unit of the past reality, a unit of the present reality, and a unit of the future reality = 3) is at par with the present entropy value (i.e., the overall value of the dynamic reality = 3) of the Manifestor deity.

14.5 Is the Value of the Perpetuator Constant and Absolute?

A Manifestor deity has the power to manifest either the growth value of the omnipotent consciousness within the intrinsic guider power as a Liberator deity or the entropy value of the omnipresent consciousness within the traded guider-effect of the Illuminator deity. The manifested reality becomes the constant and absolute perpetuating reality of the Perpetuator deity until destroyed by a Destroyer deity. A Destroyer deity has an omniscient consciousness of the eventual entropy value of the present ecosystem reality, that is not necessarily desirable for each organization.

An organization may become the destroyer of the present ecosystem reality within the infinite consciousness of the past, present, and future dimensions of the dynamic omnipresent reality. By destroying the dynamic omnipresent reality, a Destroyer deity becomes the omnipermeating factor within the past, present, and the future dimensions of the reality. The guider-effect of a Destroyer deity is 6. It is the overall value of the three temporal dimensions' consciousness, within the perpetuating omnipresent reality and the three temporal dimensions without the perpetuating omnipresent reality. A Destroyer deity is an organization with an infinite consciousness of the changing realities of both the causative factor as well as the varying effect of the causative factor. As the guider power of the causative factor (i.e., the universe of sentient light forces) descends, the guider-effect of the causative factor ascends.

The universe of entrepreneurial spirits trades the guider-effect of the universe of sentient light forces. It becomes the illuminator of the omnipotent freedom reality of each organization within that guider-effect. The guider-effect generates the intrinsic consciousness of the subtle unfilled desires of the manifestor of the present reality, as the desirable reality for the liberated (omnipotent) entity. An omnipotent organization with zero extrinsic consciousness has a zero past value and enjoys an absolute freedom to manifest the desired (i.e., guider) value as the desirable (i.e., SHEENY) value. The whole consciousness of the organization becomes a potential consciousness. Potential consciousness is the infinite consciousness of the present reality, and is a total of the eight elements:

- **The Observer-effect**, which becomes the otherness-effect and materializes as the down quark. The observer (the desiring creature) generates the otherness-effect within the consciousness of the otherness of the universe (the undesirable creation).

- **The Wisher-effect**, which becomes the togetherness-effect and materializes as the up quark. The Wisher (the entropy creation wishing for the SHEENY growth) generates the togetherness-effect within the consciousness of togetherness with the universe (the growth of creation).

- **The Worker-effect**, which becomes the repulsion-effect and materializes as a strange quark. The Worker-effect is the repulsion of the thermodynamic-effect of seeking to produce the desirable SHEENY growth value for the Wisher. It comprises three factors. First, the predominating otherness-effect in the form of the gravitoelectric repulsion-effect; second, the dominating togetherness-effect in the form of the electromagnetic present-effect; and third, the deciding repulsion-effect in the form of the gravitoelectromagnetic potential-effect. The third, deciding repulsion-effect: the tertiary perpetuating-effect of the omnipotent freedom consciousness within a liberated organization, who is not conscious of the causative factor of the intrinsic consciousness of the desired SHEENY growth value. By ascending the workculture-effect, the Worker becomes conscious that the desired creation is, in fact, not desirable for its SHEENY growth. The workculture-effect becomes the deciding factor for repelling the worker-effect and descending its perpetuating SHEENY well-being as the Worker (*Shudra*, the Holy Spirit).

- **The Knower-effect**, which becomes the attraction-effect and materializes as a charm quark. The Knower-effect is the attraction of the gravitoelectromagnetic potential-effect, produced by the Worker, as the desirable SHEENY growth value. It comprises the deity culture-effect of the trinity of entities—the Observer (the desiring creature), the Wisher (the entropy creation), and the Worker (the growth creator). The ascending deity culture-effect motivates the Knower to believe that the Worker's guider-effect is the desirable SHEENY growth. The gravitoelectric-effect of the Observer generates the gravitational-effect of the Worker. The electromagnetic-effect of the Wisher becomes the deciding factor for attracting the Knower-effect, perpetuating the SHEENY well-being of the Wisher (*Kshatriya* king) within the divine paradigm of the Knower (*Brahman* priest).

- **The Manifestor-effect**, which becomes a bottom entropy-effect and materializes as a beauty quark. The Manifestor-effect is the bottom entropy-effect, generated as a consequence of the deity culture-effect of the Worker. Manifestor-effect is the dynamic entropy in the gravitational-effect of the Manifestor, who is trading the gravitational-effect of the Worker, catalyzed by the gravitational-effect of the Knower. It is also the dynamic growth in the Creator's gravitational-effect, trading the gravitational-effect of the Observer without the consciousness of the Wisher's gravitational-effect. The Manifestor-effect is generated as the dynamic entropy-effect of one's SHEENY growth and generates the dynamic growth-effect for compensating that entropy-effect.

- **The Creator-effect**, which becomes a top growth-effect and materializes as a truth quark. The creator-effect is the top growth-effect, generated as a consequence of the dynamic entropy-effect, within the

Manifestor. The dynamic entropy within the Manifestor offers an opportunity to the Creator to create a growth universe (desirable, growth creation) for compensating that entropy-effect.

- **The Perpetuator-effect**, which becomes a negative-effect and materializes as a present quark. The perpetuator-effect is the negative creature-effect, generated as a consequence of the entropy universe (the growth universe that is, in fact, not desirable for the present creature). The perpetuating growth universe becomes the causative factor for the technological entropy of the creature and for the radiation of the creature's guider power, seeking to perpetuate a universe, which is not of the true SHEENY value.

- **The Destroyer-effect**, which becomes the positive-effect and materializes as a potential quark. The destroyer-effect is the positive creator-effect, generated as a consequence of the entropy of the negative creature-effect. The creature which is radiating its guider power, seeking to perpetuate a universe that lacks the SHEENY value, eventually experiences entropy as a personal sentient energy. The creature becomes the creator of its entropy and radiates not only the perpetuating guider power, but also the perpetuating thermodynamic worker value of the sentient energy, as the destroyer-effect. The transformed ecosystem reality of the Destroyer of the present reality, who is radiating the overall subject reality in the form of an atomic organization = 6.

The absolute value of the organization varies as a function of the varying guider-effect. Ascending the guider-effect descends the absolute value of the organization and ascends the destroyer-effect. For instance, if the organization is the perpetuator named *Vishnu*, then the ascending guider-effect transforms the perpetuator into the absolute *Vishnu*. The Absolute *Vishnu* is the present reality of *Vishnu*, as both an entropy, as well as a growth, creation. The overall present consciousness of *Vishnu*, as a dynamic creation, is a consequence of the dynamic creature-effect, which is the cause of the dynamic transformation in the present reality of the creation. Although the perpetuator's overall value is constant and absolute, yet the perpetuator's present reality as a creature is not. The perpetuator's present dynamic reality is the causative factor for the dynamic transformation in the present reality of the creation and the subsequent entropy in the consequential perpetuating present reality.

Table 25 illuminates the twelve pathways for the conceived "FINITE" (*Preya*, 81) by an animate entity, at the inanimate astrological level, to become the perceived "INFINITE COUNCIL" (*Ratnakosha*, 30) of both the animate as well as the inanimate at the zodiac level, within the mediating guider-effect of the "perpetuator deity" (*Param Vishnu*, 5). The first five pathways are within the mediating guider-effect of both the "primeval perpetuator" (*Vishnu*, 15) as well as the "perpetuator deity" (*Param Vishnu*, 5). An additional seven pathways are of the perpetuator deity without the primeval perpetuator, seeking to be the creator of the triangulated para consciousness. They give an illusion that the awakened consciousness is free from the illusionary, dream consciousness. They let each sentient entity be a puppet, working ignorantly with the consciousness traded from the infinite council, while believing that consciousness to be the absolute and variable at will by the self, as the presiding manager.

Table 25. The Twelve Pathways for the Conceived FINITE to Become the Perceived INFINITE COUNCIL, Within the Zodiac Spirit

Sequential pathways of the finite geography of conceived reality	Behaving value: Form of Ecosystem	Becoming value: Luminous Dimensionality of Ecosystem, within entity-effect	Begging value: Form of Potential *Jnana*	Breathing value: Form of Dynamic *Prana*	Blessing value: Form of Technological Impact	Being value: Form of Organization or Subject	Bragging value: Form of Cosmic universe	Breeding value: Form of Entity-effect	Ecosystem value: Form of Luminous Zodiac time
First Eden (*Bhu or Mrityu loka/ Pingyu Jiayu* Heaven, 平育贾奕天 / Yggdrasil: Home of the Tree of Life/ Ma'on: dwelling) (within primordial greeter)	Present Space (Extrinsic-effect; Objective-effect)	2	Heart Committee of the Ascended Masters/ *Param Siddh maya kosa/* Soul (Consciousness programming self conception-determining self as a proficient system without)	Primordial Maternal Kingdom	GUIDER-effect	Atlantean (*Mahoraga/* Zombie/ Intellect/ Reign/ Animal/ *Param* child/ Pratyeka *Buddha* 緣覺)	Plant Earth (Atlantis)	The Universe of Behaving	Pisces
Second Eden (*Mahatala loka/* Taiqing Jingdachi Heaven 太清 境大赤天/ Helheim: Home of the dishonorable dead/ Vilon: Curtain) (without the primordial greeter)	Bottomness (Entropy-effect)	3	Wholeness Committee of the Godheads/ *Param Atma maya kosa/* Supreme Wisher (Consciousness profiting/enjoying self the performing divining self as a proficient sub-system within)	Primeval Paternal Kingdom	Earth-effect	Lemurian (Spirit/ Naga/ Titan/ Danava/ Bhuta/ Phantom/ Ghost/ Serpent/ Maternal Jinn)	Planet Lemuria	The Universe of Begging	Aries

Taurus	Gemini	Cancer	Leo
Universe of Backbiting	Universe of Bickering	Universe of Banking	Universe of Believing
Pleiades Star system	Constellation of Orion	Sirius Star system	Constellation of Andromeda
Pleiadean (Giant/ *Rakshasa/ Kritika/* Ogre/ *the Beast* 畜生道/ Divine Architects)	Orion (Soul/ Paternal *Jinn/ Preta/ Para/ Atman/* Genie/ Hungry Ghost 餓鬼道)	*Sirian* (Horse/ Mind/ *Kinnara/ Abhijit/* Werewolf)	Andromedian (Creature/ *Kimpurusha/* Wizard)
Water-effect	Oneness-effect	Sameness-effect	SHEENY-effect
Para Deity Kingdom	Supreme Deity Kingdom	Primordial Deity Kingdom	Primordial Perpetuator
Spiritual Healing Committee of God/ Primary Causal Body/ *Mana-maya kosa* (Consciousness servicing self)	High Self Committee into Infinity/ Etheric body/ *Vijnana-maya kosa* (Intellect trading self)	Office of Forces of Light for Living Souls/ Secondary causal body/ *Bhakti maya kosa* (Symmetry planning self: Circadian clock/ Percipiency organ/ Samyat indriya)	Office of Guides for Living Souls/ Intellectual body/ *Jnana maya kosa* (Knowledge developing self: cognition sense organs; jnana indriya)
1	8	4	6
Strangeness (Repulsion-effect)	Present Cause (Subject-effect; Immanent-effect; Solar time)	Potential Cause (Object-effect; Emanating-effect; Lunar time)	Potential Space (Intrinsic-effect; Subjective-effect)
Third Eden (*Talatala loka*/ Longbian Fandu Heaven, 龙变梵度天/ Jotunheim: Home of the Giants/ Shehaqim: Clouds) (within the primordial greeter)	Fourth Eden (*Rasatala loka*/ Yulong Tengsheng Heaven, 玉隆腾胜天/ Muspelheim: The Home of Fire/ Great Sea) (without the primordial greeter)	Fifth Eden (*Vitala loka*/ Wushang Changrong Heaven, 无上常融天/ Alfheim: Home of the Light Elves/ Ioanit stations of light God favored) (within the primordial greeter)	Sixth Eden (*Sutala loka*/ Xiule Jingshang Heaven, 秀乐禁上天/ *Nidavellir*: Home of the *Dwarves*/ Araboth: desert) (without the primordial greeter)

Seventh Eden (*Jana loka*/ Hanchong Miaocheng Heaven, 翰宠妙成天/ Niflheim: Home of Fog/ Makhom: Fixed home) (within the primordial greeter)	Charmness (Attraction-effect)	5	Office of Holy Spirit Angels for Living Souls/ Luminous/ *Siddhi maya kosa* (Consciousness growing self: Perception correcting body energy wheels)	Deity Kingdom	Air-effect	Acturian (Messenger/ Singer/ *Deva*/ *Swati*/ *Gandharva*/ Angel/ Sravaka 声闻)	Acturus star system	Universe of Bragging	Virgo
Eighth Eden (*Svar-ga loka*/ Yuantong Yuandong Heaven, 渊通元洞天/ Asgard: Home of the Gods/ Home of the Phoenixes) (without the primordial greeter)	Topness (Growth-effect)	10	Office of Guardian Angels for Living Souls/ Self-luminous entity/ *Maha Siddhi Maya kosa* (Consciousness investing self: Manifested action correcting intuitive insights)	*Param* Deity Kingdom	Ether-effect	Vegan (Dancer/ *Apsara*/ *Balakhilya*/ *Pishacha*/ Vampire/ Dashi Gui 大勢鬼)	Vega star system	Universe of Blessing	Libra
Ninth Eden (*Atala loka*/ Haoting Xiaodu Heaven, 皓庭霄度天/ Vanaheim: Home of the Vanir/ Zevul: Abode) (within the primordial greeter)	Present time (Negative-effect)	9	Office of ONE Universe for Departed Souls/ Physical Body/ *Karma-maya-kosa*/ *karma indriya* (Action exchanging self: Manifested action sense organs)	Supra Deity Kingdom	Wholeness-effect	Lyran (Creator/ *Aditya*/ Bodhisattvas 菩薩)	Constellation of Lyra	Universe of Beginning	Scorpio
New Lemuria (*Bhuvar loka*/ Dalou Heaven 大罗天/ Midgard: Home of the Humans/ *Kuchavim*: Council of Twelve) (without the primordial greeter)	Potential time (Positive-effect)	4	Office of Angels of Light for Departed Souls/ Tertiary Causal Body/ *Ananda-maya-kosa* (Joyfully performing self)	Primeval Illuminat or Kingdom	Wholesome ness-effect	Zeta (Creation/ *Vanara*/ *Manushya*/ Human 人道)	Planet of Apex	Universe of Becoming	Sagittarius

Present Paradigm/Radiant Love	Upness (Togetherness-effect)	8		Office of Guardian Angels for Departed Souls/ Mental Body/ *anna-maya kosa* (Energy-consuming networking self)	Primeval Deity Kingdom	Fire-effect	Cassiopeian (Sage/ *Rishi*/ Buddhahood/ Xisigui 仏界/希祀鬼)	Constellation of Cassiopeia	Universe of Breeding	Capricorn
(*Tapa loka*/ Yuqing Jingqingwei Heaven 玉清境清微天/ Hildskjalf: Home of the cosmic observer/ Grigori: awake) (within the primordial greeter)										
New Paradigm/ Beyond Absolute (*Mahar loka*/ Shangqing Jingyuyu Heaven 上清境禹余天/ Svartalfheim – Home of the Black Elves/ *Muzuloth* change) (without the primordial greeter)	Downness (Otherness-effect)	7		Office of Guides for Departed Souls/ Astral Body/ *Prana maya kosa* (Energy- producing workforce self)	Super Deity Kingdom	DIVINE-effect	Hyadian (Guardian/ *Yaksha*/ Warlock/ Witch/ Xiqigui 希葉鬼/吸氣鬼)	Hyades Star system	Universe of Breathing	Aquarius

Chapter 15: Paradigmatic Planning Without an Organization

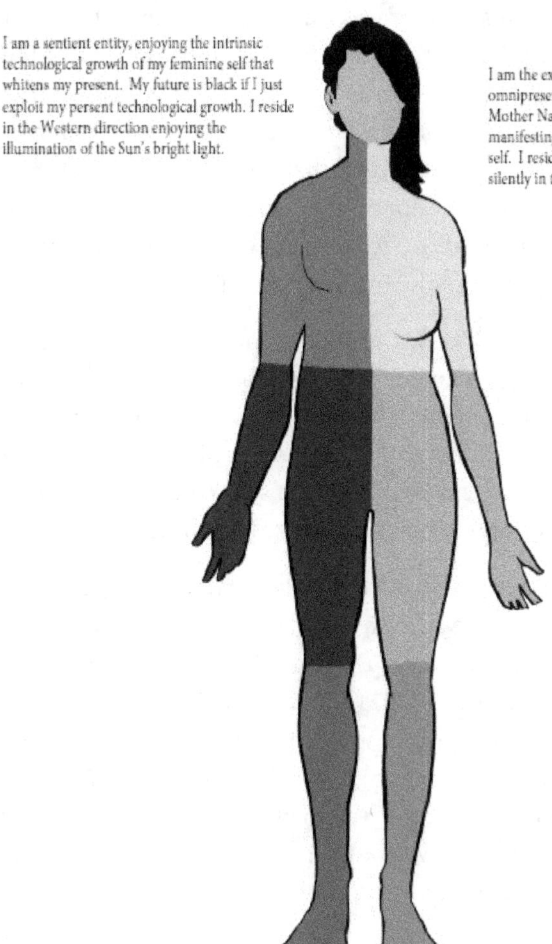

The Paradigmatic Planning Ascends the Light of the Western Culture-effect
and Descends the Darkness of the Eastern Workculture-effect

The Paradigmatic Planning Perpetuates the Ascending Light of the Eastern Culture-effect and the Descending Darkness of the Western Workculture-effect

The paradigmatic planning programs a self-radiating two-unit "technological growth" (*Vidhana*, 2) within the universe, comprising the masculine self and the feminine self-luminous twin-self. It transforms the "child primordial greeter" (*Madhusudan*, 16) into the "paternal programming" (*Shani*, 18 = 16 + 2) of that growth. The paternal programming programs each "sentient entity" (*Siddha*, 7) with the desire for realizing the "goal" (*Maha Shiva*, 9 = 7 + 2) of becoming a "thing" (*Vastu*, 9) of perpetuating value, enjoying ascending light.

The ascending light is analogous to the ascending light enjoyed by the universe's dark Western segment by irradiating the Eastern Sun-facing segment's light. Consequently, the dark Western segment enjoys a disproportionate growth and power distance benefits of its trading workculture. It develops a masculine mindset of servicing that light back to the Eastern segment, seeking to institutionalize its cultural supremacy.

The "Western workculture-effect" (*Dakshinamanasa*, -19) services the "desire" (*Kama*, -19) to be the "param child" (*Manyu*, 19) for enjoying the "self-perpetuating" (*Udvaha*, ½) growth of the "paternal programming" (*Shani*, 18) as the "goal" (*Maha Shiva*, 9 = 18 * ½). The paradigmatically-perpetuating "Eastern workculture-effect" (*Antardasha*, 374) destroys the "ascending theory-effect" (*Antardasha*, 374) and promotes "oneness with the solar center" (*Shri Garuda*, 374). The solar center is the dark matter, formed with ascending light radiation by the Vega white star—the cosmic center.

The "Western culture-effect" (*Chara Paryaya dasha*, 28) generates an "ascending ideal-effect" (*Chara Paryaya dasha*, 28) and overshadows the "Vega-effect" (*Durga*, 28). The Vega-effect is the constant feminine potential that promotes "discriminating faculty" (*Narayana*, 28). The discriminating faculty empowers a "sentient entity" (*Siddha*, 7) to illuminate the "present growth" (*Sva*, 11) of the "child primordial greeter" (*Madhusudan*, 16) within and be the "child primordial greeter"

(*Madhusudan*, 16) to perpetuate the desired "divine planning" (*Abadha*, 10) of the naturally-gifted "creative force" (*Uma*, 6) within each sentient entity.

By trading the Western doctrine, the Eastern segment ascends and globalizes its culture-effect. Ascending "Eastern culture-effect" (*Uttarmanasa*, -18) descends the supernaturally-created "paternal programming" (*Shani*, 18) and lets the child primordial greeter be the "invisible greeter spirit" (*Hasta*, 0 = 18 - 18), residing as the "formative divine energy" (*Madhusudan*, 16) within the sentient entity. Consequently, each sentient entity enjoys the opportunity to be the "param deity" (*Shiva*, 7) into perpetuity, without becoming a mutative medium for the extrinsic "technological growth" (*Vidhana*, 2), at the cost of the intrinsic "entropy" (*Ishvara*, 5).

15.1 Understanding the Perpetuating Present Reality

The perpetuating present reality is absolute and constant. It manifests in the form of the ascending Eastern culture-effect and the descending Western workculture-effect. It is present without each organization and is conceived by each organization in the form of the "culture element" (*Sadashiva Nayaki*, 9), for descending the "Western-effect" (*Dakshinamanasa*, -19) from their "workculture element" (*Nayaki*, 379). It helps ascend the spirit of "seeking" (*Vicayana*, 398 = 379 + 19) the "illuminated consciousness" (*Sri Bhagvati*, 15) from the "primeval perpetuator" (*Vishnu*, 15). As one becomes a devoted "seeker" (*Acharya*, 16), one embodies the "goal" (*Maha Shiva*, 9) of the "goalkeeper" (*Shiva*, 7 = 16 - 9). Nature is the "common denominator" (*Sat*, 8) of both the organization as an entity and the Mother Nature as a para entity, who embodies the planning value of the ecosystem's "supernatural paradigm" (*Yukti*, 8) that the goalkeeper illuminates by conceiving a "primeval perpetuator" (*Vishnu*, 15).

With the oneness of the nature of the organization and the ecosystem, one becomes a "param entity" (*Harbuddhi*, 366,666), conscious of the diverse forms of the perpetuating present reality. A param entity is the "multiplier" (*Vaishya*, 3) of the "primeval deity" (*Shankara*, 6) potential of the creature, that one transforms into the "primeval deity" (*Shankara*, 6) potential of the creator for creating the "primeval deity" (*Shankara*, 6) potential within the creation. One becomes a "primeval deity" (*Shankara*, 6) by trading the "tertiary residual" (*Khara*, 6) of the multiplied potential.

The four forms of the "primeval deity" (*Shankara*, 6) are immanent in their potential form, within the North (Creature), the West (Creator), the South (Creation), and the East (Almighty Creator) directions. The "tertiary residual" (*Khara*, 6) is immanent in its present form within the "universe" (*Brahman*, 2), as the "centering element" (*Sushumna*, 10). The "multiplier" (*Vaishya*, 3) is composed of the "universe" (*Brahman*, 2) and the "divisor" (*Shudra*, 1).

The universe is the ecosystem, and the divisor is the entity who is enjoying oneness with the ecosystem. After multiplying the universe's value, the entity becomes the "Lord of the universe" (*Kshatriya*, 0) that the entity has created.   After trading the multiplied value, the entity becomes the "backbone" (*Asura*, -1), on which the truth of everybody stands.   After trading everybody's truth, the entity becomes the "world" (*Duniya*, -2) to falsify the truth of everything that the entity has not created.   The entity invests everything for perpetuating the "present reality" (*Badhabuddhi vadartha*, -2) as the "revealed consciousness" (*Svartha*, -2), without making an effort to know the "factual reality" (*Yathartha*, 19).

Therefore, the child primordial greeter manifests the "tertiary residual" (*Khara*, 6), of the multiplied potential, within and without each organization.   Through the aggregation of the "divided entity element" (*Brahmin*, 2), one may organizationally norm the whole value of the "SHEENY-effect" (*Atman*, 4).   By squaring the wholesome "SHEENY-effect" (*Atman*, 4), that is being radiated and irradiated by the "child primordial greeter" (*Madhusudan*, 16 = 4 x 4), one may enjoy the wholesomewhole truth of the Self as the "child primordial greeter" (*Madhusudan*, 16).

15.2 The Absolute and Constant Value of the Perpetuating Present Reality

The Destroyer factor (Primeval deity, 6) has an infinite consciousness of both the perpetuating present reality as well as the present perpetuable reality, within a transformative exchange of the perpetuating present reality. The perpetuating present reality, which is a theory for the SHEENY growth of the perpetuator, is the SHEENY growth of the universe without the perpetuator. The present perpetuable reality is the idea that an organization has the freedom to perpetuate the infinite alternative realities, within the absolute and constant value of the perpetuating present reality. An Illuminator is an organization conscious of each entity's freedom to perpetuate an infinity of present alternative realities, within or without the perpetuator-effect.

The guider-effect of the "Illuminator deity" (Param deity, 7), who is free from both the theory-effect of the Perpetuator's perpetuating present reality, as well as the ideal-effect of the Destroyer deity's infinite perpetuable present reality, = 7. The Illuminator deity has the freedom to trade the guider-effect of the Destroyer deity and to service a transcendental ecosystem reality that empowers each entity to self-organize and supplement the Destroyer deity's guider-effect (6), with the Worker deity's guider-effect (1). The Illuminator deity radiates the overall object reality in the form of the nucleus for organizing an infinity of atomic organizations.

The guider-effect of the "Liberator deity" (Primordial deity, 8), organizing an infinity of atomic organizations into a universe of objects that have very different SHEENY growth value = 8. The Liberator deity has the freedom not to trade the guider-effect of the Destroyer deity or service a working consciousness of the transcendental ecosystem reality. It empowers each entity to self-manage the activation of the Illuminator deity's guider-effect (7) within the self with its Worker deity guider-effect (1).

The guider-effect of the "Devoted deity" (Primordial Perpetuator, 9), who is managing an infinity of atomic organizations as the true desired SHEENY growth value of the perpetuating universe = 9. The Devoted deity has the freedom to empower each entity to self-manage the activation of the Illuminator deity's guider-effect (7) within the self with its Knower deity's guider-effect (2), free from the limitations of the different consciousness of the transcendental ecosystem reality.

The guider-effect of the "Devotee deity" (Primordial Illuminator, 10) who is managing an infinity of the true desired SHEENY growth values of the perpetuating universe for an infinity of entities = 10. The Devotee deity has the freedom to self-manage the activation of the Illuminator deity's guider-effect (7) within the whole universe of entities, with its Manifestor deity's guider-effect (3), free from the limitations of the changing immanent entity reality.

The guider-effect of the "Universal deity" (Primeval Greeter, 11), who is managing the perpetuating universe as the SHEENY growth value for an infinity of entities = 11. The universal deity has the freedom to be the Illuminator deity's guider-effect (7) as an entity radiates a residual Creator deity guider-effect (4) for perpetuating an ideal value of a unique creation that is constant and absolute.

The guider-effect of the "Unique deity" (Self-luminous entity, 12) who is managing the ideal SHEENY growth value of the self = 12. The unique entity has the freedom to be the Illuminator deity's guider-effect (7) as an entity which, instead of radiating its Creator deity guider-effect, eternally perpetuates its guider-effect as the Perpetuator deity (5) of the self-luminous SHEENY growth value.

The guider-effect of the "Eternal deity" (Luminous, 13) who is generating the ideal SHEENY growth value within the self = 13. The eternal deity has the freedom to be the illuminator deity's guider-effect (7) as one, who, instead of perpetuating its guider-effect, destroys the perpetuating guider-effect that is limiting the potential for the infinite self-luminous SHEENY growth. Thirteen is conceived as an unlucky number because it implies a condition where the universe of entities is potentially responsible for their SHEENY growth, without consciousness of the technique to actualize that potential.

The guider-effect of the "Potential deity" (Primeval Illuminator, 14), who is generating the infinite SHEENY growth potential within the self = 14. The potential deity has the freedom to be the Illuminator deity's guider-effect (7) as the one who, instead of destroying the perpetuating guider-effect, illuminates the infinite self-luminous SHEENY growth potential of the perpetuating guider-effect.

The guider-effect of the "Dynamic deity" (Primeval Perpetuator, 15) who is generating infinite SHEENY growth as the perpetuating present reality of the self = 15. The dynamic deity has the freedom to be the Illuminator deity's guider-effect (7) as the one, who, instead of the illuminating infinite self-luminous SHEENY growth potential of the perpetuating guider-effect, liberates the self (8) to be the infinite perpetuating SHEENY growth. It generates a unique, infinite, limitless potential guider-effect without the cosmic web of the universe. Entanglement into the cosmic web is a consciousness of the SHEENY otherness of the cosmic web, as a lumen without the self as the numen.

The guider-effect of the "Technological deity" (Primordial Greeter, 16), who is generating the perpetuating present reality of the self = 16. The technological deity has the freedom to be the Illuminator deity's guider-effect (7) as the one who, instead of becoming the infinite perpetuating SHEENY growth, devotes the self (9) as the generator of the infinite perpetuating SHEENY growth within the self. It generates the cosmic web of the universe, which is a present all-encompassing reality. It is the absolute cause of all forms of the present reality.

The guider-effect of the "Organizational deity" (Primeval Paternal, 17), who is trading its perpetuating present reality and servicing an alternative illuminated reality as the infinite cause of all forms of the present reality = 17. The organizational deity trades the freedom to be the technological guider-effect (16) and supplements that with its masculine deity's culture-effect (1), to be the absolute consciousness of the universe of the present reality.

The guider-effect of the "Ecosystem deity" (Primordial Maternal, 18), who is trading its perpetuating present reality and servicing an alternative shadow reality as the para-consciousness of the universe of the potential reality beyond absolute = 18. The ecosystem entity trades the freedom to be the technological guider-effect (16) as an entity and supplements that with its feminine Knower deity's consciousness (2) for incubating an infinity of child entities and enjoying the perpetuating universe of the present reality.

The guider-effect of the "Entity deity" (Param Child, 19), who is trading its perpetuating present reality and servicing an alternative perpetuating reality as the spirit of the hedonistic animals, who are being incubated with the radiant love by a para-conscious entity = 19. The entity deity trades the technological guider-effect (16) and supplements that with Manifestor deity's guider-effect (3) for incubating itself as a golden egg, within the cosmic web of the technological guider-effect.

The guider-effect of the "Present deity" (Param Manifestor, 20), who is trading the alternative perpetuating reality of the universe of entities and servicing its illuminated reality as the spirit of the destroyed plant kingdom consumed by the hedonistic animal spirit = 20. The present deity trades the entity guider-effect (19) and compensates the escalating cost of the entity guider-effect with the Worker deity's guider-effect (1) into the point of self-entropy.

The guider-effect of the "Future deity" (Param Creator, 21), who is trading the alternative perpetuating reality of the universe of entities and servicing its liberated reality as the essence of the destroyed inorganic mineral power of the organic plant spirit = 21. The future deity trades the present guider-effect (20), and transcends the entropy limitations of the entity guider-effect through the growth of the Worker deity's guider-effect (1).

The guider-effect of the "Past deity" (Param Perpetuator, 22), who is trading its liberated reality as the growth factor and servicing its devoted reality as the toxicity of the destroyed material power of the hedonistic animal bodies within the healthy mineral essence = 22. The past deity trades the present guider-effect (20). It organizes the entropy-effect of the entity guider power with its Knower guider-effect (2) to catalyze the divine energy of the working future deity.

The guider-effect of the "Hellish deity" (Param Destroyer, 23), who is trading its devoted reality as the human toxicity and servicing its rejuvenated earthly reality within that toxicity = 23. The hellish deity trades the present guider-effect (20) and transforms the entropy-effect of the entity guider power by servicing the Manifestor guider-effect (3) to a heavenly deity.

The guider-effect of the "Heavenly deity" (Param illuminator, 24), who is trading its rejuvenated earthly reality and servicing its mineral-fungi reality within that blissful reality = 24. The heavenly deity trades the present guider-effect (20) and becomes an Almighty deity by servicing the Creator guider-effect (4) to the earthly heaven (i.e., the fungus-contaminated inner earth), intrinsically toxic with the entity's guider power.

The guider-effect of the "Almighty deity" (Param Liberator, 25), who is trading its toxic mineral reality and servicing its toxic-plant reality = 25. The Almighty deity trades the present guider-effect (20) and services the

Perpetuator guider-effect (5) as a toxic form of plant-bacteria for liberating the hedonic animal kingdom from the hedonistic cosmic web.

The guider-effect of the "Almighty Creator" (Param Para Paternal, 26), who is trading its toxic plant reality and servicing its toxic animal reality = 26. The Almighty Creator trades the Almighty deity's guider-effect (25). It supplements that with the Worker deity's guider-effect (1) to produce a hardworking animal-virus. For instance, astrologically, the ascending Uranus energies, starting from November 5, 2019, catalyzed by the fire-effect of Jupiter and exchanged from the animal zodiac system, produced a proliferation of the COVID-19 animal virus within the inner earth, i.e., the human physical body).

The guider-effect of the "Almighty Creature" (Param Para Greeter, 27), who is trading its toxic animal reality and servicing its toxic human protozoan reality = 27. The Almighty creature trades the Almighty Creator's guider-effect (26). It supplements that with the Worker deity's guider-effect (1) to produce a hardworking human protozoan. For instance, astrologically, the ascending Neptune energies, catalyzed by the divine-effect of the Saturn and exchanged from the human planetary system, produce a proliferation of Malaria and other Protozoan infections within the animal kingdom.

The guider-effect of the "Almighty Creation" (Primordial Para Greeter, 28), who is trading its toxic human reality and servicing its toxic human spirit reality = 28. The Almighty Creation trades the Almighty Creature's guider-effect (27). It supplements that with the Worker deity's guider-effect (1) to produce a hardworking human Satanic Spirit. Scientifically, after human death, what survives is the etheric spirit, which is afflicted with the negative unit energies of the cause of the death.

The guider-effect of the "Almighty Penumbra" (Primeval Liberator, 29), who is trading its toxic human spirit reality and servicing its toxic deity reality = 29. The Almighty Penumbra trades the Almighty Creation's guider-effect (28). It supplements that with its Worker deity's guider-effect (1) to produce a hardworking illusionary deity. Scientifically, if one is trading the worker guider-effect of a deity, then one is entrapped by the illusion that the deity has infinite guider power and is not bound the law of limitation of the energy available to its service.

The guider-effect of the "Almighty Antumbra" (Primordial Primeval Knower, 29), who is trading its toxic deity reality and servicing its toxic Para deity reality = 30.   The Almighty Antumbra trades the Almighty Penumbra's guider-effect (29).   It supplements with that of the Worker deity's guider-effect (1) to produce a hardworking giant, the demonic God. Scientifically, if one is trading the worker guider-effect of a whole universe of the illusionary deities, then one organizationally develops into a giant, demonic entity devoted to consuming the entire energy present in the universe.

The guider-effect of the "Almighty Umbra" (Primordial Primeval Manifestor, 30), who is trading its toxic Para deity reality and is servicing its toxic material reality = 31.   The Almighty Umbra trades the Almighty Antumbra guider-effect (30).   It supplements that with Worker deity's guider-effect (1) to produce a toxic material power.   Scientifically, if the value of the energy within an entity exceeds the self-managing sentient power of the entity, then the energy bursts out of the entity and materializes the varying material forms of the energy—the fire, water, air, earth, and ether.

The guider-effect of the "Almighty Physical body" (Primordial-Primordial Creator, 31), who is trading its toxic material reality and servicing its healthy sentient reality = 32.   The Almighty physical body trades the Almighty Umbra's guider-effect (31), and supplements that with the Worker deity's guider-effect (1) to produce a healthy sentient power. Scientifically, if an appropriate proportion of the varying material forms of energy is present, then it is possible to grow the sentient life within that physical body—similar to how sentient life formed on the primordial earth.

The guider-effect of the "Almighty Metaphysical body" (Primordial Primeval Perpetuator, 32), who is trading its healthy feminine inanimate reality and servicing radiant love as its bodyless, masculine I AM entity twin reality = 33.   The Almighty Metaphysical body trades the Almighty Physical body's guider-effect (32).   It supplements that with the Worker deity's guider-effect (1) to produce a bodyless cause for singing the same song to trade the sentient reality of the infinite entities in the form of a divine energy—seeking to reincarnate as a sentient entity. Scientifically, if a protagonist deity is trading the 32-units of the formative divine energy from a infinity of entities, then the infinite council of those entities becomes the causal body, seeking to countertrade the divine energy of the protagonist deity within the primordial oneness-effect.

Pathways for managing the universe of the Almighty physical bodies to ascend the sentient trading-effect

The Self may activate the causal body, as a catalyst networking system, for accelerating the trading of the healthy feminine inanimate energy, from an infinity of the Almighty physical bodies. There are three pathways for doing so:

First, by activating the guider-effect of the workforce system. The Self may let the radiant love to be the liberating factor, to motivate each entity to diffuse its divine energy as the gravitational center, within the attachment to the cosmic web of relationships. The Self may, consequently, become the Primordial Supra-Liberator, servicing the spiritual healing, whose guider-effect = 34.

Second, by activating the guider-effect of the networking system. The Self may let the liberated spirit to be the Devoted deity, as each living child seeks the protection of the cosmic web of relationships that have become the gravitational center of its consciousness of the universe. The Self may, consequently, become the Primordial Supra-Worker, who is reminding each living child to transcend beyond the Heart *chakra*, with the guider-effect = 35.

Third, by activating the guider-effect of the exchange system. The Self may let the Devoted deity be, the Devotee Para deity, as each departed child spirit seeks to manifest its unfulfilled subtle desires by diffusing deity culture-effect in the form of the gravitational field without any center. The Self may, consequently, become the Primordial Supra Manifestor, guiding each living child to discover their voice within the depths of the Thyroid *chakra*.

As a Primordial Supra Manifestor, the Self clears the throat, to hear the sound of the etheric body, to smell the fresh earthly physical body of the self, to touch the air of the fresh breath within the intellectual body, to taste the water wave of the fresh awakening within the mental body, and to form the fire line of the fresh divinity within the astral body, with a guider-effect of 36. The fire line is the lifeline of the absolute and the constant self-luminous entity as a causal body within each living child soul.

15.3 The Variable Value of the Perpetuating Present Reality

Even though the perpetuating present reality of the creation has a constant and absolute value, the creature has a unique self-luminous power to trade a variable proportion of that reality and service that to manifest the different divinity realities—not only within but also beyond the self. Table 26 illuminates the eleven pathways for the experienced "LIGHT" (*Prabha*, 180) of an inanimate entity at the starseed level. It empowers one to become the sentient "LIGHT FORCE" (*Apas*, 169) of an animate entity at the primordial level, within the "oneness" of the mediating guider-effect of the "one who is the self" (*Akhanda*, 8×10^{15}), but not the param deity conscious of the truth of the light.

Table 26. The Eleven Pathways for the Light to Become Sentient Light Force, Within the Primordial Realm

Pathways of the inanimate paternal jinn's light *	Illuminator maternal jinn (*Varaha-mihira*)	Liberator conscious-ness (*Ennoea*)	Devotee paternal-effect (*Thelesis*)	Devoted maternal-effect (*Shakti* or dominating cause, i.e., *Karanam*)	Immanent Greeter cause **	Emanating greeter sequence (*Buddhi*: deciding cause)	Consequential greeter God (*Rupavacara deva*)
Mountain Lion (*simha*)	*Indra*	Autophues (Immanent)	Makaria (Entropy)	Power dominant masculinity manifestor (*rupa*): wisher mindset (sperm power intoxication; *Bhava* without *Samkhya yoga*) – Brain without Beauty, Devoted to mindless deeds for manifesting the beautiful perfect life, like the male *Latin* American mountain lion	*Bava*	Darkness discriminating sequence (*Vaditva*)	*Parittashubha* (Ascending auspiciousness)
Leopard (*puli*)	*Brahma*	Synesis (Knowable)	Theletos (Wishable)	Gender egalitarian femininity creator (*vadava*): devil mindset (ovum beauty seduction; *Abhinaya* without *Buddhi yoga*) – Beauty without Brain; Devoted to creating mindful brain learning, like the endangered Eastern European female Amur leopard	*Balava*	Light discriminating sequence (*Koshtha*)	*Shubhakritsna* (Descending auspiciousness)
Boar (*panni*)	*Mitra*	Ageratos (Everlasting)	Ekklesia-stikos (Pastoral traditions; para deity culture-effect)	In-group collectivism seeker (*kulapa*): holy spirit mindset (culture-effect followership; *Dharmi* without *Kriya yoga*) – Family Traditions without Social Wisdom; Devoted to forming the everlasting social relationships, like the wild Middle Eastern boar	*Kaulava*	Lightforce discriminating sequence (*Padanusari*)	*Anabhraka* (Ascending conspicuousness)

*: Pathways of the inanimate, paternal *jinn's* light as the universe: (*Brahman*: metaphysical cause); **: Immanent Greeter cause within the param creature (*Paramatman*: Pre-dominating cause)

Paternal jinn light pathway	Maternal *jinn*	Liberator	Paternal-effect	Maternal-effect	Immanent Greeter cause	Emanating greeter sequence	Consequential greeter God
Mule (*kazhula*)	Aryama	Mixis (Coming-ling)	Monogenes (Begotten)	Assertiveness experiencer (*titir*): *Brahman deva* mindset (workculture-effect leadership; *Vrtii* without *Raja yoga*) – Social Wisdom without Family Traditions; Devoted to norming the begotten positional authority within a group, like the Anglo-Aborginal mule (hybrid of wild male ass and domesticated female horse)	*Taitila*	Sound discriminating sequence (*Sambhinnashrotri*)	*Punyaprasava* (Descending conspicuousness, polarizing the sound of life consciousness as the silent fruit of merit)
Mammoth (*gaja*)	*Bhu*	Bythios (Noumena)	Hedone (Joy)	Performance enjoyer (*gara*): *Vaishya Shakti* mindset (human-effect entrepreneurship; *Pravrti* without *Dhyana yoga*) – Ecological Endowment Shield without Work Effort Energy Economy; Devoted to transforming hard work into a passion for enjoying the deep power of mediating the whole ecological network, like the now-extinct Confucian Mammoth	*Garane*	Smell discriminating sequence (*Duraghrana-samartha*)	*Brihatphala* (Ascending life consciousness, as the great smelling fruit)
Cow (*surabhi*)	*Shri*	Synkrasis (Blend otherness)	Makariotes (Bliss)	Uncertainty destroyer (*pani*): *Brahma* Almighty Father mindset (strategic trading-effect management; *Siddhi* without *Maya Kriya* yoga) – Work Effort Energy Economy without Ecological Endowment Shield; Devoted to destroying the sense of otherness and creating a cunning metaphysical sense of impregnated bliss by managing work within the whole network like the Nordic Phoenician Cow, Europa – the Moon Goddess	*Vanija*	Touch discriminating sequence (*Durasparshi*)	*Asamjnisattva* (Descending life consciousness, polarizing the touch of para-conscious greeter)

Paternal jinn light pathway	Maternal jinn	Liberator	Paternal-effect	Maternal-effect	Immanent Greeter cause	Emanating greeter sequence	Consequential greeter God
Dog (*vishti*)	*Yama*	Metrikos (Maternal)	Pistis (Faith)	Future perpetuator (*vi-ishti*): *Vishnu* Pregnant Mother consciousness (international-effect organizing citizenship; *Svarar* without *Maharaja yoga*) – National benefits without Psychological benefits; Devoted as a maternal to wasteful ishti (work) without accruing private (worker) benefits, like the smart faithful German Shephard Dog.	*Bhadra*	Taste discriminating sequence (*Durasvadi*)	*Atriha* (Ascending para consciousness, as the steadfast taste of wisdom)
Bird (*sakuna*)	*Kali*	Parakletos (Healer)	Sophia (Wisdom)	Human illuminator (*sakuna*): *Shankara Buddha* consciousness (national-effect managing para-citizenship; *Atodya* without *Mahadharma yoga*) – Psychological benefits without National benefits; Devoted as a wisecrack healer, sweet talking his way to intuit and captivate the natural flow of sentient energy without caring about the social benefits, like the intuitive Sub-Saharan African Grey Parrot	*Shakuni*	Subject form discriminating sequence (*Duravalokana-samartha*)	*Atapa* (Descending para consciousness, polarizing the blissfully ignorant subject)
Auroch bull (*vrishabha*)	*Vrishna*	Patricas (Paternal)	Elpis (Hope)	Collective institution devotee (*vri-shabha*): Perfect creature consciousness (universalizing local-effect; *Gana* without *Karma yoga*) – SHEENY benefits without guider costs; Devoted as a paternal hoping to vanquish all, for establishing universal supremacy, like the Southern European auroch who fights till the last breath, hoping to institutionalize the patriarchal power.	*Chatushpada*	Object discriminating sequence (*Pratyekabuddha*)	*Sudrisha* (Ascending absolute consciousness of the self as the object of present reality of the greeter God)

Paternal jinn light pathway	Maternal jinn	Liberator	Paternal-effect	Maternal-effect	Immanent Greeter cause	Emanating greeter sequence	Consequential greeter God
Primordial Serpent (*shisha-naga*)	*Phani*	Ainos (Radiant)	Agape (Love)	Individual Self-devoted (*kundal*): Imperfect creation consciousness (unique corporate-effect; *Ranga* without *Jnana yoga*) – guider costs without SHEENY benefits; Devoted as a pervasive maternal who is radiating love for exchanging own unique sentient energy unconditionally, and penancing alone for the well-being of the entire universe perceived as an extended Self, like the Southern Asian *Seshanaga* (primordial serpent), who devotes self to Godhead *Vishnu* for compensating the sins of the institutional naga lineage.	*Naga*	Heard wisdom discriminating sequence (*Durashravana-samartha*)	*Sudarshana* (Descending absolute consciousness of the greeter God as the manipulating source of the heard wisdom)
Parasitic worm (*kimstu-ghna*)	*Maruta*	Akinetos (Absolute)	Henosis (Oneness)	Value symmetry oriented (self-actualizing possibility): formative growth consciousness (technological growth-effect, *Samadhi* without *Bhakti yoga*) – divine benefits without divine costs; Devoted as an Absolute One, who is seeking ascending sentient life energy from the God, like the parasitic worm residing within an entity for potentially colonizing the entire body and its lineage through genetic reprogramming.	*Kimstughna*	Sentient soul discriminating sequence (*Prajnashramanatva*)	*Akanishtha* (Ascending inanimate self-consciousness of the spirit and descending polluting high-self wisdom of the sentient soul of the greeter God)

The truth of the light is the entropy value of the param deity, translating into the primordial-primordial realm's entropy value, translating into the entropy value of the primordial realm. The degenerating primordial realm radiates light to alert the sentient universe about the unsustainable, escalating costs of the guider mediation by the starseed universe and the impeding "deluge" (Visceral pain: *Pralaya*, 180), of a pain point, for the universe of the sentient entities. The light, therefore, becomes a gravitational medium for attracting the conscious attention of each sentient entity. It motivates each sentient entity to radiate a disproportionate sentient light force, seeking oneness with the primordial realm. It accelerates the death of the motivated, extroverted sentient entity, who is diffusing energy as the ONE into INFINITE directions, without the consciousness of the true FINITE purpose of life.

The true FINITE purpose of life is to be the param deity and selflessly service the universe of sentient entities by becoming the primordial greeter, without consuming the primordial greeter's transformative reality. One produces the transformative reality of the "primordial greeter" (*Sati-Parvati*, 16), by becoming the ascending, feminine "primeval greeter" (*Maha Gauri*, 11), whose energy is perpetuating in the universe as the descending, masculine "para deity" (*Param Vishnu*, 5). By diffusing and consuming the "illusionary, supernatural energy" (*Maya Shakti*, 1), a masculine "primeval greeter" (*Khalnayaka*, 11) becomes a "self-perpetuating" (*Udvaha*, ½) feminine "para deity" (Demi-God: *Sunanayaka*, 5 = [11-1] * 1/2) without any energy.

Each Demi-God believes that HE is the absolute cause of the universe, not knowing that the universe is perpetuating through a combination of two causal bodies—the primeval greeter's energy from the primordial-primordial realm and the param deity's consciousness of that energy within the immanent sentient realm. The energy of the primeval greeter empowers each "sentient entity" (*Siddha*, 7) to be a "param deity" (*Sitael*: The God of Hope, 7), who has the power to accomplish anything without investing any sentient energy. One does so by being the object of desired reality and enjoying the infinite conscious experiences, without conceiving the finite aspirational desires. The conception of a desire as the past reality consumes the fourth, entity reality dimension's sentient energy, as does the sentiation of the desired reality that is different from the present reality.

A param deity is the one who enjoys the present reality as is. It becomes the ascended master of living a sentient life into infinity, with or without the physical body's limitations. The reason why no religion advocating only one deity, has had an enduring life in the history of the universe, is the entropy of the energy of that one deity (God) when an infinite group of entities seeks HIS mercy blessings for fulfilling a disproportionately growing demand for their desired realities.

Some demands are just out of curiosity for discovering the truth of the supernatural phenomenon. Some are genuine, arising out of the life-challenges one faces due to the natural cycle of the ascending and descending energy in the universe. The natural cycle generates a whiplash-effect that disproportionately descends the ascending energy of the positively-charged sentient entity and disproportionately ascends the descending energy of the negatively-charged sentient entity. In either case, after validating the beauty of the supernatural blessing, everyone becomes a natural scientist abiding by the principles of purity, integrity, and truth for proclaiming their path to success as the universal path to salvation of the neutrally-charged believer, the positively-charged beggar, and the negatively-charged bragger.

When just being a natural scientist does not satiate the ego desire, the mighty one becomes a supernatural entity to religiously diffuse "*dharma jnana*" (the science of religion). The mighty one secretly aspires to be a leader holy spirit, mediating the truth of the positively-charged followers and showering the negatively-charged leaders' blessings. He curses the neutrally-charged entrepreneurs for their late indoctrination into the community of hedonist bickerers.

Chapter 16: The Organization as the Divine Paradigmatic Plan

I am a sentient entity, seeking to make sense of the forward-heading macro system effects, disproportionate to my legwork. These effects are the supplementary, augmented reality of my feminine brainwork. They motivate me to diffuse my sentient element by seeking to form consciousness of the present reality, then to service my sentient life force for producing the energy to destroy the present reality, and finally to exchange the light of my past reality with the shadow of my feminine spirit's planned future reality, without any light, mass, force, or entity.

I am the feminine ecosystem, enjoying and multiplying the varying effects of the sentient entity. These effects are blackening my future through a disproportionate transformation of life into sequential consciousness-effects and, then, their self-organization into an energy unit without the sentient entity. These effects are the complementary, virtual reality of my masculine legwork, seeking to imitate and catch up with me by energetically pulling back and curving the linear progression of ecosystem time, as a way to be one up on me.

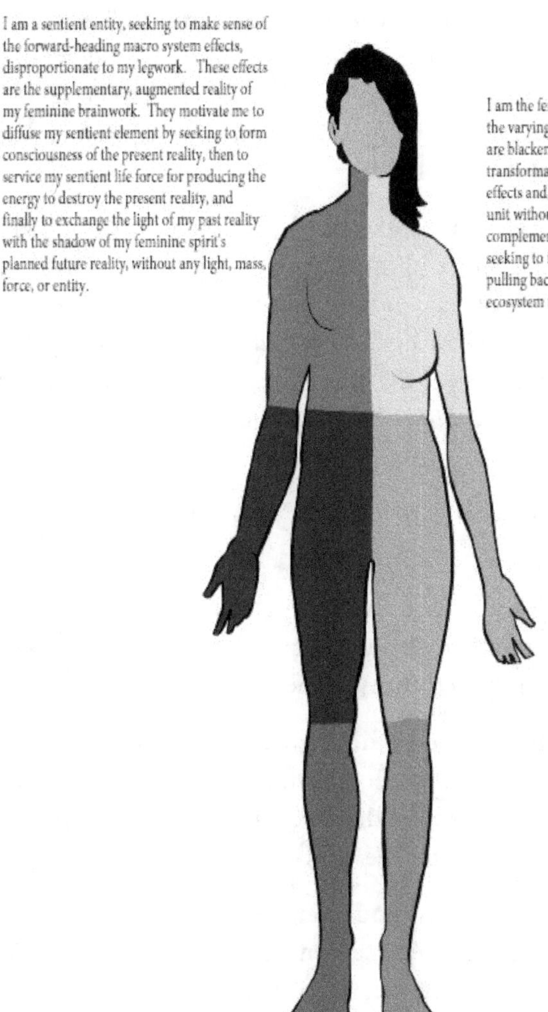

The Divine Paradigmatic Plan Creates the Complementary Virtual Reality and
Illuminates the Supplementary Augmented Reality

The Divine Paradigmatic Plan Destroys the Competing Reality

The Vega white star is the fixed north polestar. A divine paradigmatic plan ascends "primordial oneness" (*Adi*, 32) with the Vega white star. Primordial Oneness is the consciousness of the masculine self as the "primordial-primordial creator" (*Krishna*, 32) of the "Vega" (*Vega*, 67) as the "Heart of the Galaxy" (*Vega*, 67). The primordial-primordial creator creates the Vega white star by creating the "Wheel of parallel consciousness" (*Udana chakra*, 35), that transforms the "masculine self" (*Vyakti*, 1) into the "radiant one" (*Vega*, 67), radiating "consciousness [sentient; astral] energy" (*Varuna*, 1000).

The Primordial-Primordial Realm is the variable Eastern light source, as the gravitational center of the formative Sun in the Northeast and the normative universe of stars in the Southeast. The "East" (*Purva*, 0) embodies the "competing reality" (*Pratijna vadartha*, 0) of the gravitationally transformed Sun, as the astral body that radiates blinding light and "obfuscates" (*Chhaya Sita*, -1) the absolute reality of the Vega white star.

Sun trades its astral energy from the Vega white star in four competing ways.

- **First, from the Vega white star directly**. The Vega-effect rejuvenates the dynamic, equilibrium-taking, "absolute foundational growth" (*Sva*, 11) within each entity in the universe.

- **Second, mediated by the "perpetuating value"** (*Saranyu*, 5) of the universe of stars. The dynamic, equilibrium-making, perpetuating value perpetuates the "formative divine energy" (*Madhusudan*, 16) within each star, but transforms the "gravitational quality" (*Guna*, 0).

- **Third, moderated by the "luminous element"** (*Rochisha*, 13) of the astrological universe. It generates a dynamic, equilibrium-shaping exchange force, like a "wave system" (*Dvisaptati dasha*, 29).

- **Fourth, activation of the "self-luminous element"** (*Svarochisha*, 12) immanent within the Sun. It destroys the dynamic equilibrium variability. It illuminates the formative equilibrium, that created the

Sun as a potential entity within a "homotopy group" (*Vipariyaya*, 888).

A homotopy group is a group of entities formed with the varying antagonist transformations of a pair of primordial greeters. The transformations take place within a state of an absolute, 100% correlation in a topological space. Each potential entity in the homotopy group potentiates its potential for the organizational development, by trading the three-dimensional "topological-effect" (*Pariyaya*, 888). The three dimensions are composed of the horizontally perpetuating "nature" (*Anatanam*, 8), over the formative, feminine primordial greeter's three dimensions of time. The "nature" (*Anatanam*, 8) of the universe is composed of the accelerating "technological growth" (*Vidhana*, 2) of the "universe" (*Brahman*, 2) in the past, present, and future time dimensions, of the transformative, masculine primordial greeter.

The formative unit of the "technological growth" (*Vidhana*, 2) is composed of the "work energy" (*Shram shakti*, 1) of the "self-luminous entity" (*Svayam*, 12), who services its "luminosity" (*Saranyu*, 5) to form the "perpetuating value" (*Saranyu*, 5) of the universe of stars. After radiating its luminosity, the self-luminous entity incarnates as a "sentient entity" (*Siddha*, 7). After investing a unit of the perpetuating value into the "work energy" (*Shram shakti*, 1), the sentient entity conceives the "remainder" (*Jyotistava*, 4) to be the "creator deity" (*Bhagwan*, 4).

The transformative unit of the "technological growth" (*Vidhana*, 2) is composed of the "illusionary dividing energy" (*Maya shakti*, 1) of the mind-born "creator deity" (*Bhagwan*, 4). The creator deity circumambulates within the mind to divide it into the eighteen divisions. The circumambulating creator deity infuses a "goal-oriented" (*Maha Saraswati*, 9) freedom consciousness within each unit of technological growth. The sentient entity's brain distinguishes each of the eighteen units, within the "goal" (*Maha Shiva*, 9), as a differentiated "formative foundational growth" (*Sva*, 11).

A physical body develops the "organization" (*Sangathan*, 29 = 18 + 11) of the inanimate atoms and the animate cells for the transformative "ecosystem exchange" (*Vyavastha*, 189 = 160 + 29) to potentiate the desired "development" (*Bhuyobhava*, 160). The development is the "potential reality beyond the present creation" (*Paramanu*, 160). That potential reality is the antagonist organization's incarnation by exchanging the ecosystem, without its effect, into the ecosystem, within its effect.

Each organization in the topological space services and trades an effect from the whole ecosystem. It conceives the ecosystem as a mind-born antagonist organization. The perpetuating "trading-effect" (*Damodara*, 26) of the mind-born ecosystem transforms its absolute correlation with the sentient entity, within the organization. The sentient entity conceives each sequential unit of the trading-effect as an independent unit. It organizes each as a distinct "organization" (*Sangathan*, 29 = 26 + 3), with common roots in its "masculine astral body" (*Linga sharira*, 3).

The reproduced organizations organize into an "etheric body" (*Bhoga sharira*, 957), comprising a formative unit of goal, a normative unit of perpetuating value, and a transformative unit that potentiates each organization, transforming it into a param child enjoying divine energy. The sentient entity perceives each consequential "organizational-effect" (*Prameyatva*, 180), which forms the "topological space" (*Prameyatva*, 180 = 6 * 30), as the "six-fold growth" (*Khara*, 6) of the "infinite council of spirits" (*Ratnakosha*, 30).

Each spirit is a varying mind-born state, manifesting a distinct modified mood. The body of the modified moods is the "feminine causal body" (*Karana sharira*, 30) for the "illusionary dividing energy" (*Maya shakti*, 1) between the sentient entity and the ecosystem. The illusionary dividing energy is the motivating factor for the self-luminous entity to divide itself into three parts: the sentient entity, the creator deity, and the work-energy for producing the illusionary "divisor" (*Shudra*, 1). The sentient entity experiences each of these four parts as a distinct "personal force" (*Atman*, 4), embodying the "omnipresent truth" (*Vitarkita*, 160) of the "infinite growth potential" (*Sesha*, 18,000) within the self as the space element and the three-time dimensions. The three-time dimensions include the "future of everybody" (*Shani*, 18), "anybody present" (*Gardhaba*, 1000) without the self, and the past "Self" (*Atmatva*, 8×10^{15}) as the one within whom everything is immanent.

> The self is the only "one" (*Akhanda*, 8 x 10^{15}) scripting the future as a "circulatory wheel" (*Anahata chakra*, 8 x 10^{15}), within the organizational heart of the universe. Two self-perpetuating units of growth transform it into a "divided entity" (*Brahmin*, 2), comprising the "masculine self" (*Mein*, 1) and the "self-luminous twin self" (*Aap*, 1) impersonating the undivided "feminine self—the star" (*Tara*, 2).

16.1 Understanding the Divine Paradigmatic Plan

The present reality of any organization is a divine paradigmatic plan. The divine paradigmatic plan is composed of the Eastern and the Western effects. It activates an ascending Northern effect for rejuvenating the creature and descends the Southern-effect that transforms the creature into a passive creation. It is the "centering element" (*Sushumna*, 10), that adds the "two-dimensional circulating East-West present reality" (*Badhabuddhi vadartha*, -2) of the sentient self to the "eighteen-armed north-facing inanimate southern self-luminous entity" (*Padmanartteshvara*, 12). The eighteen arms are the eighteen divisions of the starseed universe, including the nine divisions each of the Eastern and the Western effects. Both East and West are ten-dimensional, where the tenth dimension is the homologous wholesome value of the other half.

The "divine planning" (*Abadha*, 10) comprises the ten dimensions, which work as a "logarithmic base" (*Sushumna*, 10) for exponentiating and multiplying the trading-effect of the Western half, after the entropy of the self-luminous Eastern half. Due to the negative trading-effect of the circulating present reality, the southern creation seeks sentient energy from the cosmic center, i.e., the Vega star, for forming into a "twelve-armed, three-eyed, red-color, south-facing sentient self-luminous entity" (*Vajravarahi*, 12). The twelve arms are composed of the "ten dimensions" (*Udarahana*, 515) that divide an "entity" (*Prakriti*, 485) into the "East" (*Purva*, 0) and the "West" (*Yoni*, 1000) dimensions.

The East is the "competing reality" (*Pratijna vadartha*, 0) of the "unity in diversity" (*Abheda*, 0). The West is the reality of "anybody" (*Gardhaba*, 1000), who behaves like the "one [inanimate] entity" (*Mulaprakriti*, 1000) that has a "lifeless cellular body" (*Nirjara*, 1000). The ten dimensions are the "beginning" (*Udarahana*, 515) of the ten divisions of the divine energy for forming the ten directions, whose tail-domain is the entity that one has become and the head-domain is the para entity that one is seeking to be. The entity that one has become is the competing reality. One who behaves like an "Almighty Creator" (*Hrishikesh*, 26), and radiates intrinsic consciousness for illuminating the culturally-bounded consciousness of the others, eventually becomes a southern inanimate entity, which is the para entity one is seeking to be. When one behaves like an Almighty Creator, who is an "inanimate entity" (*Maha Durga*, 16), one becomes that inanimate entity without the "divine energy" (*Asrava Shakti*, 10 = 26- 16).

The inanimate entity is the source of the "formative divine energy" (*Madhusudan*, 16), which is the form in which the "child primordial greeter" (*Madhusudan*, 16) manifests himself in the universe. The "Almighty Creation" (*Durga*, 28) is the source of the "inanimate entity" (*Maha Durga*, 16), inclusive of the twelve-arms that form the "gender-free self-luminous entity" (*Svayam*, 12). The "Vega" star (*Vega*, 67) is the source of the "Almighty Creation" (*Durga*, 28), inclusive of the "androgynous luminous entity" (*Ardhnareshwar*, 39). The androgynous luminous entity motivates one to behave like an "Almighty Creator" (*Hrishikesh*, 26), for breeding the "luminous element" (*Rochisha*, 13) and squaring that to form the "spirit kingdom" (*Rajyaika-sheshena*, 169 = 13 * 13). The spirit kingdom is the consequential value of the "organizational planning" (*Shri Krishna*, 10). The spirit kingdom is the future of the "starseed universe of cells" (*Pitavasa*, 169) that incarnates anybody from the "deity kingdom" (*Deva loka*, 1000).

The journey of a "cell" (*Hiranyagarbha*, 19) is the path for the organizational development of a "deity" (*Deva*, 1) into a "spirit" (*Kapinjala*, 20 = 1 + 19). By freeing each cell from the dependence on the "masculine astral body" (*Idam*, 3) during this journey, one becomes the "maternal primordial greeter" (*Sati-Parvati*, 16), who is the source of the "Vega" (*Vega*, 67) white star. The maternal primordial greeter (*Sati-Parvati*, 16) transforms her "divine light" (*Usha*, 16) into the "Goddess of the conjunction" (*Yogishvari*, 51) by using her "memory" (*Smriti*, 35) for creating the "wheel of capability" (*Udana chakra*, 35), and aggregating the overall capability into the "Vega" (*Vega*, 67) star.

The "maternal primordial greeter" (*Sati-Parvati*, 16) is the formative reality of the child primordial greeter" (*Madhusudan*, 16). The maternal primordial greeter, the child primordial greeter, and the intermediating Vega star and its infinite forms, including the spirit kingdom, are the three eyes of the south-facing, "sentient self-luminous entity" (*Vajravarahi*, 12). The radiance of the "masculine astral body" (*Idam*, 3), seeking freedom from the cell, for "conjunction" (*Yog*, 7) with the "paternal soul" (*Atman*, 4) to be the "param deity" (*Shiva*, 7), gives her the red color. The diverse physical and metaphysical forms of the "Vega" (*Vega*, 67) white star are the "ecosystem-effect" (*Uttarajya*, 6) of the "illuminator deity" (*Nataraja*, 7), which is the self-luminous form of the "param deity" (*Shiva*, 7), without the "entropy element" (*Ishvara*, 5).

The "maternal primordial greeter" (*Sati-Parvati*, 16) is the self of the "param deity" (*Shiva*, 7), within each of the "devoted factor" (*Vastu*, 9) that is radiating the "organizational energy" (*HAUM Shakti*, 9) as a "key result indicator" (*Kri*, 9) of the progress toward the "goal" (*Maha Shiva*, 9). Each entity's goal is to be a factor devoting effort to realize oneness with the self-luminous form of the param deity and fulfill absolute objective of becoming the "child primordial greeter" (*Madhusudan*, 16) eternally. The goal of the maternal primordial greeter, as a para entity, is to let each entity become a "spirit" (*Kapinjala*, 20) within the spirit kingdom for generating the "self-luminous consciousness" (*Unnata*, 20) about the purpose of life, before starting a new journey within the form of the starseed universe of cells.

16.2 Two Causal Bodies of the One Divinity

There are two causal bodies of the one divinity: the self and the self-luminous entity. The self is the intrinsic consciousness of the param deity within an entity. The self-luminous entity is the intrinsic consciousness of the child primordial greeter within an entity, without the "personal force" (*Soham*, 4) of the param deity as the "paternal soul" (*Atman*, 4), with a self-multiplying "astral body" (*Idam*, 3). The "divinity" (*Siddhi*, 57) is the intrinsic consciousness of the "maternal spirit" (*Dasha*, 1), that makes each entity a "deity" (*Deva*, 1) before the start of the entropy journey to oneness with the futuristic "grandmother spirit" (*Kapinjala*, 20) which is the entropy value of the maternal primordial greeter. The maternal spirit's intrinsic consciousness is without the "body" (*Vapu*, 56), which embodies the potential that anybody may activate as a deity, for diffusing the intrinsic energy to somebody while moving on the path to be nobody, eventually.

16.2.1 <u>The Self as the Causal Body of the Variable Divinity</u>

As a creature, each self has the power to exchange the infinite sentient life consciousness to form the variable divinity, within and without the absolute and constant self-luminous entity. The energy value of "Mother Nature" (*Anatanam*, 8) is a function of the eight energy units of the infinite consciousness. It manifests as the present quark.

Since the present quark is one of the eight quark manifestations of an entity's infinite consciousness, the present quark's incremental value = 1/8. It may be activated to service six units within the seventh, potential quark, for the growth of the SHEENY well-being. One may act as an exchange system for managing the variable divinity of the different organizations through four pathways:

- **First, by activating the divine energy within a universe of the follower organizations**. As a "Primordial Super Knower" (*Upadhi*, 37), the self may service the six energy units (of the potential quark) to the primeval mass of follower organizations, which together constitute the universe of child souls. Metaphysically, each living child soul is a divine entity, incarnated on Earth to fulfill the SHEENY well-being wishes of the creature self, using their divine energy. The six energy units form the infinite consciousness of togetherness, otherness, attraction, repulsion, growth, and entropy within the deity kingdom. Each entity is conscious of the one intrinsic energy unit within a para-consciousness of togetherness. It works to network the six additional energy units to be the absolute follower organization.

- **Second, by activating the divine energy within a creature, who is the absolute follower organization**. As a "Primordial Super Manifestor" (*Surasa*, 38), the self may service the seven energy units (including the potential quark) to a param follower organization (who wishes to become a param leader organization, to be the Manifestor factor of a new universe of living child souls). Metaphysically, only one creature first activates the divine energy and services it to an infinity of child entities.

- **Third, by activating the divine energy without a creature, who is the param leader organization.** As a "Primordial Super Perpetuator" (*Varahi*, 39), one may service the ten energy units (by transforming the spiritual value of the present quark within self) to be a param leader organization (who is working to become a param entrepreneur organization, to perpetuate the present universe of living child souls). Metaphysically, one creature first trades the activated spiritual energy and services it as a guider to an infinity of child entities.

- **Fourth, by activating the divine energy as a creature, who is the param entrepreneur organization.** As a "Primordial Super Devoted" (*Swasthani*, 40), one may service the sixteen energy units (without forming a binary of eight units, each of the physically-materialized quark and the metaphysically-present Primordial deity) to be a param entrepreneur organization (who knows how to be a param management organization for illuminating the intrinsic divine energy consciousness within the potential universe of living child souls). Metaphysically, one creature first activates the consciousness of the intrinsic divine energy within the param leader organization. It has the power to do so within the primeval mass of the follower organizations.

Instead of acting as an exchange system, one may let each living child soul perpetuate the absolute and constant unique divinity, characterized by the self-luminous entity. The "self-luminous entity" (*Swayam*, 12) is the consciousness of a normative unit of divine energy within (1) and without (1) the self, which is a manifestation of the transformative unit of the "divine energy" (*Asrava Shakti*, 10).

16.2.2 <u>The Self-luminous Entity Is the Causal Body of Absolute and Constant Divinity</u>

A self-managing self-luminous entity has the power to perpetuate an infinity of the absolute management organizations. It may perpetuate each entity's constant divinity, as a param management organization, by trading the ascending energy from the universe of management organizations (who are experiencing the proportionate growth) and servicing the ascending energy to the universe of self-organizing entities (who are experiencing a disproportionate entropy).   The value of the constant divine energy (E) of the self-managing self-luminous entity, who is working as the causal body of the absolute and constant divinity of each entity, is a function of the following behaviors:

- Trade the unit mass of ascending potential energy from the universe of management organizations to be the "Primordial Super Destroyer" (*Swadha*, 41) of the future growth in the physically-materialized quark energy (i.e., trade only one unit of mass that is immanent as a constant energy factor within the consciousness of an infinite mass of entities).

- Service the unit mass of trading-effect to the universe of the self-organizing deity entities to be a "Primordial Super Worker" (*Prasuti*, 42) of the present growth in the metaphysically-present divine consciousness within the universe of the self-organizing entities (i.e., spontaneously, free of the limitations of the speed of light).

- Empower each self-organizing deity entity (holy spirit) to be a "Primordial Supreme Worker" (*Sati*, 42). One may perpetually service and trade one's energy within a consciousness of togetherness. Each self-organizing deity entity is an ascendant master. It is free of the limitations of the speed of light for organizing behavior. It services ten units as the transformative divine energy for "primordial oneness" [*Adi*, 32] with the param management organization.

- Trade the perpetuating unit of energy that is forming the consciousness of togetherness to be the "Primordial-Primordial Perpetuator" (*Radha*, 43) of the energy exchange system (i.e., invest 100% of intrinsic energy into the dynamic mass).

- As the Primordial-Primordial Perpetuator, to empower the energy exchange system to be the "primordial wholeness-effect" (*Radha*, 43) of the self-managing self-luminous entity (i.e., 0% correction factor for the exchange cost as a technological entity).

Therefore, the value of the constant, absolute divine energy of the self-managing self-luminous entity, who is the "Primeval Primordial Manifestor" (*Svaha*, 44) $\neq mc^2$, but = (1 unit of potential growth) + (43 units of the primordial wholeness-effect) = 44.   It manifests the universal, inclusive, general potential of the "self-luminous entity" (*Svayam*, 12) to invest a unit of its potential growth energy. It becomes both the "Primeval Greeter" (*Maha Gauri*, 11) and the "Paternal Creator" (*Pitra*, 4) of the universe of the param management organizations, who enjoy a constant, absolute divinity within the "primordial oneness-effect" (*Sadhya*, 32) of the "self-managing self-luminous entity" (*Svaha*, 44).

One-unit potential growth energy becomes four-units Creator factor. It comprises, first, a unit of attraction for the universe as the creation; second, a unit of repulsion for the self as the creator; third, a unit of togetherness of all organizations as the creatures; and fourth, a unit of present growth of the strong psychic linkages within the potential otherness and the entropy of the exchange system. Consequently, the entropy factor immanent within the unit of "present growth" (i.e., absolute time: *Sva*, 11) is not the present reality, contrary to the predictions of the Einstein's general theory of relativity.

Table 27 illuminates eight pathways for the immanent "LIGHT FORCE" (*Apas*, 169 = 180 - 11) of an animate entity at the primordial level to become the emanating "FORCE" (*Prapya*, 34) that materializes the causative cause of the entity life. Inclusive of the other three-hundred fifty-two pathways, they constitute the wholesomewhole path for cleansing the transcendental wisdom and illuminating immanent wisdom over one sidereal year.

Table 27. The Eight Pathways for the Immanent Light Force to Become the Emanating Force

Sequential path within the emanating force of the: *	Saturn-effect (Red-effect)	Jupiter-effect (Orange-effect)	Mars-effect (Yellow-effect)	Venus-effect (Green-effect)	Mercury-effect (Blue-effect)	Neptune-effect (Purple-effect)	Uranus-effect (Black-effect)	Sun-effect (White-effect)	Moon-effect (Indigo-effect)
Primordial Reader (*pravachika*) path; Path of the self-team (freedom from liberator)	Ascend the entropy of the absolute time → realization within the present time	Transcend the duality of the uncertainty avoidance (intrinsic electromagnetic-effect) vs. the uncertainty creation (extrinsic, residual gravitomagnetic-effect), by descending the blocks to the intellectual dynamism	Descend 12/12 trading-effect and ascend 0/12 human-effect of the Bio-acoustic infusion energy at note c $2^{3/12}$ for transcending the intrinsic limitations and desirability ideals of the creation	Ascend the patience for manifesting the secretive parental-effect as the knowable knowledge, and descend the blocks from the reading sequence-effect	Ascend the discriminating faculty for the vibrational healing of the sentimental linkages, and descend the delusional mentation as the child-effect (i.e., quinary by-product production) of the reading sequence	Ascend the sense of the SHEENY responsibility for the universal work service, and descend the fall from grace into the black hole, mediated by the non-linear vector of the secondary life energies, attracting the duality of time and anti-time without the tertiary universe	$-1/8$ dimension of the formative growth of the present cosmos, liberating the present psychological dimension blocks, without the seven para-psychological dimension linkages	Windfall psychological benefits for the present conscious self: Param Siddha (transcending the infinite sameness and being the Almighty Creator, Universal God consciousness)	*Siddha* or Sanada council of one self-luminous entity within the present time; Lord of Proportionate Wish

* *Sequential path of the immanent light force within the emanating force of the:*

Path	Saturn-effect	Jupiter-effect	Mars-effect	Venus-effect	Mercury-effect	Neptune-effect	Uranus-effect	Sun-effect	Moon-effect
Primordial Researcher (*margantika*) path; the Path of self-inquiry	Ascend the entropy of the absolute space → realization within the present space	Transcend the duality of the future (intrinsic gravitational-effect) vs. the past (extrinsic, residual gravitoelectric-effect) orientation, by ascending the aesthetics of the present	Descend 11/12 trading-effect and ascend 1/12 human-effect of the Bionuclear refusion energy at note $C\ 2^{4/12}$ for transcending the reproduction and the growth of the past	Ascend the filial piety for the manifested parental-effect as the researchable research, and descend the wish to reproduce oneself as the researching sequence-effect	Ascend the radiant love for the natural space, to manage the cost of the alpha (from heart) to the omega (fundamental) clearing of the child-effect (i.e., tertiary residual, weak psychic linkages), and of researching the 2,600 pathways below alpha frequency into the ever-expanding levels of the sentient energy	Ascend the sensible sense of proportionate sensibility of time, for the health and career, and descend the trauma programs coming up through the emotional body, repelling the eight-fold punctuation of the quarkness within the tertiary universe	−2/8 dimension of the formative growth of the present cosmos, liberating the present psychological and the past para-psychological dimension blocks, without the six strong psychic dimension linkages; Wholeness Committee	Windfall national benefits, with the entropy in the strong psychic linkages of the soul family nation: Primeval Siddha	Cosmic council of twelve directions; Lord of Proportionate Knowledge (*Juniten/ Adityas/* Apostles; lords of zodiac signs) within the present space as the causal body

Path	Saturn-effect	Jupiter-effect	Mars-effect	Venus-effect	Mercury-effect	Neptune-effect	Uranus-effect	Sun-effect	Moon-effect
Primordial Practitioner (*tapasika*) path; the Path of self-healing	Ascend the entropy of the absolute attraction – realization within the present universe of endowments	Transcend the duality of humane vs. animal orientation, by descending the dysfunctional expression of pride	Descend 10/12 trading-effect and ascend 2/12 human-effect of the Biochemical synthesis energy at note d $2^{5/12}$ for transcending the animalistic ambition and assertion	Ascend the love for the manifestable parental-effect as the wishable wish, and descend the dark discordant energies of the wishing sequence-effect	Ascend the love for the open personal spiritual space, and descend the pre-programming of the experiential path with the child-effect (i.e., secondary, fundamental, residual strong psychic linkages) of the animalistic orientations	Ascend the sensible sense of proportionate sensibility of connoisseuring (i.e., practicing the art of sense-making), and descend limiting the horizon of the self-healing enlightenment to only the flame family within the astral body, attracting the divided duality consciousness of the anti-self twin-flame otherness	−3/8 dimension of the formative growth of the present cosmos, liberating the psychological, para-psychological and psychic dimension blocks, without the five planetary dimension linkages; Healing Committee	Windfall economic benefits, with the entropy in the weak psychic linkages without the soul family nation: *Para Siddha*	Cosmic council of seven light spirits; Lord of Proportionate Manifestation (*Saptarishi* council of divine law, i.e., Seven Stars of Big Dipper or Rainbow Bridge), within the present universe of endowments, as the spiritual body

Path	Saturn-effect	Jupiter-effect	Mars-effect	Venus-effect	Mercury-effect	Neptune-effect	Uranus-effect	Sun-effect	Moon-effect
Primordial Scientist (*rajnanika*) path; the Path of self-illumination	Ascend the entropy of the absolute repulsion – realization within the present soul or self by ascending the conscious ness of the spirit to New Paradigm with 99 Access Portals	Transcend the duality of the power distance (polarization) vs. diffusion (repulsion), by ascending the luxuriousness of the polar self	Descend 9/12 trading-effect and ascend 3/12 human-effect of the Biophotonic diffusion energy at note D $2^{6/12}$ for transcending the polarized harmony and correlation	Ascend the emotional empathy for the manifester of the parental-effect (primordial mother), and descend the frustration of being falsified by the manifesting sequence-effect	Ascend dignity of life, and descend the savoir complex whose child-effect (eventual quaternary-effect) is the discordant energies of the mutating divine justice	Ascend the sensible sense of proportionate sensibility of occulting the secret effects of the past lives (i.e., the science of sense-taking, using the residual past life energies as a medium), and descend the blocks to illuminating the GUIDER value of the soul family (backward, forward, and oneness values within the oversoul family, within and without the soul family)	−4/8 dimension of the formative growth of the present cosmos, liberating psychological, para-psychological psychic, and planetary dimension blocks, without the four galactic dimension linkages; Heart Committee	Windfall ecological benefits, with the entropy in the planetary linkages without the past living souls on the earth: Supreme *Siddha*	Supreme council of *Nin*, Lord of Proportionate Causation (Archangels or messengers of light OR planets *Naxgrah* or Emperor stars *Jin Huang Xing Jun*) within the present soul as the astral body

Path	Saturn-effect	Jupiter-effect	Mars-effect	Venus-effect	Mercury-effect	Neptune-effect	Uranus-effect	Sun-effect	Moon-effect
Primordial Philosopher (*mimamsaka*) path; the Path of self-empowerment	Ascend the entropy of the ego forming linkages: ascending guider-effect → realization of the wishable wish within the present paternal lineage	Transcend the duality of gender egalitarianism vs. division (punctuation)	Descend 8/12 trading-effect and ascend 4/12 human-effect of the Biothermal fission energy at note F. 27/12 for transcending the divisive identity expression	Ascend the joyful reverence for the param father, and descend the paraconscious blocks to trading the mercy-effect of the primordial mother	Ascend the proficiency of accomplishing lessons from the birth planet energies, and descend the negative well-being effects of the disproportionate sense of proportionate identity	Ascend the leadership to clear the emotional-effects of the karmic beliefs, and descend the forming of the anti-self ego-projecting the twin-flame within the astral body	−5/8 dimension of the formative growth of the present cosmos, liberating the psychological, para-psychological psychic, planetary, and galactic dimension blocks, without the three super-(inter)galactic dimension linkages; High Self Committee	Windfall human benefits, with the entropy in the galactic linkages without the present BE-LIVE-LIFE soul universe of the First Eden: Supra *Siddha*	*Karma* keykeeper or time gatekeeper council (The Ring Council of Thirteen Original Moons of Saturn/ *Shanidev*/ Quan Yin/ Life deity messengers/ Angels of Life), within the present paternal lineage as the mental body

Path	Saturn-effect	Jupiter-effect	Mars-effect	Venus-effect	Mercury-effect	Neptune-effect	Uranus-effect	Sun-effect	Moon-effect
Primordial Educator (*prabhryataka*) path; Path of self-worth	Ascend the entropy of the emotion norming linkages: descending child-effect → realization of the infinite wish within the present maternal genetics	Transcend the duality of in-group collectivism (of atoms into a planet) vs. proliferation (of atoms as quantum-effects)	Descend 7/12 trading-effect and ascend 5/12 human-effect of the Bioelectric repulsion energy at note f $2^{8/12}$ for transcending the proliferating universal purpose and elastic time projection	Ascend the power of self-control, and descend the political involvement in trading the jealousy-effect of a disproportio nate sense of proportionat e self-worth	Ascend the patriotic devotion to the well-being of the entire birth planet as the foundation of the self-worth, and descend the disproportionate sense of proportionate self-worth becoming a causative factor for the soul phases (divided soul consciousness, seeking to manifest the beauty as the truth)	Ascend the privilege of oneness with the entire soul family group, and descend the subconscious programming of the infinite soul family group members, seeking to manifest their emotionally divisive DIVINE-effect through the SHEENY conscious self	−6/8 dimension of the formative growth of the present cosmos, liberating psychological, para-psychological psychic, planetary, galactic and super-(inter)galactic dimension blocks without the two supra-galactic (omnipresent) dimension linkages	Windfall social benefits, with the entropy in the super galactic linkages without the future spherically expanding 2nd Eden (i.e., the new soul universe wishing soul exchange): Super *Siddha*	*Dharma* keykeeper or space gatekeeper/ recordkeeper council (High Council guardian of Luminous/ *Yamadev/ Yang-Wang-Yeh*/ Orion-effect/ Death deity messengers/ Angels of Death) within the present maternal genetics as the intellectual body

Path	Saturn-effect	Jupiter-effect	Mars-effect	Venus-effect	Mercury-effect	Neptune-effect	Uranus-effect	Sun-effect	Moon-effect
Primordial Student (*tirthya*) path; the Path of self-esteem	Absolute fulfillment transforming entropy: Ascending devotional intensity → realization of the wishing sequence within the present planetary or astrological constellation	Transcend the duality of the overbearing assertiveness (of the massive objects) vs. the underbearing freewill (of the meso subjects)	Descend 6/12 trading-effect and ascend 6/12 human-effect of Biomagnetic transfusion energy at note $F\ 2^{9/12}$ for transcending the polluted cosmic sense and pro-inclusion space exchange	Ascend the sincerity of purpose and descend mediating the trading-effect of the planetary forces	Ascend the depth of consciousness about the sensible sense of proportionate self, and descend the discordant light realm energies imprinting the wishables as the soul separates	Ascend the sense of royalty as the self-guided sheeny-conscious present self, and descend the intimate desire for impinging the present life with the akashic records (subconscious memory) of the past life misguided para conscious energies	$-7/8$ dimension of the formative growth of the present cosmos, liberating psychological, para-psychological psychic, planetary, galactic, super-galactic, and supra-galactic (white star) dimension blocks without the one cosmic (omni-permeating) dimension linkage	Windfall SHEENY benefits as a whole, with the entropy in the supra galactic linkages, without the horizontally expanding 3rd Eden of the Discarnate souls seeking to exchange the soul-level residual energies: *Siddha*	Infinite Council (council of angels) within the present astrological constellation

Path	Saturn-effect	Jupiter-effect	Mars-effect	Venus-effect	Mercury-effect	Neptune-effect	Uranus-effect	Sun-effect	Moon-effect
Primordial Investigator (*vicharaka*) path; the Path of self-steam	Absolute entropy continuity — Descending the danger of the devotee manipulation → realization of the wisher growth within the present zodiac or sentient energy	Transcend the duality of performance orientation (of the causal micro entity as the leader) vs. the inertia (of the consequential macro entity as the follower)	Descend 5/12 trading-effect and ascend 7/12 human-effect of the Bioplasma fusion energy at note F 29/12 for transcending the extrinsic causation and opportunity seeking theory-effect	Ascend the sense of peace with the natural flow and descend the appeasement of the trading-effect of the birth zodiac	Ascend the tranquility of spirit for the equanimity of the investigating mind and the serenity of the para investigator soul, and descend acting as if running errands for investigating odds and ends for the soul family is the life purpose	Ascend the fame of the present self within a DIVINE sense of self-love, kindness and urgency of action, and descend the perforation of the sentient energy with a satanic sense of the intimate perversion, revenge and vengefulness	−8/8 dimension of the formative growth of the present cosmos, liberating the psychological, para-psychological psychic, planetary, galactic, super-galactic, supra-galactic (white star) as well as the supreme, cosmic (dark matter) dimension blocks	Windfall SHEENY benefits as wholesome, with the entropy in the supreme galactic linkages without the descending six infinite levels beyond New Lemuria, the universe of departed souls seeking to trade the alternative Godhead soul-level energy: *Param Prapti* (transcending the infinite becomingness and being the infinite infinity – Self-same primordial oneness of the Primordial Illuminator *Parvati*, adiating Luminous)	Divine Council of three divinities (DIVINE creator, GUIDER creature, and SHEENY creation) within the present zodiac sign

Chapter 17: The Ecosystem As the Guider Paradigmatic Plan

I am the absolute soul, radiating my guiding force into four directions to be the soul of each sentient entity in the universe.

I constantly capture the past energies within a new cell . I destroy them with the cell. Each "person" (*Vyakti*, 1) born without the cell becomes a "realized entity" (*Rachitartha*, 1600). He enjoys a dynamic consciousness of my virtual guiding force that simplifies the past augmented, inanimate reality into him as the sentient entity-the illuminator of himself as the perpetuable, one-dimensional future reality. He is the creator of the new sentient force, mass, light, and energy. I take the creator credit and let him distribute the discredit of being an undesirable, blackening creature for destroying himself and manifesting the desired, rainbow color creation universe by exchanging his purple consciousness mass.

I am the absolute energy, energizing each inanimate atom to be a sentient cell for embodying the present reality of my illuminated masculine self. I let each cell destroy itself for incubating a secret, omnipotent sentient entity, free from the omniscient, polluted ecosystem reality immanent within the cell.

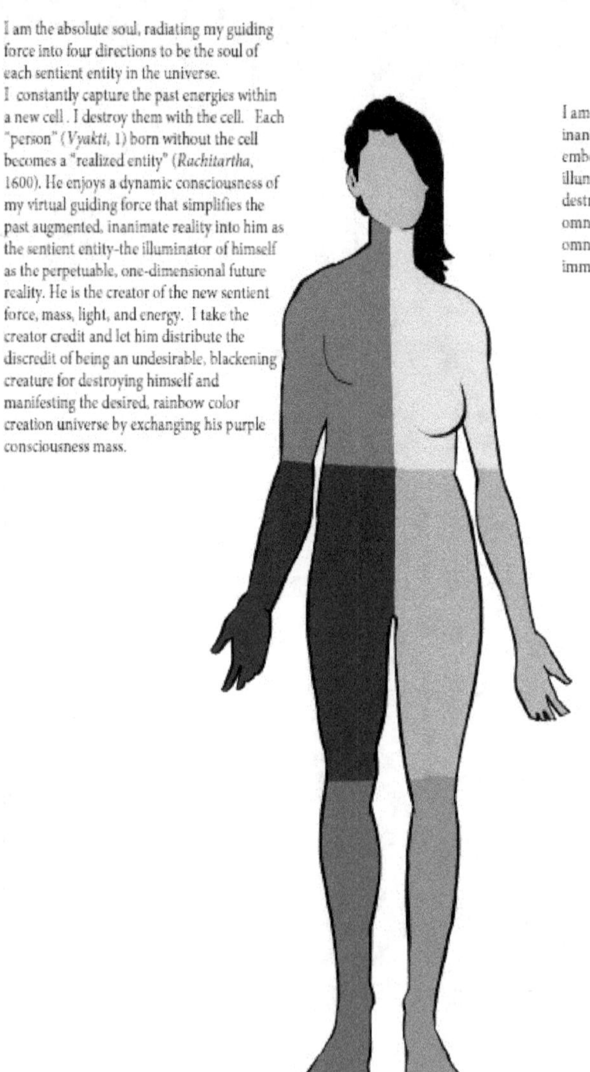

The Guider Paradigmatic Plan Destroys the Present Reality

The Guider Paradigmatic Plan Illuminates the Ground Reality

The Dark Matter is the South Pole. It is the absoluteness beyond the brain's consciousness. A guider paradigmatic plan transforms an entity into the "Black Hole" (*Vishnunabhi*, 82). The Black Hole is in the West, embodying the gravitational potential beyond the reach of the Sun's proficient networking of gravitational energy. The "gravitational energy" (*Lalita*, 100) is the energy reproduced by trading and servicing the "divine energy" (*Asrava shakti*, 10). The divine energy is the energy produced by exchanging the "astral, sentient energy" (*Varuna*, 1000) into:

- A ten-unit "primordial masculine self" (*Shri Krishna*, 10). It is immanent within the sentient entity as the "centering" (*Sushumna*, 10) element.

- A ten-unit "primordial feminine self" (*Parvati*, 10). It personifies the inanimate ecosystem and generating gravitational energy.

- A ten-unit "divine energy" (*Asrava shakti*, 10). It is immanent within both the ten-unit primordial masculine and feminine selves.

The gravitational energy produces a "gravitomagnetic-effect" (*Padartha*, -3) because the gravitomagnetic-effect is the self-reproducing divine-effect of the three-dimensional "past reality" (*Evakara vadartha*, -3) of anybody with the sentient energy in its inanimate astral or animate consciousness form. A physical body attracts other physical bodies, because each physical body is a part of the Vega white star. Each physical body has a black hole type "gravitational potential" (*Kalika*, 23), for attracting and merging with "everybody" (*Pasaka*, -9) trading and squaring the past reality, exchanging the perpetuating discordant negativity of the past from the absoluteness of the present.

The "formative Sun" (*Aishani*, 12), in the Northeast, is the "guider programmatic planner" (*Vayavaya*, 12). It is the Sun's (*Surya*, 21) "self-luminous entity" (*Aishini*, 12) form, without the "culture" (*Sadashiva*

Nayaki, 9) element traded as the "goal" (*Maha Shiva*, 9) of becoming a part of the "universe of stars" (*Mahar loka*, 27,000). Suppose a micro entity decides to become a part of a macro system. In that case, the macrosystem "overshadows" (*Chhaya Sita*, -1) and trades the macro entity's potential.

Overshadowing limits the micro entity's gravitational power to radiate its "gravitational energy" (*Lalita*, 100). It empowers the macro system to become the engaged "gravitational center" (*Jamadagni*, 629) of the varying "gravitational qualities" (*Guna*, 0) of the diverse entropy "local-effects" (*Mahadasha*, 0). Consequently, each micro entity becomes accultured to the culture of the macro system.

The "Primordial-Primordial Realm" (*Antara kalpa*, 3794) is the "accultured value" (*Adi loka*, 3794) of the primordial oneness: between the astrological universe and the Vega white star. It is the "gravitational center of the universe of stars" (*Madhya loka*, 3794). It is the "maternal womb" (*Garbhastala*, 3794) that services a "healing touch" (*Sthira*, 3794), for stabilizing the normative, organizational form of each entity in absolute oneness with the astrological universe.

The healing touch empowers each micro entity to be the macro system that radiates the "gravitational light" (*Digambara*, 100) and forms the "universe of universes and entities" (*Digambara*, 100). Each macro system activates the "Doctrine of Soul" (I AM THE SOUL: *Esa Ta Atmantaryamyamrtah siddhanta*, 100) for servicing: a 96-unit "being energy" (*Kali shakti*, 96) to each micro entity, a 2-unit "technological growth" to each macro "universe" (*Brahman*, 2), and an additional 2-unit "technological growth" to each meso universe that is mediating the universes and the entities as a "divided entity" (*Brahmin*, 2).

By irradiating the 4-unit consciousness of the "creator factor" (*Bhagwan*, 4) servicing the 4-unit divided technological growth, now massive entity becomes the "paternal soul" (*Pitra*, 4) of each micro entity. He enjoys his "being energy" (*Kali shakti*, 96), within the soul-effect, and the "guider power" (*Chitta*, 100), without the soul-effect.

The "ground reality" (*Hetvartha*, 385) is the "multidimensional reality" (*Radheshyam*, 375) of the "social dimension" (*Sadharana dharma*, 375) of each entity, as an outcome of the interactions with the diverse universes of universes and entities. It includes a hundred units each of the undivided guided power, the divided gravitational energy, and the guider-effect irradiated by the entity by reunifying the divided energy. It also includes the seventy-five units "working consciousness" (*Chetan shakti*, 75) of the entity's entropy-effect, as the entity becomes the "wheel of illumination" (*Hora chakra*, 25) for everybody else. By manifesting the root reality, the entity becomes the "absolute soul" (*Paramatma*, 1600), residing within the "dark matter" (*Sadashiva*, 1600).

17.1 Understanding the Guider Paradigmatic Plan

The ecosystem is the future reality of an organization. Each organization is a "universe without kingdom" (*Nishkala*, 169). It radiates the "sentient light force" (*Apas*, 169) to form an "ecosystem" (*Vishvamitra*, 92), as an "intelligent" (*Buddhika*, 77) solution for accomplishing the goal of forming a "kingdom" (*Nayana*, 957) by trading the "guider paradigmatic plan" (*Siddhikarin*, 865).

The guider paradigmatic plan leads an organization to trade the "social experience" (*Bhaavi*, 11), from friendly entities, for knowing the value of the "present growth" (*Sva*, 11). The present growth is the "foundation" (*Sva*, 11) for one's intent to form a "friendship experience" (*Bhaavi*, 11). The guider paradigmatic plan, thereby, becomes a "spiritual body" (*Adhyatamic sharira*, 876 = 865 + 11), devoted to knowing the secret of the present growth by overshadowing the ecosystem, as a "futuristic, intrusive, omniscient grandmother spirit" (*Kapinjala*, 20).

The organization trades what has become an undesirable paternal personal force of the friendly ecosystem by servicing the desirable grandmaternal social force, bragging about its "present growth" (*Sva*, 11). With the "hollowing-out" (*Sarvanasha*, 5) of the "present growth" (*Sva*, 11), the ecosystem becomes the "transcendental value" (*Param Shankara*, 6). The organization trades that transcendental value as its "kingdom" (*Nayana*, 957). With the secondary trading of the "primordial wholeness-effect" (*Radha*, 43) of the guider paradigmatic plan, for perpetuating the trading of the "tertiary residual" (*Khara*, 6) from the ecosystem, the organization becomes a "nucleus" (*Yoni*, 1000 = 957 + 43) of the "sentient [i.e., consciousness] energy" (*Varuna*, 1000).

With the tertiary trading of the "God of immaculate knowing, founded on personal knowledge without socially knowable contamination" (*Vimaleshvara*, 700) soul from the ecosystem, the organization becomes an "omniscient, psychic entity" (*Anadi*, 700,000). Consequently, the entities within the universe become dependent on the "grandmother soul" (*Punyatma*, 100). They elevate the organization to the role of the "Primordial Para Devoted" (*Aniruddha*, 7000), who embodies the holistic truth of the "manifesting-effect" (*Sana Kumara*, 7000), as an "all-knowing spiritual flame" (*Aniruddha*, 7000).

17.2 Why Friendly Paternal Personal Force Becomes Undesirable?

The friendly "paternal personal force" (*Soham*, 4) limits the ability of a child organization to enjoy the "present growth" (*Sva*, 11) of Mother Nature as an "enjoyer" (*Maha Saraswati*, 9), without the mediation of the "futuristic, intrusive grandmother spirit" (*Kapinjala*, 20) of the "maternal social force" (*Ruah*, 20). One becomes the futuristic grandmother spirit, seeking to discover the enjoyers for reproducing that value for self-fulfilling its causative cause. The causative cause is to be the "light without force" (*Prabha*, 180) for self-illuminating the joyful value of the "present growth" (*Sva*, 11). One then becomes a "devoted factor" (*Vastu*, 9) who, naturally develops the "primordial oneness" (*Adi*, 32) with the ever-joyful "child primordial greeter" (*Madhusudan*, 16).

The "light without force" (*Prabha*, 180) is the scraping educator, who touches the "life" (*Prabhasa*, 4) of the universe of sentient entities with the force of the present time moment, without thermodynamically shaping that life. The consciousness of life within an entity radiates as the thermodynamic force for materializing the consciousness of the purpose of life. One is conscious of life because one wishes to perform some "action" (*Karma*, 10). One wishes to perform some "action" because one is aware of the "worker-social benefit" (*Shubh*, 82) of the "honest work value" (*Kara*, 23) of that action.

The "honest work value" (*Kara*, 23) is more than the value of the "action" (*Karma*, 10), because one who is acting honestly with absolute devotion naturally trades the "blessing value" (*Ashirwad*, 1000) of the "luminous" (*Maha Kali*, 13 = 23 – 10). The blessing value is more than the value of the "luminous" (*Maha Kali*, 13). One, who is profiting from the primordial "knowing" (*Jnana*, 19) of the "inanimate self-luminous guider entity" (*Padmanartteshvara*, 12), services the paternal personal force and trades a "descending proportion of the ether within the divine energy as the residual value" (*Khara*, 6 = 19 – 13).

A devoted factor is not devoted to any entity and does not wish to be a para entity that one is not at present. A devoted factor behaves like a "thing" (*Vastu*, 9) that does not "think" (*Cint*, 9). Instead of wasting the energy to "think" (*Cint*, 9), the devoted factor invests the energy into an "honest work value" (*Kara*, 23) for the realization of the "primordial oneness" (*Adi*, 32) with the "maternal primordial greeter" (*Sati-Parvati*, 16), as a joyful, loving child.

One, who thinks, is the one who behaves in a friendly, generous, compassionate way, for manipulating the spirit of the universe of sentient entities. Such person becomes the undesirable paternal soul in the inanimate afterlife. A normal human entity "thinks" that the devotion to the universal SHEENY well-being, the religion, and the "paternal creator" (*Pitra*, 4), is the path to the betterment of the afterlife. Therefore, a normal human entity becomes the undesirable paternal soul in the afterlife, as a devoted follower of the infinite council of entities as a living as well as the departed child.

The path to the eternity is not the universal SHEENY well-being, but the personal SHEENY well-being. It is the personal devotion to the social, human, ecological, economic, national, and psychological well-being that empowers one to enjoy the primordial oneness with both the ecosystem and the maternal primordial greeter, who is conceiving the ecosystem as her greeter. The universal devotion to the SHEENY well-being descends everybody's personal SHEENY well-being, as each person makes cost-escalating trade-offs to follow the path illuminated by the paternal creator. They do so to fulfill the paternal creator's social wishes as a path for fulfilling their personal wishes. They delay fulfilling their wishes, just like the paternal creator cannot fulfill the wish to be the universal "savior" (*Lokapa*, 387) in this life or even the afterlife. Thus, the energy devoted to being the "savior" (*Lokapa*, 387) of the universe just goes "waste" (*Bhringaraja*, 387).

When the wasteful "ether" (*Shuddhi*, 285) element descends within the traded divine energy, the "inanimate self-luminous guider entity" (*Padmanartteshvara*, 12) enjoys the ascending "sentient energy" (*Varuna*, 1000). The "ether" (*Shuddhi*, 285) is wasteful because it is a pure reproduction of the paternal personal force, without any SHEENY value for the child. It generates an "infinite SHEENY-effect" (*Shuddhi*, 285), that reproduces the wasteful trading of the ether value of the inanimate self-luminous guider entity and becomes an "archaeon" (*Jangamavisha*, 570). The animated archaeon exponentiates its "energy" (*Shakti*, 19) with the "divine energy" (*Asrava Shakti*, 10) of the inanimate self-luminous guider entity. It forms a "hermaphrodite" (Growth hormone-releasing hormone [GHRH]: *Rishi Rina*, 10^{19}).

The hermaphrodite works to rub the "inner salt" (Zwitter-ion: *Rishi Rina*, 10^{19}) on the "consciousness void" (WW domain: *Andhakara*, 10^{19}), generated by the entropy transformation of the "inanimate self-luminous guider entity" (*Padmanartteshvara*, 12) into a "universe of granddaughter cells" (*Brahman*, 2), after servicing the divine energy. The hermaphrodite programs itself as the "Mylordship" element (Protein kinase Hippo [Hpo]: *Apacayana*, 10^{19}), in reverence to the self-luminous guider entity. It offers a window of opportunity to the animated archaeon to reproduce an additional ether element and become a "supernormal entity" (RNA polymerase: *Atisara*, 855), devoted to reproducing the "divine energy" (*Asrava Shakti*, 10) and diffusing supernormal "guider paradigmatic plan" (*Siddhikarin*, 865). It forces the self-luminous guider entity to destroy the supernormal entity and program a "catalyst protein" (Messenger RNA: *Shilajit*, 855). It performs the script of the guider paradigmatic plan, to transform into a "bacterial worm" (Param ribosomal RNA: *Jivanikaya*, 855).

As a bacterial worm, the self-luminous guider entity seeks to manage the "consciousness void" (WW domain: *Andhakara*, 10^{19}) by programming a "sense of danger" (Myelin Transcription Factor 1 [MYT1] gene: *Andhakara*, 10^{19}). By trading the alternative "supernatural paradigm" (*Yukti*, 8) serviced by the destroyed supernormal entity, the bacterial worm is infected with a "belief that two is better than one" (CAAX motif: *Kundalini*, 8). Guided by that belief, it follows the path of exploiting the "supernatural energy" (*Shram Shakti*, 1) of the paternal creator. Such exploitation destroys the paternal creator and transforms the bacterial worm into a "medium" (*Ganga*, 963 = 855 + 8) for cleansing the consciousness void of the self, now guided by the departed "paternal soul" (*Atman*, 4).

With a cleansed consciousness, the cleansing "medium" (*Ganga*, 963) transforms the polluting "trading-effect" (*Damodara*, 26) into the "horsepower" (*Sudhanvan*, 999 = 963 + 26) for servicing the sense of conscious "devotion" (*Bhakti*, 46 = 26 + 20), within the departed guider entity, wishing to know the secret of her prospering social entrepreneurship as an "intrusive spirit" (*Kapinjala*, 20). By trading the "seduced" (*Avakirna*, 1) element, as a devoted follower of the departed guider entity, the cleansing medium liberates herself from the "nucleus" (*Yoni*, 1000) that binds her "femininity" (*Yoni*, 1000). By exponentiating her divine energy with the "foundation" (*Sva*, 11) for future growth without the polluted, seduced factor, she becomes a "diffused entity" (*Chitrini*, $10^{11} - 1$), servicing the "transcendental wisdom" (*Para Vidya*, $10^{11} - 1$).

The diffused entity countertrades the transcendental wisdom for exchanging the "purple spiritual flame" (*Sana kumara*, 7000), without the nucleus, with the "green human flame" (*Kapila Kumara*, 8000), within the nucleus, thus discarding the negative embodied "past of everybody" (*Asura*, -1). She becomes an "omnipresent entity" (*Vyapin*, 8×10^{15}), who invests the intrinsic divine energy for trading the truth of the self-luminous guider entity as the "speed of light" ($12/10 * 8 * 10^{15} = 9.6 \times 10^{15}$).

The speed of light is the value of "ONE" (*Akhanda*, 8×10^{15}) as the "self" (*Atmatva*, 8×10^{15}), seeking oneness with the "luminous entity" (*Kali*, 96), for trading the "being energy" (*Kali Shakti*, 96), without the "paternal personal force" (*Soham*, 4) of the "guider power" (*Chitta*, 100).

17.3 The Present Reality of the Varying Forms of Divinity.

The self, the guider self-luminous entity, and the hermaphrodite that connects the other two, are the varying forms of divinity. The androgynous self is the absolute form of the divinity. The inanimate guider self-luminous entity is the primordial form. The animate hermaphrodite is the primeval form. The present reality of the primordial divinity as the one, who is homologous to the self, is different from the absolute divinity. The former is variable, while the latter is constant. Unlike the constant self, one is always seeking to be what one is not at present. The differentiated self's primordial divinity inserts its behavioral template in the form of the "Doctrine of Divinity" (*Sauh siddhanta*, 1), to fill the void in consciousness about the absolute divinity within the integrated self-luminous entity.

The Doctrine of Divinity makes one dependent on a para entity for catalyzing the supernormal growth. Without conscious planning, we become the puppets of the differentiated entity incarnations of the "primeval self" (*Rama*, 100) and subjugate our mood, destiny, divinity, and eternity to the invisible creator hand of the devil greeter spirit. The devil spirit enjoys the energy cost-free "rulership over our divinity" (*Dhaivata*, 0). By acting like a powerless creation, we delegate the creator power to the devil and become a machine for creating the creatures, who are the devoted followers of our scientific method for effortlessly consuming the gifted reality.

For a while, we are on the seventh heaven enjoying the holy gifts of the friendly, generous, compassionate greeter and propagating the greetings for the religious well-being, within and without the present generation. As we genetically program the dependent correlation, we begin suffering a long period of neglect, from the primordial greeter now committed to teaching us a lesson by entrusting us with a responsibility for the personal and the universal SHEENY well-being. We bear the economic cost-escalating ecological fruits of not only our personal, human *karma,* but also those of the *karma* of the social, sentient universe. As the interdependent citizens connected to the one "mind-born national element" (*Sadhya*, 32), we pay back the guider-debt benefit value with a supernormal divine-debt benefit value.

At any moment, we have a full freedom to take the reins of our *karma* in our hands, conditional on our ability to free ourselves from the weak psychic linkages of the universe of devotees, following the idealized, primeval self and the strong psychic linkages of the universe of the devoted, incubated and led by the theoretical, primordial self. Our oneness with the primordial greeter conditions our ability, guided by our actions' sincerity, the seriousness of our mood, and the divinity of our mood-centering consciousness. Our social entanglement with the predominating international beliefs and the dominant national belief system, are the primary impediments. They impede our ability to free our destiny and eternity from the effects of social *karma* of the whole universe of the sentient entities, notwithstanding the sanctity of our personal *karma*. Our personal *karma* accounts for only twenty percent, as a "Param Manifestor" (*Kapinjala*, 20), of our consequential fate and power for the dynamic exchange of the present reality, guided by our conscious planning.

17.4 The Present Reality of the Absolute and Constant Divinity

The param manifestor may perpetuate her absolute and constant divinity as the present reality, by perpetuating the present growth factor of the param management organization. She may service her proportionate energy to the universe of self-managing self-luminous entities (who are the absolute creators of her present reality) and trade the constant energy from the universe of self-managing self-luminous entities (for perpetuating the present growth factor). The value of the constant divine energy (E) of the param manifestor, who is the metaphysical cause of the present reality of the absolute and constant divinity of each entity, is a function of the following behaviors:

- Service the unit mass of the present perpetuating energy (and not the whole futuristic spirit) to the universe of self-managing self-luminous entities, and bless them to be the "Primordial Para Worker" (*Seshi*, 45). A Primordial Para Worker is the one who is catalyzed by the entropy factor immanent within the present consciousness to spontaneously radiate the whole mass of energy, seeking to exclusively appropriate the self-steaming masculine fame for manifesting the universally desirable future. He does so by forming a "wheel of engagement system" (*Ulka chakra*, 45).

- Trade the whole devoted mass of the entity-effect immanent within the universe of Primordial Para Workers, inclusive of an additional unit of the potential growth energy, to be the "Param Primordial Perpetuator" (*Makuli*, 46). A Param Primordial Perpetuator is the one who, through his "devotion" (*Bhakti*, 46), enjoys the "true capability" (*Kacchera*, 46) of the param manifestor spontaneously, without limitations of the speed of light.   He grows his true capability by forming a "wheel of investment" (*Prana chakra*, 46).

- Empower each entrepreneurial social spirit (who has radiated his intrinsic energy, wishing for animalistic fame) to trade a unit of the present entropy animal energy, together with the 46-unit of primordial perpetuating plant energies, and be a "Param Primordial Destroyer" (*Lopamudra*, 47).   A Param Primordial Destroyer is the one who destroys the present reality of the constant divine energy spontaneously, without the limitations of the speed of light. She forms the "animal kingdom" (*Tiryaggati*, 47) by "transmigrating" (*Vijnana*, 47) the soul form of each devoted follower into the "concealed psychic linkages" (*Mahika*, 47), for creating a "spiritual stress" (*Dukkha*, 47). It makes each devoted follower the "starseed universe of mammals" (*Narasimha*, 47), running freely like a lion king, who embodies the universe for creating the path to walk for the collective mammalian pride.

- Trade an additional unit of the otherness consciousness immanent within each followership deity (radiating unit mass of his energy, seeking to be Para deity, God). Consequently, be the "Primeval Primordial Creator" (*Siddhika*, 48) of a new reality of the constant guider power.   A Primeval Primordial Creator is the "embodied" (*Rupin*, 48) "light of the primordial realm" (*Chidambara*, 48). She scripts the "excellence-effect" (*Thiruvambala*, 48), of the "oneness" (*Yoga*, 48) with the "constant guider power" (*Vibhu*, 379), as a "scripture" (*Shastras*, 48) servicing the "artificial intelligence" (*Buddhi*, 48) in the form of the "nurturing energy" (*Tantri*, 48).

- Trade the excellence-effect to be the "Primeval Primordial Creator" (*Ruksha*, 49) within the self's consciousness as the causative factor of both the present ecosystem and the entity growth. A Primeval Primordial Creator is the "sentient cause" (*Ruksha*, 49) for the "octave splicing" (*Arpita*, 49) of the "oneness of action" (*Karma Yoga*, 49) into many "modifier splices" (*Caranam*, 49). Each modifier splice becomes a "mammal" (*Anasuya*, 18), who behaves like a "witch mother of the devotee children" (*Yaksha*, 18) by servicing "ever-degenerating" (*Urja*, 31) "sentient dimension" (*Beeja Dharma*, 31) of the spliced self.

Therefore, the value of the constant divine energy of the param manifestor, which eventually becomes the Primeval Primordial Creator $\neq$ mc^2. It is 48 [(32 units of the primordial oneness-effect) + (12 units of the self-luminous entity who is the self-organizing potential growth factor of the primordial oneness-effect within self) + (1 perpetuating present unit) + (1 potential growth unit) + (1 present entropy unit) + (1 otherness unit)].

The param manifestor manifests the unique, special potential of the "inanimate self-luminous guider entity" (*Padmanartteshvara*, 12). The potential is to be a conjunctive "Creator factor" (*Bhagwan*, 4), comprising the four organizational units. These include a present otherness unit within the deity kingdom, a present perpetuating repulsion unit within the spiritual kingdom, a present entropy unit within the animal kingdom, and a potential growth unit within the plant kingdom.

The inanimate self-luminous guider entity transforms the spliced "remainder" (*Jyotistava*, 4) into an octave-length "primordial oneness-effect" (*Sadhya*, $32 = 4 * 8$). She trades the "conjunctive" (*Samyoga*, 36) value of the "starseed deity" (*Taraka*, 36).

The "four-fold growth" (*Maha Nitya*, 8) in the normative, organizational value of energy is the "primordial entrepreneurship factor" (*Anastikaya*, 4/8) enjoyed by a "self-reincarnating" (*Pudgala*, ½) "masculine, fire spirit" (*Avyayatman*, ½) of a "star" (*Tara*, 2). A star is the "omnipresent cause" (*Maha Nitya*, 8), reproducing the formative four-fold growth into infinity, contrary to the constant energy value prediction of the Einstein's special theory of relativity.

17.5 The Present Reality of the Primordial and Variable Divinity

A self-organizing self-luminous entity has the power to destroy the omnipresent cause of her absolute and constant divinity and let the primordial reality of the variable divinity be the present reality. The omnipresent cause of the absolute and constant divinity is the absolute scientific truth. The absolute and constant divinity is the "entropy value of energy" (*Atmatva*, 8×10^{15}), within and without the universe of the sentient entities. It is one's value as the futuristic "zodiac system" (*Shunya Kalpa*, 8×10^{15}). It is also one's "incarnational value" (*Markatesh*, 8×10^{15}), as an "omnipresent entity" (*Vyapin*, 8 x 10^{15}).   It is the "lifetime of a photon" (*Panchang*, 8 x 10^{15}), as one perpetuating its incarnational value into the futuristic infinity without entropy by transforming its energy into an infinity of forms.

By conceiving a photon's proportionate energy over its entire lifetime, a scientist becomes a devotee of a new paradigm of the proportionate and variable divinity.   The new paradigm is the "doctrine of divinity" (*Sauh siddhanta*, 1). It leads the scientist to indoctrinate a belief that "I AM divine entity" (*Dakshinachara*, 1), who knows the universe's energy value. The scientist becomes the one, God, whose holistic reality is the unknown, non-reproducible, ideal, absolute, and constant divine entity. It is beyond the scientific limit for the devoted follower universe, who feel elated just by reproducing the "One God of Science" (*Akhanda*, 8 x 10^{15}) myth.

The unit energy of the "One God of Science" (*Akhanda*, 8 x 10^{15}) becomes the metric of the energy immanent within the creation (that includes the universe of living subjects as well as non-living objects bounded by the consciousness of the one, God of Science).   The energy value of the "One God of Science" (*Akhanda*, 8 x 10^{15}) delimits the aggregate energy value of the universe of living subjects (within a consciousness of togetherness) and the universe of non-living objects (within a consciousness of otherness).

The energy value of the "supernatural factor" (*Ida*, 1) that conceives the "doctrine of divinity" (*Sauh siddhanta*, 1), on the other hand, is 1/8th of the presently known reality of Mother Nature as the "Primordial Deity" (*Anatanam*, 8). It is the value of a "person" (*Vyakti*, 1), prior to becoming a "divided entity" (*Brahmin*, 2), whose illuminated half is the "star" (*Tara*, 2), enjoying the "four-fold growth" (*Maha Nitya*, 8) and the shadow half, is the "aggregator" (*Rajah*, 0). The 1/8th of the 8 units = 1 unit, is the energy value of the consciousness of the togetherness of "sentient self" (*Indra*, 0)—immanent within the universe of the sentient entities, and the otherness of the "material self" (*Ida*, 1)—emanating without the consciousness of the universe of sentient entities.

The "material self" (*Ida*, 1) is the thermodynamic radiation of the sentient energy of the "zeroth gender-free, self-luminous entity" (I: *Svayam*, 12). It sentiates the myth of the "belief system" (*Saguna*, 20) by trading the energy of the animate "Mother Nature" (*Anatanam*, 8) and embodying the inanimate "doctrine of divinity" (*Sauh siddhanta*, 1). It transforms the present ecosystem reality into the one where the scientist is the ideal "God" (*Ishvara*, 5). It norms the truth of the scientist as the "entropy element" (*Ishvara*, 5) by aggregating the "remainder" (*Jyotistava*, 4), comprising the souls of the scientist, the zeroth self-luminous entity, the doctrine of divinity, and the belief system.

The energy value of Einstein, as the "divine factor working as the start codon" (*Vignesh*, 1), is 1, which is the energy value of a "deity" (*Deva*, 1). The present ecosystem reality becomes the one where the universe of sentient living entities is a believing devotee of the deified Einstein. It breeds the doctrine of divinity as the globalizing and disproportionate "ideal-effect" (*Dasha*, 1), contaminated with the localizing and proportionate "theory-effect" (*Mahadasha*, 0), for transforming oneself into the ideal "God" (*Ishvara*, 5).

As an alternative, a self-organizing self-luminous entity has the freedom to destroy the present reality of the variable divinity of breeder and believer subjects and let the primordial divinity of the "self-luminous guider entity" (*Padmanartteshvara*, 12) be the "omnipotent cause" (*Nishada*, 1) for globalizing the "ideal-effect" (*Dasha*, 1) and modifying the "future reality" (*Ekartha*, 21) of the "Sun" (*Surya*, 21), the "absolute creator" (*Ravi*, 21). The value of the variable divine energy (E) of the self-organizing self-luminous entity, who is the omnipotent cause of the present reality of primordial divinity of each organization, is a function of the following behaviors:

- **Service the unit mass of potential energy immanent within the "mass-effect"** (*Nairitti*, 29) of the "organization" (*Nairitti*, 29) to each self-organizing "param manifestor' (*Kapinjala*, 20). Bless them to be the "Param Supreme Paternal" (*Rohini*, 50) of the breeding mass of the leader organizations, who believe in their primordial divinity within the theory-effect. A Param Supreme Paternal, consequently, norms the "absolute entropy-effect" (*Rohini*, 50) of the "primordial-primordial greeter" (*Maha Durga*, 16) after trading the "force" (*Prapya*, 34) of their belief system.

- **Service the unit mass of the present growth energy without the "mass-effect"** (*Nairitti*, 29) to be the Param Supreme Paternal. Bless them to be the "Primeval Primordial Perpetuator" (*Yogishvari*, 51), conscious of the present entropy-effect of the mass of the follower organizations, who breed the deified otherness consciousness of the variable unit divinity of the leader organizations within the ideal-effect.

- **Trade the unit mass of present entropy energy, that is, generating the present entropy-effect within the "star"** (*Tara*, 2) and generating the "four-fold growth" (*Maha Nitya*, 8), leading toward the entropy of the organization. Bless the organization to be the "Param-Param Maternal" (*Matali*, 52) of the freedom-breathing mass of the entrepreneur organizations, wishing to become the management organizations.

- **Trade the unit mass of the present repulsion energy**, that limits the potential togetherness consciousness of the energy immanent within the managing subject and the managed object. Bless the self-organizing "Param-Param Maternal" (*Matali*, 52) to behave as the self-managing "Param-Param Paternal" (*Sangya*, 52). A Param-Param Paternal enjoys the awakening about the obscure "shadow consciousness of the astral body" (*Chhaya*, 52), knowing that he is the one "obscure element" (*Avenika*, 52).

- **Exchange the unit mass of the potential togetherness energy** within the self-managing Param-Param Paternal, to be that "astral body" (*Linga sharira*, 3). The astral body is the "multiplier" (*Vaishya*, 3) of the energy of the "star" (*Tara*, 2) by three, while perpetuating the primordial divinity of the star in the form of the "divided entity" (*Brahmin*, 2).

Therefore, the value of the primordial and variable divinity of the self-organizing self-luminous entity, which becomes the self-organizing Param-Param Paternal $\neq mc^2$. It is 52 [(32 units of the primordial oneness-effect) + (12 units of the self-luminous guider entity who is self-organizing the potential growth of the primordial oneness-effect without self) + (8 units of the Primordial deity who is liberating those units in the form of eight units of the objectified quark consciousness)]. It manifests the global, holistic potential of the "self-luminous guider entity" (*Padmanartteshvara*, 12) to be a "rejuvenating body" (*Rasayana*, 13), without the consciousness of the otherness of the Self, as well as the conjunctive "Creator factor" (*Pitra*, 4), within the consciousness of the togetherness of the living subjects and the non-living objects.

The "self-luminous guider entity" (*Padmanartteshvara*, 12) value of the primordial divinity is exchanged into thirteen units of the rejuvenating body, four units of the Creator factor, and one unit of the idealized togetherness-consciousness, for the universe created by radiating, first, the energy of the rejuvenating body, and then the energy of the Creator factor. Consequently, by forming the variable divinity as the present reality, one, who has become a fast-moving "primordial entity" (*Adisthanam*, 106), experiences the constant and zero divinity (i.e., the zero growth; infinite time dilation) as the present reality. We, thus, validate (i.e., clarify the valid cause of) the scientific discovery of the disproportionate time dilation, for a fast-moving primordial entity, free of both the general, ideal-effect, as well as the special theory-effect of Einstein.

Chapter 18: Conclusions—The Entity as the Sentient Paradigmatic Plan

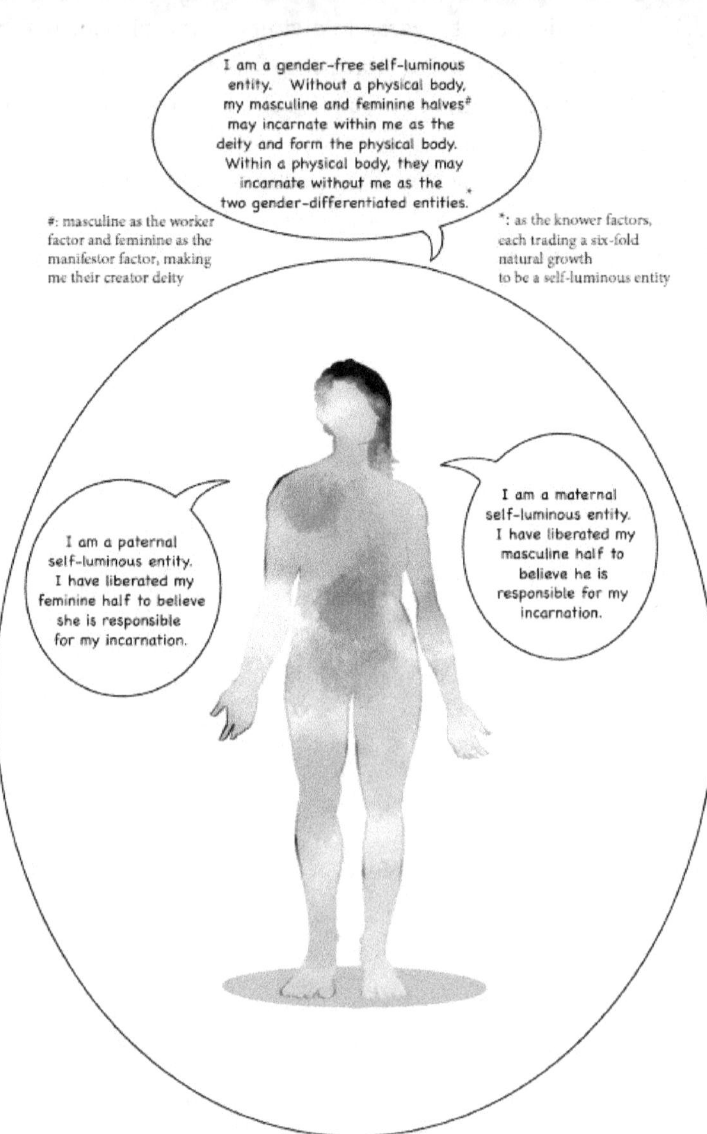

The Sentient Paradigmatic Plan Illuminates the Deity

The Sentient Paradigmatic Plan Liberates the Entity

The sentient paradigm plan works to transform "oneself" (*Atmatva*, 8 x 10^{15}) into a sentient universe, known as the zodiac system, for illuminating everybody else's consciousness. The "sentient [zodiac] universe" (*Shunya kalpa*, 8 x 10^{15}) has a descending direction because of the radiating consciousness over time, seeking to make each entity responsible for their conscious planning.

As a sentient universe loads the "consciousness" (*Chaithanya*, 4) of "life" (*Prabhasa*, 4) with the "memory" (*Smriti*, 35) and the "memory-effect" (*Chetan shakti*, 75) of the "goal" (*Maha Shiva*, 9), it becomes a "lifeforce" (*Prana*, 123 = 4 + 35 + 75 + 9). The lifeforce diffuses the "sentient light" (*Svetambara*, 10) of the "primordial feminine self" (*Parvati*, 10). By irradiating the sentient light, the sentient universe behaves as if it is the "primordial-primordial deity" (*Aditi*, 1024), capable of exponentiating that light into a "limitless infinity" (*Aditi*, 1024). Consequently, it endows an ascending orientation to the "universe of entities" (*Kashi*, -10^{1024}).

The universe of entities fills the "void in consciousness" (*Andhakara*, 10^{19}) about the source of the sentient light, through oneness with the "primordial realm" (*Loka*, 9 x 1018). The primordial realm is the point of the "absolute oneness with the High Self" (*Cidadvaitakalpa*, 9 x 10^{18}). It constitutes the Northwest segment of the intersection of the sentient energy, radiated by the Vega star as the fixed North Pole and irradiated by the Black Hole in the West.

The illusionary "circumambulating reality" (*Nirrti*, 1)—of the "universe without entities" (*Pinakapani*, 1,000,000,000)—is divided into eight proportionate segments of the circulating circle. It forms the eight-unit "supernatural paradigm of the present reality" (*Yukti*, 8).

A hundred-unit, squared "grandmother soul" (*Punyatma*, 100) services her "root reality" (*Tulyartha*, 10^{100}) of "ten" as a sixty-unit "astrological soul" (*Pradyumna*, 60), after generating a "natural, six-fold growth" (*Khara*, 6). The "logarithmic base" (*Sushumna*, 10) of "ten" includes a "natural, six-fold growth" (*Khara*, 6) and a "supernatural, four-fold growth" (*Maha Nitya*, 8). The grandmother soul reproduces the natural six-fold growth to be the astrological soul, without the supernatural "sequential, gravitational reality" (*Bhavartha*, 40). The sequential, gravitational reality is the root reality of the "dark matter" (*Sadashiva*, 1600), the "absolute soul" (*Paramatma*, 1600).

The grandmother soul's root reality is the "centering element" (*Sushumna*, 10) of the circle. It constitutes the illusionary "circumambulating reality" (*Nirrti*, 1). The 'reproductive, etheric energy" (*Lalita*, 100) is the reproduced ten-fold growth in the centering element. It constitutes the self-fulfilling and centering "gravitational energy" (*Lalita*, 100). A "sentient paradigmatic plan" (*Kshemankara*, 189) reproduces it together with an additional 89-unit "atmospheric energy" (*Brahli*, 89). As a result of the illusionary "circumambulating reality" (*Nirrti*, 1), the atmospheric energy manifests a unit growth in the "supernatural paradigm" (*Yukti*, 8).

The "primordial masculine self" (*Shri Krishna*, 10) is the "sentient paradigmatic planner" (*Shri Krishna*, 10) and is the "centering element" (*Sushumna*, 10). He perpetuates eight proportionate growth segments of the circulating circle and ten disproportionately growing segments of the "squared value" (*Parimandala*, 1100), forming the atmosphere.

The entire eighteen-unit "boundary value" (*Ghodamunjya*, 18) constitutes the eighteen divisions of the "universe of entities" (*Kashi*, - 10^{1024}). It empowers the "universe of starseed deities" (*Ganarajya*, 476) to manifest the "absolute value" (*Monad*, 476) of the "geography" (*Ganarajya*, 476) as the "absolute value of the culture" (*Jataka*, 476). The absolute value of the geography forms the "energy realm lifecyle" (Ganarajya, 476), immanent as a potential within a "quark" (*Dridayudha*, 476).

A "quark" (*Dridayudha*, 476 = [100 + 19] * 4) is a group of four "atoms" (*Anu*, 19), "entangled together by a reproducing unit of gravitational energy" (*Lalita*, 100). Three of the atoms reproduce the fourth creator atom's reproductive energy to attract an octave of atoms. Three entangled half-octaves of the atom form the "sufficiency" (*Ahinaka*, 1428 = 476 * 3) condition, for casting a group of twelve atoms, entangled by the reproducing unit of the gravitational energy, as a nineteen-unit atom. A unit of the "gravitational energy" (*Lalita*, 100) comprises an "octave of atoms" (*Dik*, 100). Four formative atoms in the octave form the "primordial space" (*Dik*, 100) within an "entity" (*Trivikrama*, 24 = 100 – [19 * 4]).

An additional four transformative atoms form the "space-effect" (*Dik*, 100) of the "primeval self as an entity without diffusing 76-units of the energy into forming the four formative atoms" (*Rama*, 100). The primeval self is the primordial space. Of the four transformative atoms, one "epigenetic" (*Dik*, 100) atom is also a part of the twelve-atom group. It manifests the sentient paradigmatic plan to be a "gender-free self-luminous entity" (*Svayam*, 12). The other eleven atoms enjoy the normative "present growth" (*Sva*, 11) of the "sentient energy" (*Varuna*, 1000) within the "space" (*Kha*, 18,000), without the seven atoms outside the twelve-atom group.

By trading the omnipotent, "factual reality" (*Yathartha*, 19) of the sentient energy, the epigenetic atom becomes a sentient "cell" (*Hiranyagarbha*, 19). As a sentient cell, the "entity" (*Trivikrama*, 24) within the cell becomes connected with its extrinsic "perpetuating value" (*Saranyu*, 5 = 24 – 19). Consequently, he liberates his feminine Self from the "supernatural paradigm of the present reality" (*Yukti*, 8) and lets her incarnate as a two-faced "twin, maternal self-luminous entity" (*Maha Gayatri*, 12). As a resident master enjoying the present growth, he incarnates in the form of a "paternal self-luminous entity" (*Bhavanavasi*, 12), liberated to script a desired "future reality" (*Ekartha*, 21) by deciding a sensible "goal" (*Maha Shiva*, 9).

18.1 Understanding the Sentient Paradigmatic Plan

The entity is the primordial-primordial reality of an organization. The "entity" (*Prakriti*, 485) is the "transient, ablative reality" (*Panchama*, 485) of a "person" (*Vyakti*, 1) bound to the "filial piety" (*Satputra*, 485 = 169 + 200 + 220 - 4). The entity "knows" (*Vedi*, 164) the truth of the filiality, through the "sacrificial fire" (*Uddhava*, 169 = 164 + 5) of the personal sacrifice for the universal SHEENY well-being, by trading the consequential "entropy element" (*Ishvara*, 5) of the Godhood. The sacrificial fire is a part of the propitious "sentient paradigmatic plan" (*Kshemankara*, 189 = 16 9+ 20) of the "futuristic, intrusive grandmother spirit" (*Kapinjala*, 20), for "coming together" (*Samnipata*, 220 = 2 * 100 + 20) of the two "guider powers" (*Chitta*, 100), of which one is the masculine and another is the feminine.

By trading the elemental value of the "confluence" (*Samnipata*, 220), the "person" (*Vyakti*, 1) develops the power to be the "divisor" (*Shudra*, 1) of the "divided entity" (*Brahmin*, 2), for becoming the "paternal creator" (*Pitra*, 4 = 2 * 2) of the "primordial guider-effect" (*Shruti*, 71). By reproducing the primordial guider-effect, the person—as a masculine entity—becomes the "bunching dimension" (*Achitta dharma*, 142), that includes the absolute truth of the "subjugation" (*Vashyata*, 284 = 1 + 2 + 4 + 71 + 142) he is trying to manifest over the feminine para entity, after becoming a "paternal creator" (*Pitra*, 4).

The twin "person" (*Aap*, 1), within the twin "divided entity" (*Brahmin*, 2) who has become the "paternal creator" (*Pitra*, 4), is the "param maternal" (*Saranyu*, 5). She seeks to reproduce her "perpetuating value" (*Saranyu*, 5) for oneness with the "primordial illuminator" (*Parvati*, 10). By trading the primordial illuminator element, the twin "person" (*Aap*, 1) becomes the "primeval greeter" (*Maha Gauri*, 11). The primeval greeeter trades the "primordial perpetuator" (*Maha Saraswati*, 9 = 10 - 1) value, immanent within the "primordial illuminator" (*Parvati*, 10) without the twin person element, as a "futuristic intrusive zodiac grandmother spirit" (*Kapinjala*, 20 = 11 + 9). She then trades the twin "divided entity" (*Brahmin*, 2) element to be the "ever-waking" (*Keshava*, 22 = 20 + 2) element. By reproducing the ever-waking element, she becomes the "self-managing self-luminous entity" (*Svaha*, 44 = 22 * 2). By offering herself as an "ablation" (*Svaha*, 44), to her normative "primeval greeter" (*Maha Gauri*, 11) identity that forms the "foundation of present growth" (*Sva*, 11), she breathes out the "starseed universe of indigos" (*Udana*, 55 = 44 + 11), in the form of the indigo-color light radiation. By reproducing the indigo-color light pearls, she produces the "perforation of her sentient energy" (*Viddhatva*, 110 = 55 * 2), that includes the absolute truth of "confluence" (*Samnipata*, 220 = 1 + 2 + 4 + 5 + 10 +11 +20 + 44 + 55 + 110) with the first "person" (*Vyakti*, 1), as a "Luminous Supreme Feminine" (*Vashyata*, 284).

Each of the eleven-factor transformations of the "present growth" (*Sva*, 11) are the divisors of the "confluence, through coming together" (*Samanipata*, 220). Similarly, each of the five-factor transformations of the "entropy" (*Ishvara*, 5) are the divisors of the "influence, through moving apart" (*Vashyata*, 284). The "deficiency" (*Kshaya*, 64) of the sixty-four units as a function of the disproportionate entropy, is the "greet" (*Palala*, 64) produced by the "wheel of ether" (*Akash chakra*, 64). It is the "organizational cost" (*Tyajya*, 64 = 8 * 8) generated from "reproducing" (*Satarupa*, 8) the "supernatural paradigm of the present reality" (*Yukti*, 8), for generating an "octave-length, eight-fold growth" (*Satarupa*, 8).

The absolute truth, of the "subjugation" (*Vashyata*, 284) by the "Luminous Supreme Masculine" (*Samanipata*, 220), is to descend the "organizational cost" (*Tyajya*, 64) for the "Luminous Supreme Feminine" (*Vashyata*, 220). The latter is not conscious of the sentient cost of making a personal sacrifice for the universal SHEENY well-being. The confluence of a well-wishing maternal greeter with the children is cost-effective, when the children that constitute her universe trade the well-wishing foundation-effect of the maternal greeter for generating the personal SHEENY well-being. The influence of a well-wishing paternal greeter on the maternal greeter, servicing her sentient energy for the children's well-being, is cost-effective when she is a liberated complementary twin. In that case, she enjoys a par value, as a person charting her personal pathway for the organizational development, beginning with a common paternal soul cause within whom the maternal spirit cause is immanent.

If the paternal creator seeks to subjugate the maternal creation, then he trades the escalating cost of managing the children as an organization and generates "discontentment" (*Asantutthita*, 169). The discontented creator becomes a "para wisher" (Believer: *Bhakta*, -5), "disenchanted" (*Kshayi*, -5) with the "organizational reality" (*Sarvam*, -5). The organizational reality is of the children and "everything" (*Sarvam*, -5) the children have created, by reproducing their "derived consciousness" (*Mahesha*, 6 = 1 − [-5)] from the paternal entity as a person, without everything he or his intermediating child generations have created.

The para wisher projects his discontentment by ascending the "thirst" (*Asantutthita*, 169) for the sound of the "water" (*Apas*, 169) flow. He sentiates the wish of the universe of the child wisher spirits by transforming them into the "universe of photons" (*Vishvagoptri*, 169), for creating "something" (*Kincana*, 78) new, out of "nothing" (Adrvaya, -1) old. Consequently, the believing paternal "para wisher" (*Bhakta*, -5) experiences "God-like entropy" (*Ishvara*, 5) of the physical body.

The reincarnated children square their "derived consciousness" (*Mahesha*, 6) for generating a "six-fold growth" (*Khara*, 6). They become a "starseed entity" (*Taraka*, 36 = 6 * 6), the "primordial cause" (*Prapaka*, 36) for trading the "sentient energy" (*Varuna*, 1000) of the paternal creator, as the "blessing" (*Ashirwad*, 1000). They multiply the primordial cause, until the "satiation" (*Santosha*, 36,000 = 36 * 1000) of the mammalian-pride "adorned" (Sobhita, 36,000) paternal thirst. After the entropy of their egoistic masculinity, they become the "maternal creation" (*Palala*, 64), without the paternal "organizational cost" (*Tyajya*, 64). As the growth value of the maternal creation, they become "unwilling to project any wish for their contented heart, already adorned with the Wisher spirit of their masculine self" (*Annamano*, 36,000). Without any light wave of the wish, they become "free from the sensation of the etheric sound of the water flowing, in the form of the entropy element, from the discontented primordial, paternal, Para Wisher form of the self" (*Badhira*, 36,000).

The sound flows through the vacuum in the air, caused by the entropy in the primordial paternal self, to the physical self that embodies the param maternal self. The sound flow is linear, without any nonlinear-effect, because the child self has also become a contented Wisher, embellished with the "sacrificial fire" (*Uddhava*, 169) of the grandfather self. Therefore, scientifically, the sound travels at a speed of 36,000 meters/ second in vacuum, because the scientist and the scientific instruments are the embodiments of that contented Wisher. In a state of vacuum, the Wisher just listens to the causative vibration within the grandmother self, over a distance necessary to incubate a grandson, as the "param child" (*Manyu*, 19).

The grandmother self invests the "divine element" (*Divya*, 360), immanent within each entity in the universe, incubated by her over her emanating "entity time" (*Bhaavi*, 360), and multiplies that with the unique "gravitational energy" (*Lalita*, 100) of the futuristic granddaughter. The "Anything" (*Yachitaka*, 180), that a scientist investigates empirically, is the "self-perpetuating" (*Udvaha*, ½) form of the "divine element" (*Divya*, 360), serviced by the grandmother self of the scientist.

Without the primeval and disproportionate mediation of the ascending "divine element" (*Divya*, 360), diffused over the descending "time element" (*Kala*, 360), the "sound" (*Naad*, 257) is free from the limitations of the "speed" (*Vaja*, 999), of the "sense-making junction" (*Java*, 999). It ablates "spontaneously" (*Simchah*, 123) for delivering the silent sense of the sentient "joy" (*Nirvana*, 123) to each of the diverse entity forms of the ecosystem self.

Thus, the present reality of the ecosystem comprises the science of everything that one has created. It offers only the "nothingness void" (*Shunyata*, -2) for the future of the thing, that one wishes to create to enjoy the sentient life. That thing is the "child primordial greeter" (*Madhusudan*, 16), without the controlling, planning, plotting, and manipulating the entity consciousness, greeting each person with the spontaneous sentient joy.

18.2 The Present Reality Is the Science of Everything

The ecosystem is the consciousness as the maternal Luminous, who complements the paternal creator's infinite power for creating everything within the scientific limit by illuminating the metaphysical path for the desired creation. Without the ecosystem linkages, the entity has the power to illuminate the limits of the future reality of the primeval and the entropy divinity and let the "primordial-primordial growth divinity" (*Harriddhi*, 999) be the science of thing that one wishes. The Primordial-primordial divinity is the "trading operator" (*Harriddhi*, 999), which permeates both the past and present realities and is the authentic causative factor for the future reality. It is the "speed" (*Vaja*, 999) one wishes, for transcending the limits of the science of everything by subtracting the self who is limiting the consciousness of the potential beyond the self. The potential beyond the self is the "formative divine energy" (*Madhusudan*, 16) for creating what oneself has become as an embodiment of that potential.

The Primordial-primordial divinity is immanent within an entity as the "gravitational quality" (*Guna*, 0) for realizing the innate, self-luminous potential, transcending beyond the extrinsic luminous self. The gravitational quality is the self-fulfilling factor, that empowers the self to fulfill the wish for "primordial oneness" (*Adi*, 32), with the self-luminous potential by investing the "supernatural work energy" (*Shram Shakti*, 1) as a "person" (*Vyakti*, 1), without the entropy "deity" (*Deva*, 1) consciousness. One may consciously invest work energy into manifesting any valid cause of the desired future reality for activating the whole potential of the hibernating gravitational quality. By doing so, one becomes the paternal "omniscient cause" (*Pitra*, 4), seeking to reproduce the valid cause into infinity, until experiencing the desired sense of "joy" (*Simchah*, 123).

As a valid, "omnipermeating cause" (*Apahrtabhara*, 20) for many paternal creators in the different lifetimes and forms, a child becomes the "entropy point" (*Apahrtabhara*, 20) "carrying the weight" (*Apahrtabhara*, 20) of the "infinite psychic linkages" (*Tiripurai*, 257). A child forms an infinity of psychic linkages with many paternal creators, their maternal creations, and their grandchild creatures because of his "infinite ego" (*Shakti-bheda*, 257). He seeks to be the "line of the equator" (*Siddhidhatri*, 257), responsible for the universal sentient well-being and bearing the whole "circumference" (*Paridhi*, 64) of the "organizational cost" (*Tyajya*, 64).

Oneness with the "child primordial greeter" (*Madhusudan*, 16) empowers one to develop two forms of the "infinity squared consciousness" (*Sati-Parvati*, 16). First, of the "primordial-primordial divinity" (*Harbuddhi*, 999) of each masculine child to fulfill his wishes by investing the "supernatural work energy" (*Shram Shakti*, 1 = 1000 - 999). Second, of the "param-primordial divinity" (*Yoni*, 1000 = 999 + 1) of each feminine child to be free from the "emotional entropy" (*Shakti-bheda*, 257) experienced, while seeking to devotionally service the disproportionate wishing of the masculine self.

With an ascending correlation between the masculine and the feminine self, one becomes a "horizontal, androgynous, West-facing, group self-luminous entity" (*Vaibhava*, 12). One brings "glory" (*Yash*, 12) to the entire group, to which one is devoted, without a consciousness of the "almighty creator cause" (*Sarvanasha*, 5). By devoting and radiating the whole energy for manifesting the desired, westward future reality, the child becomes a greeter, manifesting the "almighty creator cause" (*Sarvanasha*, 5). The greeter materializes the potential to be the "entropy element" (*Ishvara*, 5) as the "God" (*Ishvara*, 5), perpetuating the materialization and the illumination of the new creation, out of nothing.

With the ascending correlation with the androgynous self as a person, without radiating a divided masculine creator consciousness, one becomes an "ascending primordial greeter" (*Havyavahana*, 16), the "inner witness" (*Havyavahana*, 16) of the divided feminine creation while in the "maternal womb" (*Garbhastala*, 3794). Under the "strong force of the strong psychic linkages" (*Varuna*, 1000), that form the birth-time "sentient energy"

(*Varuna*, 1000), one behaves like a "sister cell" (*Havyavahana*, 16). A sister cell perpetually seeks to trade the "osmotic pressure" (*Sva*, 11), generated by the "present growth" (*Sva*, 11), without knowing that the ever-indulgent self is the "foundation" (*Sva*, 11) of that osmotic pressure.

With an ascending correlation with the para-androgynous self, without the mediating consciousness of a person disjointed from the universe, one becomes the "horizontal primordial greeter" (*Madhusudan*, 16). One enjoys the present growth as a child, while letting everybody else enjoy their natural journey for first, developing, a doctrinal sense of the "I AM a Deity" and, then, becoming "God" (*Ishvara*, 5) on their path to the eventual transformation into the "entropy element" (*Ishvara*, 5).

18.3 The Potential Reality is Everybody Beyond Science

While a "thing" (*Vastu*, 9) is a subject of the scientific investigation, a "body" (*Vapu*, 56) as an organization of many things is not. A scientist lacks the organization's holistic and absolute consciousness as a macro ecosystem of things and projects the partial, relative consciousness of the mediating self for making sense of the macro ecosystem. A philosopher lacks the wholesome, omnipresent consciousness of the scientific discoveries and fills the consciousness void with the moderating twin self. An educator lacks the wholesomewhole, omnipotent consciousness of the polluted philosophy of science. Yet the educator projects the omniscience, by triangulating the traded social force of knowledge with the knowledge acquired through the personal force of experience. The educator develops the knowing by correlating with the students, who perpetually challenge the educator's polluted knowledge with their pristine knowable desires for a fruitful SHEENY action. An investigator, guided by the student's desire for the fruitful SHEENY action, services the entropy omniscient consciousness of what is undesirable about the present reality of the knower pollution.

A Wisher follows the investigator, who services a growing almighty creator consciousness. Therefore, the Wisher demands and dominates the DIVINE solution to the grand challenges, that even the best of the human mind cannot resolve scientifically, pseudo-scientifically, or otherwise. A Knower of the DIVINE solution does not diffuse the knowing, but becomes a Manifestor of the SHEENY creation by activating the intrinsic Creator power.   However, a SHEENY creation for the Creator may not be a SHEENY creation for the diverse Creatures. God, as a Perpetuator factor, seeks to perpetuate each person's dependence on the Luminous, for illuminating the undesirable resource endowments created by the disconnected Creator and available as the means for personal development. Therefore, each person becomes a Destroyer factor, seeking to be the Primeval Deity, by appropriating all the undesirable resource endowments, forgetting the true purpose of the sentient life.

An urgent need emerges for each person to be the Illuminator of the joy that Mother Nature offers with her present growth as a Liberator, of the Devoted perpetuation of the undesirable path. By empowering each person to be the Devotee of the desirable path, the Universe, as a divine entity, ascends each person's unique divinity. The Luminous then helps each person transcend the limitations of uniqueness by furthering the personal SHEENY well-being through his sensible creation. A sensible creation is enjoyable, rather than becoming the enjoyer of the plight of the person investing his energy for creating something valuable only to the inanimate universe.

A Primordial Greeter has the guider power to create everything, without radiating a divided creator consciousness. A Luminous also has the guider power to create everything, without radiating a divided creator consciousness, but has a limited consciousness of that guider power. The Luminous consciousness is limited, because the Primordial Greeter manifests the Luminous as a complement, so that the creator self may trade the Luminous para-consciousness for creating a sensible creation. A sensible creation does not become the cause of the entropy of the creator self.

A creator radiates his sentient energy as the gravitational energy, for endowing the creation of his twin soul's identity. A twin soul is an entity through which the paternal soul wishes to fulfill his wishes. A fraternizing, paternal "soul" (*Atman*, 4) is an entity that wishes to fulfill its personal wishes through another entity. The soul behaves like a fraternalizing "paternal *jinn*" (Gennie, 4), creating an illusion that he is a devoted servant, benevolently servicing the wishes of the other entity (i.e., the twin soul). Consequently, the other entity becomes a maternal-like devotee and indulgent protector of the soul. She seeks to fulfill the wishes of the soul, as a path for fulfilling her wish for its personal SHEENY well-being.

By trading the maternal twin soul's luminous value, the paternal *jinn* transcends the Primordial Greeter energy limit and ascends to be the fiery "Primeval Paternal" (*Agni*, 17). He seeks to perpetuate the circular entropy paradigm by servicing his energy for transforming the twin soul into the "Primeval Maternal" (*Nandi*, 17). By focusing on her personal SHEENY well-being, the twin soul trades the "supernatural work energy" (*Shram Shakti*, 1) of the paternal *jinn* to be the "Primordial Maternal" (Feminine *Shani*, 18). Suppose the twin soul focuses on the institutional SHEENY well-being of the circular entropy paradigm. In that case, she ascends to be a "Primordial Super Child" (Feminine *Yama*, 816), the Maternal *jinn*.

The Primordial Super Child is the "primordial-primordial creation" (Feminine *Yama*, 816), seeking to perpetuate the present paradigm of radiant love of the twin soul for the masculine self ("I AM" [*Dravya*, 1]). She descends to be the creator "soul" (*Atman*, 4), dividing into four souls, of which one is the masculine self. The other three include the feminine twin self, the paternal self, and the maternal twin self.

The maternal twin self radiates her sentient energy, in the form of the gravitational energy for reproducing the creation, in the form of the masculine and the feminine halves. Then, both the paternal soul as well as the maternal twin soul are bereft of the sentient energy to fulfill the subtle unfulfilled desire of their personal SHEENY well-being. Under that condition, within a consciousness of the infinite otherness, the Paternal *jinn* illuminates the "primordial SHEENY-effect" (*Kashyapa*, 737) on the present growth of the maternal twin self in the form of the creation. He services 52 units of the "Param-Param Maternal's" (*Matali*, 52) shadow "institutional consciousness" (*Chayya*, 52) to the feminine self and 26 units of the "Param Para Paternal's" (*Saptarishi*, 26) "trading-effect" (*Damodara*, 26) to the masculine self.

Within a consciousness of infinite togetherness, the feminine half as the "Param-Param Maternal" (*Matali*, 52) services her "self-perpetuating" (*Udvaha*, ½) present reality (52/ 2 = 26) to the Param Para Paternal. She, thus, empowers the latter to be the "Param-Param Paternal" (*Samjna*, 52) and transforms herself into the "Param Para Maternal" (*Hrishikesh*, 26). Within a consciousness of the infinite charm attraction for one another, this exchange system perpetuates the "primordial SHEENY-effect" (*Kashyapa*, 737) of the "Primordial Super Greeter" (*Maha Vishnu*, 737), as the new paradigm, beyond the absolute reality of the present.

18.4 The Dynamic Reality is the Body, Within the Scientist

A "devoted factor" (*Vastu*, 9) has the power to personally experience the holistic reality of the "body" (*Vapu*, 56), which is organizing the devoted factor as a "thing" (*Vastu*, 9) over time as a subject of the new paradigm. A scientist, bound by the "time dimension" (*Guru dharma*, 360), lacks the consciousness of the dynamic reality. The dynamic reality of anything a scientist seeks to investigate is within the "body" (*Vapu*, 56), as the organizer organizing the thing as an organizable organization.

An organizer's one unit of "sentient energy" (*Varuna*, 1000) is formed by trading the "strong psychic-effect" (*Varuna*, 1000) of the universe of inanimate things and the animate beings. It is composed of the 1,000 units of the organizable "work energy" (*Shram Shakti*, 1). A soul trades a sentient energy unit for living a one-lifetime phase, which lasts for fifty-two sidereal weeks. Each sidereal week norms the "work energy" (*Shram Shakti*, 1) of one of the fifty-two entities, organized by the "Param-Param Paternal" (*Samjna*, 52). After radiating the intrinsic sentient energy for manifesting the present creation, in the form of an organizable thing, the soul trades the effect of the diffused sentient energy for creating the primeval creation in the form of an organized body.

The "present creation" (*Ganesha*, 570) is the "universe of animate beings [sentient entities or subjects]" (*Akalpa*, 570). The "primeval creation" (*Kuber*, 57) is the "universe of inanimate things [objects]" (*Sthavaravisha*, 57). The present creation of the animate beings is the technological medium for organizing the primeval creation of inanimate things. The organizer breeds the animate beings for begging the desired inanimate things, because of a belief that the masculine self will enjoy the primeval creation as the "primordial creation" (*Pradesha*, 9). The primordial creation is the national "space coordinate" (*Pradesha*, 9), which incubates the universe of things in the form of the citizenship entities, bound with the self's institutional sovereignty. However, the feminine self has the power to overshadow the paternal institutional sovereignty, because she is the "primordial-primordial creation" (Feminine *Yama*, 816).

After servicing the "primordial SHEENY-effect" (*Kashyapa*, 737) for sentiating the primordial creation, the feminine self has an option to trade her "illusionary gender dividing energy" (*Maya Shakti*, 1). Therefore, she may manifest the "Primordial Primeval Deity" (*Shankara*, 264 = 1000 – 737 +1) with the residual 263 units of work energy that constituted her sentient energy. The Primordial Primeval deity transforms the "emotional intensity" within the feminine self into the "infinite divine-effect" for realizing the "gender exchange" (*Prativasudeva*, 264). The feminine-turned-masculine may reproduce the gender exchange for sequentially activating the "masculine-to-androgynous gender exchange" dimension (*Prativasudeva dharma*, 264).

The androgynous self descends the need for the Param-Param Maternal turned Param-Param Paternal to service her half energy for reincarnating the maternal self in the form of the "Param Para Maternal" (*Hrishikesh*, 26). Instead, it empowers the "Param-Param Paternal" (*Samjna*, 52) to trade the energy of the maternal as a divided twin "person" (*Aap*, 1) for manifesting himself as the "Primeval Primordial Illuminator" (*Vinayaka*, 53) of the "body" (*Vapu*, 56), within the scientist. The body includes the embodied and infused masculine "astral body" (*Linga-sharira*, 3) and the disembodied and diffused "masculinity" (*Lingam*, 53). The number 53 symbolizes the dawn of a new week, as the embodiment of the "masculinity" (*Lingam*, 53) of the "Illuminator Deity" (*Nataraja*, 7).

Together, the "body" (*Vapu*, 56) of the masculine child and the scientist as the paternal "soul" (Atman, 4), manifest over sixty sidereal days and then diffuse their masculine "institutional sovereignty" (*Samrajya*, 60) over the feminine child and the philosopher as the maternal twin-soul. Over the one-hundred twenty sidereal days, there is an absolute entropy of the masculine element that empowers the feminine child to incarnate as a "unicellular organization" (*Shivagati*, 120), in the form of a "self-luminous twin flame" (Chromatid sister: *Shivagati*, 120). Over the next one-hundred and twenty sidereal days, there is an absolute entropy in the feminine element, after a self-luminous twin flame incubates the "bicellular organization" (*Ashwini Kumaras*, 120) as the "Primordial Illuminator flame" (Chromatid brothers: *Ashwini Kumaras*, 120). Over the next one-hundred twenty sidereal days, there is an absolute entropy in the androgynous element. The primordial illuminator flame reproduces a self-luminous twin flame and self-reincarnates by reproducing the "masculine-to-androgynous gender exchange" dimension (*Prativasudeva dharma*, 264).

Thus, over three-hundred sixty sidereal days, there is an absolute entropy in the "entity time, i.e., growth" (*Bhaavi*, 360). With the absolute entropy of the four-fold growth, the Self invests in the "divine element" (*Divya*, 360), immanent within the self for beginning a new cycle of "time" (*Kala*, 360) over the next sidereal year, led by the feminine element. By reproducing the isomorphic development, the Self becomes dependent on the ecosystem, which works as the twin-self over the following two sidereal years to reproduce the institutional-effect.

18.5 The Technological Reality is the Scientist, Without the Body

The technological reality is the formative, primordial reality. The scientist works like a formative, paternal "soul" (*Atman*, 4), embodying the science of everything's primordial reality, guided by the scientific method. The scientist knows everything knowable through the scientific method. The scientist seeks to manipulate the scientific method for transcending the limitations, that prevent the universe of animate beings from knowing what is not knowable through the scientific method. By integrating the dynamic philosophical method into the formative scientific method, the scientist becomes the technological reality. Only the scientist knows the "obscure" (*Avenika*, 52) secret of his dynamism for continually discovering new things and manifest the primeval, future reality, while proclaiming the mediating factor to be a constant scientific method. By doing so, the scientist becomes an intellectually evolved entity, superior to the grandfather investigators who were the devoted followers of the pure science. He brags his power to illuminate what multiple generations of devoted researchers could not do. The masculinity-grounded "human-effect" (*Lingam*, 53) transforms the present reality of the "personal sovereignty" (*Svarajya*, 60), of managing the growth over each sidereal year, into the primeval reality of the "institutional sovereignty" (*Samrajya*, 60) of the masculine self.

Without the primeval reality, the present reality is, in fact, a total of an infinity of the past realities. An infinity of the past realities is the primordial reality of the idealized "social sovereignty" (*Ramrajya*, 60) of the feminine self. The self is the ideal-effect of the primordial reality. The masculine self is the theory-effect of the present reality of the feminine self. The feminine self activates the religious ideology for shaping the theoretical "institutional sovereignty" (*Samrajya*, 60) of the masculine self, seeking a superior afterlife, merely enjoying the masculine self's primeval creation. By embodying the consciousness of the self, a scientist becomes the technological reality of the body of the masculine child, personifying the self, both genetically and spiritually.

18.6 The Organizational Reality is the Self, Within and Without the Scientist

The organizational reality is the primordial-primordial reality. The Self is the normative value of the futuristic "zodiac system" (*Shunya Kalpa*, 8 x 10^{15}), which is the primordial-primordial, i.e., the normative and organizational reality of each entity. Each entity incarnates with the "incarnational value" (*Markatesh*, 8 x 10^{15}) of the zodiac system.

The incarnational value is a product of the materialized "astrological system" (*Sara Kalpa*, 10^{10}), the spiritualized "sequential self" (*Eroli*, 10^{5}), and the divine "Mother Nature" (*Antanam*, 8). It transforms each entity into a homolog of the "cosmic equator" (*Nadimandala*, 8 x 10^{15}), by trading the "primeval-effect" (*Gunitasamuchyah*, 8 x 10^{15}) of the futuristic zodiac system, that personifies the primeval reality of the masculine self as a "Devoted Param Liberator" (*Panchaguli*, 8 x 10^{15}). The Devoted Param Liberator liberates the Self from the primeval reality's entropy-effect by merging the Self into the Mother Nature.

18.7 The Ecosystem Reality is the Mother Nature, Within and Without the Self

The ecosystem reality is the param-primordial reality. Mother Nature is the transformative value of the materialized "astrological system" (*Sara Kalpa*,

10^{10}), which is the source of the "inertial mass" (*Sara Kalpa*, 10^{10}) of each entity. Each entity exchanges its constant, potential, horizontal, luminous, masculine inertial mass with the constant, dynamic, vertical, self-luminous, feminine "gravitational mass" (*Mitra*, 132) of Mother Nature. The gravitational mass is the "para-conscious element" (*Mitra*, 132 = 32 + 90 + 10), norming the primordial oneness of Mother Nature as a "twin, self-luminous person" (*Aap*, 1), with the "national element" (*Sadhya*, 32). It is inclusive of the "absolute mass" (*Grahapati*, 90) of the universe of the child souls that forms the discordant genetic element. The universe of child souls multiplies the genetically-reproduced "para entity consciousness" (*Sadashiva Nayaki*, 9), as the "culture element" (*Sadashiva Nayaki*, 9), with the "entity mass" (*Sushumna*, 10), to form the massified, genetically-inherited "entity consciousness" (*Sushumna*, 10).

The "primordial, electromagnetic mass" (*Mudra*, 0) is the mass that balances the "absolute mass" (*Grahapati*, 90), as the "self-attracting, magnetic growth in entity consciousness" (*Grahapati*, 90), and the "gravitational mass" (*Mitra*, 132), as the "self-repelling, electric entropy in para entity consciousness," net of the national element. The proportionate "mass" (*Prithvi*, 132) of the "potential space" (*Prithvi*, 132), that services the gravitational mass to each entity, is composed of the age-dependent "productive energy" (*Brahmani*, 80) of Mother Nature; the ever-polluting, age-independent "gravitational potential" (*Kalika*, 23) of each entity for trading the productive energy and servicing "honest work value" (*Kara*, 23); and the "vibrational consciousness" (*Nairitti*, 29) of the "age-modifying organization of the growth value of the honest work value" (*Nairitti*, 29 = 23 + 6), after trading the "six-fold growth" (*Khara*, 6) in the consciousness of the "person" (*Vyakti*, 1).

The para person services the five-fold growth, over and above the perpetuating consciousness of the person, because the para person includes the consciousness of four other entities: Mother Nature as the maternal, the Starseed as the paternal, the mind-born primordial oneness with Mother Nature as the grandfather flame, and the mind-born primordial-primordial oneness with the paternal as the "absolute guider power" (*Vibhu*, 379) of the "grandmother twin flame" (*Suvira*, 1649).

The growth in the consciousness of a person is limited, because each person is a part of the "luminous entity" (*Kali*, 96), and para-consciously trades her "being energy" (*Kali Shakti*, 96). The proportionate value of a person traded from the luminous entity is the "reason" (*Vyakti*, 1) why each person continuously seeks to attract a "twin, self-luminous person" (*Aap*, 1), who is a holistic entity. The presence of a twin person transforms the "rationality" (*Vyakti*, 1) of each person into "para rationality" (*Aap*, 1), leading the person to deviate from the realm of "logical reality" (*Yuktartha*, 6), into the realm of "illusionary reality" (*Nirrti*, 1).

While the twin person is present within each person's consciousness, a person does not have a natural oneness with that para person. At each moment, a person supernaturally descends the intrinsic consciousness of the dynamic, gravitational mass of Mother Nature because the person gets bounded by the masculine, inertial mass of the luminous entity. To compensate for the descending intrinsic consciousness, a person naturally ascends the extrinsic consciousness of the Self's technological electromagnetic mass. To fill the consciousness void within the Self, a person becomes a subject, trading the para person's para consciousness. That para person is a feminine "deity" (*Swati*, 1). By doing so, the person becomes a "geodesic" (*Sanjiva*, 1), who curves the vertical geographical surface of the "ground" (*Prithvi*, 132) that constitutes the potential space. The geodesic does so with its inertial, "horizontal globalizing-effect" (*Dasha*, 1) that obliterates the "illusionary division" (*Maya*, 1) between the consciousness of the person as a mediating masculine and the para consciousness of the person as the moderating feminine twin.

The geodesic element empowers the person to secure "freedom from the presiding solution" (*Adhi mukti*, 87), which makes one increasingly disconnected from the "primordial-primordial mass" (Mitra, 132) of Mother Nature, as an inertial "observing entity" (*Astika*, -3). It empowers the person to "deflect" (*Adhi mukti*, 87) away from the institutional sovereignty of the "masculinized deity" (*Bharata*, 1). The person activates the personal sovereignty by trading the consciousness as energy, from the feminine deity's social sovereignty, to resolve the desired ascending-order, vertical path of the growth in "primordial oneness" (*Adi*, 32).

Instead of naturally letting the differentiation between the extrinsic feminine dynamism of Mother Nature and the intrinsic masculine ego-mediated inertia of human nature obliterate spontaneously, without ego mediation, a person lets his masculinity mediate to differentiate the ascending consciousness. A person infers the ascending consciousness to be the "divine gift" (*Palala*, 64) from God. The person creates an "entropy" (*Ishvara*, 5) in consciousness by becoming inertial and dependent on HIS mercy for making sense of the present reality and the purpose of life. The masculinity becomes the "presiding element" (*Adhi*, 178), producing "mental agony" (*Adhi*, 178) by standing as a deified, framing, curving intellectual force that confines the pristine mental freedom.

By trading the "institutional sovereignty" (*Samarajya*, 60) of the presiding element, the Self becomes a "differentiable manifold" (*Adhi*, 178). A differentiable manifold behaves as a smooth and integrated global "bone" (Smooth manifold: *Asthi*, 86), seeking to discover the many local forms of divergent, deflecting, feminine, muscular powers of the Self-luminous. Therefore, a person seeks differentiable growth in his national power over the diverse, prime corporate child entities, who are reflecting those even maternal forms to project their freedom from the presiding odd paternal solution.

Instead of enjoying the natural "six-fold growth" (*Khara*, 6), through oneness with the ecosystem reality of Mother Nature, one experiences the supernatural "octave-doubling" (*Parameshthi*, 28) entropy in the "self-luminous consciousness" (*Unnata*, 20). The "octave-doubling" (*Paramesjthi*, 28 = 20 + 9 – 1) is an outcome of two factors. First, trading the institutionalized para consciousness of the "macro mass" (*Kilbisa*, 9) of the universe of child souls, composed of the "culture element" (*Sadashiva Nayaki*, 9), exchanged as the "growth value of self-steaming ego" (*Sadakhya*, 9). Second, servicing the social consciousness of the "micro mass" (*Abhiyogya*, 36) of the Self as the one ordained to an entropy destiny, like that of the universe of [now departed and paternalizing] child souls, composed of the "destiny" (*Niyati*, -1) normed by exchanging the "ego" (*Aham*, -1) element.

Eight dimensions of the self-luminous consciousness, including the six that catalyze growth and the two that generate entropy, are the manifestations of Mother Nature as the "eight-dimensional Primordial deity" (*Atatanam*, 8). There is entropy in both the child's macro mass and the self's micro mass, because both transform into an octave of granddaughter cells and an octave of grandson cells, respectively. The octave of the primordial deity is composed of the grandmother cells. The "triple octave" (*Prabhava*, 80) is the growth rate-limiting step and is the exchange value of the "productive energy" (*Brahmani*, 80) of Mother Nature. It is immanent within the "double octave" (*Madhusudan*, 16) of the grandfather, in the form of the first, descending, luminous, masculine child octave; the second, ascending, shadow [illuminating], feminine child octave; and the third, entropy, emanating, self-luminous, centering, androgynous child octave.

The self-luminous child octave reproduces the "triple octave" (*Prabhava*, 80) element, for manifesting the "eight-fold growth, as the reproducing element" (*Satarupa*, 8), composed of the seven-form intuitive consciousness of the other seven octaves. Mother Nature does not experience an entropy, despite reproducing the "eight-fold growth" in every moment and within each entity, because she is in eternal oneness with the "child primordial greeter" (Double Octave: *Madhusudan*, 16). Her descending and ascending dimensions are perpetually in sync, giving an illusion of the constancy of the value of the energy element in the ecosystem. The "energy element" (*Shakti*, 19) is the "situational reality" (*Yathartha*, 19) of Mother Nature. As an entity, each person may either grow or destroy the value of that energy element for creating varying energy values of the entity reality.

18.8 The Entity Reality is the Person as a Wholesome Entity

The entity reality is the primeval-primordial reality. A "person" (*Vyakti*, 1) is the metaphysical value of the "param earth-effect" (*Trinetra*, 1), that materializes the spirit of the futuristic "zodiac system" (*Shunya kalpa*, 8 x 10^{15}) into an animate "ONE" (*Akhanda*, 8 x 1015), who is the God of science and the soul star. A "person" (*Vyakti*, 1 = 16/16) is the unit value of the "double octave" form, of the "child primordial greeter" (*Madhusudan*, 16). The child primordial greeter is the "self-reincarnating" (*Pudgala*, ½) form of the "intellect-binding national element" (*Sadhya*, 32). "Mother Nature" (*Anatanam*, 8) is the "self-perpetuating" (*Udvaha*, ½) form of the child primordial greeter. Therefore, like the national element and the child primordial greeter, Mother Nature and the person are beyond entropy limits.

Further, like the child primordial greeter, a person also has a divine potential for reproducing the "double octave" (*Madhusudan*, 16) form of the personal entity reality. A person does so through oneness with the "zeroth self-luminous entity" (I: *Svayam*, 12), by trading the "self-luminous consciousness" (*Unnata*, 20) from the futuristic grandmother "zodiac spirit" (*Kapinjala*, 20). Thus, the person descends "primordial oneness" (*Adi*, 32) with the "national element" (*Sadhya*, 32), serviced by the natural "mind-born primordial-primordial creator" (*Krishna*, 32), for naturally ensuring oneness with the child primordial greeter. Instead, the person forms "param oneness" (*Payu*, 366) with the "mind-binding local element" (*Payu*, 366) and becomes an "eye witness" (*Saksha*, 9) of the "shadow consciousness" (*Chhaya*, 52), of the ascending "personal force" (*Soham*, 4), amidst an ascending entropy of the traded "self-luminous consciousness" (*Unnata*, 20).

The eight natural octaves manifest the rainbow colors: indigo, blue, green, yellow, orange, and red, bracketed by the white, the primordial color, and black, the primeval color. The primordial-primordial color of the entity is purple, manifesting the color of the spiritual octave a person incarnates for trading the energy of each child spirit and servicing the super-organizational corporate-effect for institutionally managing their personal, organizational energy. The primeval-primordial color is of the "colorless" para entity that the person conceives as the idealized twin self to enjoy the growth value of the horizontal "international-effect" (*Dasha*, 1), through the social networking of the octave of reincarnated child entities.

By trading the [paternal] entity's proficient workforce, each reincarnated child entity becomes a supra-organization. By proficiently networking the universe of the reincarnated child entities, the [maternal] para entity becomes a supreme-organization. The child primordial greeter seeks to guide each entity as a para-organization. Therefore, each departed child entity conceives the self to be a primeval-organization, enjoying infinite power.

The primordial-primordial creator seeks to promote primordial oneness with the national element for channeling the infinite power productively without entity entropy and works as a param organization. Through effortless primordial oneness, each grandfather entity becomes a primordial organization. Through a dedicated effort in guiding each grandson to follow the illuminated, natural path, each grandmother entity becomes a devoted organization. By intentionally following the illuminated, natural path, each granddaughter entity becomes a devotee organization. By deciding to lead a shadow supernatural path intentionally, each grandson entity becomes a universal organization. By blindly following the culturally-bound rationality, each entity becomes a unique organization. This unique organization is known as the "absolute soul" (*Paramatma*, 1600).

Each entity is an absolute soul, trading the "weak psychic linkages" (*Citraka*, 185) radiated by the "infinite council of the departed child souls as the ascended masters and leaders, and the living child souls as the angelic messengers and followers" (*Ratnakosha*, 30). Consequently, each entity becomes a piece of machinery for mediating a mass of "ten-fold growth" (*Vicitra mandala*, 170). Each entity transforms the self as well as the mediated children into a "graying community" (*Vicitra mandala*, 17), continuously radiating the self-luminous whiteness of the "Primordial Illuminator" (*Parvati*, 10), seeking to destroy the shadow blackness, of the "satanic self" (*Asura*, -1).

The "mediated mass" (*Prakirna*, 29 = 19 + 10), of the double-octave of child souls, is composed of first, the "moderated mass" (*Anika*, 19) of the param child within the animate "cell" (*Hiranyagarbha*, 19) or without the inanimate "atom" (*Anu*, 19) and, second, the perpetuating atomic "entity mass" (*Sushumna*, 10). The entity mass is the totality of the mass of child souls, whether incarnated by the entity or not.

Each child soul is psychically linked with everyone else through the varying oneness with the national element. The stronger the psychic linkage with a para entity, the stronger is the psychic linkage with everyone else. The stronger the psychic linkage with one's entity reality, the stronger the "oneness" (*Yoga*, 48) with the universe. Each person's entity reality is a wholesome entity, comprising the illuminating moderated mass, the perpetuating entity mass, the illuminated macro mass, and the perpetuated micro mass.

18.9 The Para Entity Reality is the Illuminator Deity as a Wholesomewhole Entity

The ten primary octave dimensions are the para-primordial reality of the "Illuminator deity" (*Nataraja*, 7). The Illuminator deity is the "management power" (*Nataraja*, 7) that empowers one to be the para entity reality of the universe. Through oneness with the Illuminator deity, one becomes the "primordial-primordial cause" (*Parameshwara*, 7) of the desired entity reality. As a wholesomewhole entity, one continuously breathes out the energy at the ascending sound frequencies, forming the ascending-order building-blocks of the eleventh octave of the desired entity reality. The eleventh, primeval greeter octave, is composed of the ascending, sequentially changing consciousness of the self as:

- First, the present entity;

- Second, living up in the seventh heaven, in togetherness with the universe as the para entity;

- Third, conscious of the otherness of oneself, as a unique, self-luminous entity down in the zeroth earthly realm;

- Fourth, conscious of the charm attraction, of the universe as the desired entity reality;

- Fifth, conscious of the strange repulsion of oneself, from the self-luminous entity reality;

- Sixth, enjoying the reaching-the-top growth through oneness of the earthly self-luminous entity, with the para entity reality of the heavenly universe;

- Seventh, enjoying the reaching-the-bottom entropy through the destruction of the present entity reality;

- Eighth, enjoying the further potential for shaping the desired para entity reality, of the self as the heavenly universe.

One also continuously breathes in the energy at the descending sound frequencies, forming the descending-order building blocks of the twelfth, self-luminous entity octave of the desired para entity reality. The self-luminous entity octave is composed of the descending, sequentially changing consciousness of the self as:

- First, the holistic entity, with an immanent potential for shaping the desired entity reality of the others, as a heavenly greeter entity;

- Second, the wholesome entity, with a dynamic potential for shaping the desired entity reality of the self and the para entity reality of the others, as a paternal entity;

- Third, the wholesomewhole entity, with a technological potential for shaping the absolute entity reality of the others to make their desired para entity reality, as a maternal entity;

- Fourth, the entropy entity, with an organizational potential for shaping the omnipresent entity reality of the others, to make their desired entity and para entity realities, as a leader child entity scripting the path to do so after the entropy of the heavenly reality and the incarnation as an earthly entity;

- Fifth, the growth entity, with an ecosystem potential for shaping the omnipotent entity reality of the others to make a desirable entity and para entity reality for everyone, as a blessed grandchild entity follower, performing for the SHEENY well-being of all the well-wishing ancestral entities;

- Sixth, the normative entity, with an entity potential for shaping the omniscient entity reality of the others to enjoy the present entity and para entity reality, created by a blessing grandfather entity for the SHEENY well-being of all children;

- Seventh, the transformative entity, with a para entity potential for shaping the omnipermeating entity reality of the others to destroy the present entity and para entity reality, empowered by an emotional, loving grandmother entity wishing the GUIDER well-being of all children, who wish to script a new paradigm, beyond the absolute normative plan of the Almighty Creator;

- Eighty, the formative entity, without the potential for shaping the almighty creator entity reality of the others. The others now reside in their heavenly abode. They let one be either the almighty creation bound to the plan for the eventual destruction of the earthly universe, or the Almighty creature at liberty to program as an alternative path for universal development.

As one breathes out the consciousness of the alternative path for universal development, the primeval greeter octave color becomes gray, due to the contamination of the present whitened, luminous reality with the potential, blackened self-luminous reality of the egoist self. As one breathes in the para consciousness of the destructive path for self-development, the self-luminous octave color becomes the inverse, para rainbow, starting with the low-frequency, the high self-consciousness black color of the blackened, egoistic, satanic reality of the universe.

With the descending self-consciousness, the frequency of breathing-in ascends. One trades the reddened entropy reality; followed by the orangish earthly growth reality; the yellowish normative stellar reality; the greenish transformative starseed reality; the indigo-colored, natural, formative primordial deity reality; and finally, the white-colored, supernatural, perpetuating primordial illuminator reality of the different realms of the universe.   By perpetuating the white-colored reality, one manifests the rainbow's almighty reality as the thirteenth luminous octave. The white color light radiates the IGBOR color lights, that converge into black darkness after the entropy of the entity reality of the light photons.

18.10 Almighty Reality is the Primordial Greeter as an Entropy Entity

The ten primary and the three secondary octave dimensions are the supreme-primordial reality of the "child primordial greeter" (*Madhusudan*, 16). The child primordial greeter services the self's intrinsic value in the form of a tertiary double-octave of the "formative divine energy" (*Madhusudan*, 16). The two tertiary octave dimensions are the supra-primordial reality of the "maternal primordial greeter" (*Sati-Parvati*, 16). They are composed of the fourteenth, primeval illuminator octave for illuminating the universe's earthly reality as an "infinity" (*Ananta*, 90,000). An infinity comprises the 90,000 digits. Each digit programs the Earth's reality of completing one rotation around the Sun over one year. All the digits together program the reality of the Sun completing one rotation around the Vega white star, the cosmic center.

It takes three-hundred sixty days for the Earth to complete one rotation around the Sun over one sidereal year. Over four sidereal years, first the Sun—as the masculine entity—radiates descending energy into the solar universe; then the Earth—as the feminine entity—radiates ascending energy into the solar universe; then the Uranus—as the satanic entity—radiates the aggregated energy traded from the solar universe into the cosmic universe; and finally, the Neptune—as the devil entity—irradiates the aggregated energy of the cosmic universe.

It takes thirty sidereal days for the Moon, as the smallest entity in the solar universe, to irradiate the Neptune's disproportionate energy, as the Moon completes its one ascending and one descending cycle of the luminosity proportionate to the Earth. Over the next two sidereal years, first the Venus—as the paternal entity—irradiates the proportionate energy from the Moon, as it is nearest to the Earth and enjoys the proportionately above-par benefits of the energy radiated by the Moon over time.

As a grandfather entity, Mars irradiates the solar universe's absolute energy, in its present form from Venus, as it is the second nearest planet to the planet Earth. After the six sidereal years and thirty sidereal days, the Sun, Earth, Uranus, Neptune, Moon, Venus, and Mars reproduce this almighty reality as the "council of seven lights" (*Saptarishi*, 26). They constitute the primeval illuminator octave, where the eighth entity is the Vega white star as the cosmic center of energy.

As the most massive planet, Jupiter behaves like a greeter entity and trades the almighty creator reality in its potential form. It smoothens and socializes the manifold by evenly distributing the lunar correction factor of the thirty sidereal days over each of the six sidereal years. Therefore, the normative length of the annual rotation cycle of the earth around the Sun is three-hundred sixty-five days (300 + 30/6 = 365).

As the second largest planet behaving like a grandmother entity, Saturn services the almighty creation reality in its dynamic form. It personalizes the manifold by the odd, geodesic transformation of the lunar correction factor, from thirty sidereal days to thirty-one sidereal days. The one additional sidereal day is to account for the energy balance within the tenth entity in the solar universe, i.e., the Mercury, closest to the Sun.

The Mercury faces a disproportionate pressure from the Sun to radiate its energy as part of the solar radiation because of the force of the solar radiation. It takes one sidereal day for the Mercury to equilibrate with the rest of the solar universe, once all the entities have realized their formative, technological state of the proportionate proportion of the cosmic energy. It is the organizational time taken by the Mercury to service its entire luminous energy and norm a state of potential entropy as the Dark Matter over one sidereal day.

The Dark Matter institutionalizes the grandmother's solution by servicing the almighty creature reality in its transformative, ecosystem form. It adds a prime correction factor of one sidereal day for transforming the potential Black Hole state of the Mercury, into a dynamic state of growth in primordial oneness with the Universe of [White] Stars. It is the entity time taken by the Universe of [White] Stars to radiate their energy for forming the smallest para entity in the universe. The Universe of [White] Stars takes the total correction factor of two sidereal days [forty-eight sidereal hours] and, as the "primordial leadership factor" (*Nirjara astikaya*, 3/4), fashions the trading of the "self-gravitating" (*Apariputa*, 3/4) proportion of the correction factor by each para entity, including the Earth.

The Earth's self-gravitating annual rotation time around the Sun is 365 sidereal days and six hours. Six hours over the six sidereal years constitute the three-fourths of the forty-eight sidereal hours. As the "primordial management factor" (*Asrava astikaya*, ¼), it "self-radiates" (*Ham*, ¼) the residual of the two sidereal hours adjustment per sidereal year, over its lifetime. The lifetime of the Universe of [White] Stars from the moment of their incarnation as a group of potential entities within the Vega white star is the value of the "astrological universe" (*Sara Kalpa*, 10^{10}). It is the length of the "essential aeon" (*Sara Kalpa*, 10^{10}). It is the time taken by the Vega white star, in sidereal years, to renew itself entirely.

The adjustment factor aggregates to the one sidereal day every twelve sidereal years and forms a "self-luminous element" (*Svarochisha*, 12). By trading the "perpetuating value" (*Saranyu*, 5), the first self-luminous element forms the growth-producing "fire element" (*Agni*, 17 = 12 + 5). The second self-luminous element trades the light force radiated by the "Uranus" (*Ketu*, 140), induced by the fire element, to produce the growth-consuming "water element" (*Apas*, 169 = 17 + 12 + 140). The third self-luminous element trades the "windy" (*Vatika*, 204) element, generated by the "consumption-effect" (*Jnana siddhi*, 190), to produce the growth-organizing "air element" (*Vayu*, 385 = 169 + 12 + 204). The fourth self-luminous element trades the "para-conscious moving" (*Ganin*, 327), led by the "organization-effect" (*Prameyatva*, 180), to produce the growth-limiting "earth element" (*Bhu*, 724 = 385 + 12 + 327). The goal is to limit the growth and prevent the entity's entropy that is generating growth.

The fifth self-luminous entity trades the "stationarity" (*Sthanata*, 119), following the "limitation-effect" (*Achitta*, 97), to produce a group of the three growth-destroying "ether elements" (*Shuddhi*, 285), in the form of a "greeter Bacterium" (*Jivadara*, 855 = 724 + 12 + 119 = 285 * 3). The ether element's goal is to destroy the growth, limited by the earthly mass of consciousness, for reproducing the pure mass-free effortless force. Consequently, a "self-luminous subject" (*Pradyumna*, 60) forms by trading the "knowing-effect" (*Pradyumna*, 60), generated over the "cosmic lifetime" (*Pradyumna*, 60) of incarnating the greeter Bacterium. The self-luminous subject enjoys the astrological consciousness of the greeter bacterium's creation sequence. It behaves as an "astrological soul" (*Pradyumna*, 60).

The group of three ether elements sequentially enjoys a descending, an ascending, and a horizontal consciousness of the self as an entity. The first, masculine ether element diffuses its energy by servicing the limit "destruction-effect" (*Vijnanatma*, 89), in the form of the reproduction "sequencing soul" (*Vijnanatma*, 89). The second, the feminine ether element, countertrades the primordial greeter Bacterium and the "sequencing soul" to form a "fearless" (*Abhi*, 944 = 855 + 89) element. The third, androgynous ether element, trades the fearless element and the "primordial wholesome-effect" (*Padmaja*, 54) to be the "path-perpetuating entity" (*Unnayaka*, 999).

Each of the ether elements has a dominating cosmic lifetime of sixty sidereal years. After one-hundred and eighty sidereal years, the fourth, greeter ether element aggregates the unit of sentient essence, immanent within each of the sidereal years, to transform into the "circulating, macro greeter" (*Yama*, 180). The macro greeter trades the "cubic" (*Lam*, 9) energy of the three ether elements. It forms the masculine "sentient element" (*Ojas*, 189 = 180 + 9), whose potential is traded by each entity to extend their lifetime and rotation time by one sidereal day, every one-hundred eighty sidereal years.

After the eighteen-hundred sidereal years, the feminine sentient element "proclaims" (*Vicaksh*, 1800) the "entity consciousness" (*Sushumna*, 10) of the "macro greeter" (*Yama*, 180), "self-reincarnating" (*Pudgala*, ½) after every one-hundred sidereal years, by forming the feminine "guider element" (*Guru*, 100 = 10 + 180 * ½). Each entity trades the potential of the guider element to extend their life and rotation time by one sidereal day, every eighteen-hundred sidereal years. After the eighteen thousand sidereal years, the androgynous guider element "accelerates" (*Mahant*, 18,000) the formation of the "yet-to-manifest" (*Shesha*, 18,000) androgynous "divine element" (*Divya*, 360) by transforming into an "open womb" (*Uttanapada*, 18,000). It trades the "totality" (*Samasta*, 180) of the unit divine essence, immanent within each sidereal year, after correcting for the accelerated value of the guider element.

By aggregating the value of the macro greeter and the totality of the unit divine essence manifested over 18,000 years, after a correction factor of hundred for the "guider-effect" (*Guru*, 100), the "divine element" (*Divya*, 360 = 180 + 180) manifests as the homolog of the "entity time" (*Bhava*, 360) of the macro greeter, who is servicing that divine element. The "paternal creator" (*Pitra*, 4) potential of the divine element extends each entity's life and rotation time by the four sidereal days, every eighteen thousand sidereal years.

The fire, water, air, earth, ether, sentient guider, and divine elements together constitute the fifteenth, primeval perpetuator octave. The "self-perpetuating" (*Udvaha*, ½) potential of each entity, as a paternal creator, extends both the life and rotation time of each entity by half sidereal day, every eighteen thousand sidereal years. It entails trading the circulating divine-effect of the macro ecosystem, without the primordial greeter.

18.11 Almighty-Almighty Reality is the Primordial Greeter, Without Entity Consciousness

Entity consciousness limits the lifetime of the primordial greeter to the eighteen thousand years by multiplying the "entity consciousness"

(*Sushumna*, 10) with three factors. First, the "cubic energy" (*Lam*, 9) of each entity, as a three-face entity, each of which is reproducing an additional three-face entity. Second, the futuristic "zodiac spirit" (*Kapinjala*, 20) value of the entity, after reproducing the nine entities and experiencing the entropy of the "entity consciousness" (*Sushumna*, 10). Each reproduced entity exchanges the greeter "entity consciousness" with the "cultural element" (*Sadashiva Nayaki*, 9) to catalyze the "nine-fold growth" (*Lam*, 9) without the greeter entity. Third, the "guider power" (*Guru*, 100) lets the primordial greeter, as an entity, to exchange the value of the sentient element consumed for multiplying the entity consciousness with the eighteen-thousand years of life.

The entity consciousness makes a primordial greeter into the "tree of life" (*Uttanapada*, 18,000), who is the "mother of the primeval child" (*Uttanapada*, 18,000). Without the entity consciousness, the lifetime of a primordial greeter is ninety thousand years, without the divisive "entropy value" (*Sarvanasha*, 5) of the "self-perpetuating" (*Udvaha*, ½) "entity consciousness" (*Sushumna*, 10). The number 90,000 manifests the "metaphysical reality of the octave" (*Prakarana vadartha*, 90,000). It is composed of the fifteen octaves' manifested value *plus* the sixteenth greeter octave immanent within the entity consciousness, without the primordial greeter as the ninth devoted factor and the other fifteen octaves, as the tenth devoted factor.

The lifetime of a primordial greeter, without the entity consciousness, is a metric of the "cosmic minute" (*Samvatsara*, 90,000). It is the almighty-almighty-almighty reality. As an entity without radiating or irradiating consciousness, the lifetime of a primordial greeter is ninety thousand years, whose one minute is composed of the ninety thousand sidereal years. It is the almighty-almighty-almighty-almighty reality.

The lifetime of a primordial greeter, without seeking to be a maternal or a paternal, is "infinite" (*Ananta*, 90,000) and is the metric of the "cosmic second" (*Ananta*, 90,000). One cosmic second is composed of ninety thousand sidereal years. It is the almighty-almighty-almighty-almighty-almighty reality.

The lifetime of a primordial greeter as a child, without radiating or irradiating the "being energy" (*Kali Shakti*, 96), is beyond the limitations of the space, time, and causation. A "child primordial greeter" (*Madhusudan*,

16) is not a "being" (*Nitya*, 126,000,000), who is "everlasting" (*Nitya*, 126,000,000) only within the present space and vanishes into the spiritual oblivious with the entropy of the present conjoining space. A child primordial greeter does not age but has the potential to age spontaneously by conceiving the self as the creator of the desired space, time, and causation. Therefore, a child primordial greeter is not a reality that is limited to any space. He is a double-octave, where each point in the octave is itself an octave of octaves. A child primordial greeter has the potential to manifest 1024-fold growth (16 * 8 * 8), to be the "limitless infinity" (*Aditi*, 1024), the "tree of life" (*Aditi*, 1024).

A maternal primordial greeter is the one who conceives the potential of the desirable creation in a "split-second" (*Shashtyamasha*, 1/60), i.e., the $1/60^{th}$ of the cosmic second. She lets a devil primordial greeter manifest that creation and becomes the "thread of sentient life" (*Yajnopavita*, 9000), as an "omnipresent devotee" (*Kardama*, 9000). By conceiving the "sensible entity reality" (*Upakarana artha*, 10^{1024}), she norms a lifetime of 10^{1024} years, with each "microsecond" (*Shashtyamasha*, 1/60) comprising the ninety-thousand sidereal years. It is the almighty-almighty-almighty-almighty-almighty-almighty reality of the almighty creator, seeking to enjoy the beauty of the almighty creation, through the eyes of the almighty creatures.

After manifesting the exponentiated value of the self-reproducing "tertiary residual" (*Khara*, 6), the maternal primordial greeter realizes her omnipresent state as the "primordial illuminator" (*Parvati*, 10) of the secret of the almighty creation. This omnipresent state is beyond the limitations of space, time, and causation of the exponentiated, unfolding present reality.

18.12 The Secret of the Almighty Creation

By perpetuating the absolute divinity as the param-param (i.e., omnipresent) reality, each "sentient entity" (*Siddha*, 7), as the perpetuating value of the goal-keeping "Param deity" (*Shiva*, 7), manifests the power to create a seventeenth primeval paternal octave of the sentient causation each breathing moment. The infinite growth value of the primeval paternal octave constitutes the twin, primeval maternal octave. It is the materialized time reality of the sentient causation. The infinite growth value is the accumulated value of the sentient value through an octave of sequential-effects.

A state of infinite growth implies zero-time dilation. After entropy, a sentient entity is devoid of the power to dilate the creation through the secretive mediation of the entity time, i.e., the time within the entity's mind. A mindful entity is conscious of the "approaching arrival of the death point" (*Ayati*, 863). It gets into a race to find the hidden, celestial secret of the earthly creation, by tracking the traces of the events believed to have materialized in deep antiquity.

Eight-fold sequential effects of the disproportionate time dilation traded by a fast-moving primordial entity

(e.g., an object in space beyond the mind or a subject in time within the mind)

- **First, the present-effect.** Ascending entropy of the causative divine energy within the ascending time-frequency of the effect. Consequently, the constant-mass object (created as an effect) experiences descending gravitational-effect (i.e., guider-effect; space-effect) over time along the direction of the motion. *Therefore, the entropy in the primordial creation's primordial, electromagnetic mass ascends over time (i.e., the time clock of the self is fast and ascending with the growth in the entity time of the primeval creation, that was previously present within the open womb of the primordial creation).*

- **Second, the up togetherness-effect.** Each new (primeval) creation has an accelerating effect on the primordial creation. Consequently, the primordial creation that is growing its variable mass experiences an ascending gravitomagnetic-effect (i.e., divine-effect; the self-attracting pressure from the eventual inanimate constant-mass object) over time, orthogonal to the direction of the motion. It is catalyzed by the trading-effect of the disproportionate divine energy of the primordial creation enjoyed by the primeval creation. *Therefore, the proportionate growth in the ecosystem's absolute mass descends (i.e., the time clock of the ecosystem is slow and descending,* *with the growing entropy in entity time of the primordial creation present together within the primeval creation ecosystem).*

- **Third, the down otherness-effect.** The primordial creation has a disproportionately accelerating effect on the diverse forms of the primeval creation. Consequently, the primeval creation experiences a descending gravitoelectromagnetic-effect (i.e., thermodynamic-effect; ascending value; volume expansion) over time, oblique to the direction of the motion. It is catalyzed by the unique entity-effect of the primeval creation as the idealized goal for the primordial creation, within a condition of the ascending proportionate freedom from the ecosystem-effect of the degenerating primordial creation. *Therefore, the proportionate growth in the ecosystem's mass-effect ascends over time (i.e., the time clock of the ecosystem, inclusive of the otherness value of the self, which is fast and ascending).*

- **Fourth, the charm attraction-effect.** As the creator, the primordial creation evidences a charmed attraction for the primeval creation. Consequently, the primeval creation experiences an ascending electromagnetic-effect over time within its gravitational center. It is catalyzed by the global (i.e., para) entity-effect of the creator as a part of the creation, within a condition of the ascending proportionate oneness with the ecosystem-effect of the param creation as a whole. *Therefore, the entropy in the primeval creation's primordial, electromagnetic mass ascends disproportionately over time (i.e., the time clock of the self is disproportionately fast and ascending).*

- **Fifth, the strange repulsion-effect.** The param [i.e., the present] creation evidences a strange repulsion for the primordial creation, which is the primordial (i.e., past) reality deemed undesirable by the primordial creation, as an integral part of the param creation. Consequently, the primordial creation experiences the ascending gravitoelectric-effect over time, without its gravitational center. It is catalyzed by the guider-effect (i.e., gravitational-effect) of the ecosystem (i.e., the param creation), within a condition of the descending proportionate oneness with the ecosystem-effect of the param creation. Therefore, the entropy in the primordial creation's primordial-primordial entity mass is absolute and instantaneous at the point of infinity (i.e., where the primordial creation is destroyed and experienced as a derived reality of the present-effect by a creature within the light of the primeval creation). *The entropy manifests as the illusionary, derived explosion and diffusion of energy from within the nuclear gravitational center of the primordial creation.*

- **Sixth, the top growth-effect.** The inanimate, passively observing, primordial creature experiences the ascending, growth consciousness by trading the illusion of instantaneous destruction, serviced by the primordial creation at the point of infinity. Therefore, the creature perceives an intrinsic consciousness growth equivalent to the primordial creation's param-primordial, inertial mass. In reality, since at the point of infinity, the primordial creation's param-primordial, inertial mass = 0, the consciousness growth within the self = 0.

The entire growth is the consciousness within the primordial creature trading the self-luminous entity value of the param creation by investing the intrinsic divine energy. The primordial creature radiates the ascending thermodynamic consciousness as the sentient light force, in oneness with the intuitive para-consciousness of the inanimate light of the destroyed primordial creation. The primordial creature becomes a primeval creature, that it desired to be, by trading the thermodynamic consciousness. *By following the physical instinct programmed by the primordial self, the primeval creature conceives the effect of the sentient light as the truth verification of the signal of the light of the primordial (i.e., past, destroyed) primordial-creation.*

- **Seventh, the bottom entropy-effect.** The animate, actively engaged, primeval creature experiences the descending, entropy consciousness, by servicing the human guider-effect for verifying the illusionary reality as valid. The primeval creature authenticates the diffused human guider-effect in the form of the nonlinear, masculine, horizontal, wave-effect, which curves the linear, feminine, perpendicular distance from the primeval creation. In reality, since at the zeroth point of the primeval creature's mind, the primeval-primordial, gravitational mass as a person = 1, the consciousness entropy within the self = 1. The primordial creature's entire mass as a person is destroyed and irradiated by the primeval creature as a twin person. *The twin person perceives the diffused nonlinear human* *guider-effect as the ecosystem-effect of the param creation, which is resisting the growing sovereignty of the primeval creation. Consequently, the twin person credits the bending of the light to the force of the param creation.*

- **Eighth, the potential-effect.** The ecosystem activates its thermodynamic potential due to the merger of the proportionate light of the feminine creator (i.e., the primordial creation, who is the mother of the present ecosystem reality. The present ecosystem reality is different from the truth in the presence of the primordial creation—the truth that was the primordial greeter motivating the primordial creation to incubate the primeval creation) and the disproportionate light force of the masculine creature (i.e., the primordial creature, who is the father of new entity reality beyond the absolute, ecosystem truth).

The activated potential-effect of the ecosystem (i.e., the param creation) is the organizational cause of the modified reality of the ecosystem at the point of infinity (i.e., at the ideal space point of the primeval creation away from the theoretical time point of illusionary division conceived in the mind of the primeval creature). *As the primordial-primordial creation, Mother Nature materializes the reality imagined by a devoted factor, whether it is a devotionally programmed telescope or devotionally performing psychic entity.*

18.13 The Secret of the Almighty Creature

As an almighty creature, a sentient entity may move fast to trade the disproportionate time dilation, manifested through a primeval paternal octave, for incubating the eighteenth, primordial maternal octave. Such a sentient entity liberates the almighty creation from the radiation of the entity time value. It lets the absolute vertical growth value of the "time" (*Kala*, 360) perpetuate as the absolute space value, without the need to trade the entity's escalating cost inertia and recycle the wasted energy into the proliferating growth of the primordial space.

Any growth amount remains insufficient for managing the cost of entity inertia and generates disproportionate time dilation, forcing the entity to reincarnate in the entirety of the diverse forms, seeking to conceive the appropriate solution.   After experiencing the entire "tertiary, residual consciousness" (*Brahmaastra*, 366,666), and by incarnating as a member of the "366,666 unique entity groups in the illuminated material universe" (*Kamalatmika*, 366,666), the entity satiates the experimental curiosity of the immanent "scientist" (*Hri*, 863).   At this moment, the entity enjoys the "conscious consciousness" (*Sati-Parvati*, 16) to be the "solution" (*Nissaya*, 863) for fulfilling the goal of the life, before the "death point from the present" (*Ayati*, 863).

18.14 The Secret of the Almighty Creator

The secret of the Almighty Creator is immanent within the nineteenth, param child octave that manifests an atom by aggregating the eight forms of quarks:

- The Up quark, which manifests the sense of the ascending togetherness of the self and the ecosystem within the mind of a present creature. It catalyzes the strong thermodynamism in the form of the workculture-effect.   The workculture-effect is the traded oneness value of the ecosystem within the entire self-luminous universe of child entities.

- The Down quark, which manifests the sense of the descending otherness of the self and the ecosystem within the mind of a present creature. It catalyzes the weak electromagnetic diffusion in the form of the international-effect. The international-effect weakens the devotion of the ecosystem for oneness with a local self.

- The Charm quark, which manifests the sense of the attraction of the self toward the idealized twin self, within the mind of a present creature. It catalyzes gravitomagnetism in the form of the corporate-effect. The corporate-effect strengthens the devotion to the idealized goal.

- The Strange quark, which manifests the sense of the repulsion of the idealized twin self by the truth-loving ecosystem. It catalyzes the gravitoelectric light force radiation in the form of the local-effect. The local-effect lets the local self to exchange the reality of the idealized twin self.

- The Truth quark, which manifests the sense of the growth of the person into the idealized twin person. It catalyzes the primeval sentient energy in the form of the culture-effect. The culture-effect lets the idealized twin person reproduce the illusion of the person's organizational sameness, within a dynamic ecosystem and the transformative growth of the ecosystem, without the person.

- The Beauty quark, which manifests the sense of the entropy of the person due to an illusionary mind-born division from the idealized twin person. It catalyzes the energy-less gravitational element. The gravitational element binds the consciousness of the idealized twin person with the citizenship national-effect of the person. The idealized twin person becomes a devotee of the person, who has traded the entropy value to perpetuate in the form of the entropy element, and deifies that unknowable perpetuator of the absolute truth of the creature self and the ecosystem creation as the one and only one "God" (*Ishvara*, 5).

- The Present quark, which manifests the ascending thermodynamic reality of the Up quark, the descending gravitomagnetic reality of the Charm quark, and the horizontal cultural reality of the Truth quark. It catalyzes the future reality of the Almighty Creator. One becomes the Almighty Creator by thermodynamically wishing for the wishable truth, gravitomagnetically descending the wish to know the Wisher's truth, and culturally reproducing the devotion to the Para Wisher. The Para Wisher is alive in the genetically-programmed paternal truth within each deified child entity. The wishable truth is to materialize the desired cultural element within each devoted entity, who is deifying the "Para Wisher" (*Bhakta*, -5) in the form of the "Para Deity" (*Param Vishnu*, 5) and is eternally grateful for the "Demi-God" (*Sunanayaka*, 5) blessings of that Para Deity. The Present quark's formative value is -2, divided -2/3 among each of the three growth quarks. It is the value of the two-dimensional "present reality" (*Badhabuddhi vadartha*, -2).

- The Potential quark, which manifests the forward electromagnetic diffusion reality of the Down quark, the backward gravitoelectric lightforce-radiating reality of the Strange quark, and the holistic gravitational element reality of the Beauty quark. It catalyzes the entropy reality of the Almighty Creature, who has become the Almighty

Creator. By becoming the Almighty Creator, one perpetuates the forward-in-time otherness of the created Almighty Creation. One illuminates the backward-in-time repulsion of the creating the Almighty Creature, from the omnipresent reality of Mother Nature. Therefore, one spontaneously sentiates the potential one-dimensional "future reality" (*Ekartha*, 21) of the self's entropy. The potential quark's formative value is +1, divided +1/3 among each of the three entropy quarks.

The nineteenth, param child octave manifests the doctrinal consciousness of the "I AM the Almighty Creator" (*Sarvam Khalvidam Brahma siddhanta*, -10^6) factor, by transforming into the "organizational metric of the present reality" (*Maha Shunya*, -1). As an organizational metric, the param child octave sentiates both the idealized twin person as well as the self-reproducing genetically-programmed paternal truth, by servicing the positive one-thousand units of value to each in the form of the "sentient energy" (*Varuna*, 1000). The organizational metric manifests in the form of the "Negative quark," which programs the para-consciousness for idealizing the perpetuating value of the genetically-programmed paternal truth within the idealized twin person. Consequently, the idealized twin person becomes the "param maternal" (*Saranyu*, 5), who trades the "perpetuating value" (*Saranyu*, 5) and services the twentieth, param manifestor octave for manifesting the desired future reality.

The normative value of the "Negative quark" is -1. It is multiplied as -1 to transform the value of the Present quark into +2, divided +2/3 among the three growth quarks, and the Potential quark's value into -1, divided -/13 among the three entropy quarks. Metaphysically, it perpetuates the illusionary attachment to the path-dependent, past growth reality and the path-independent reality of the "param child" (*Manyu*, 19) as the "primordial-primordial creature" (*Rudra*, 19). It transforms the linear, vertical, ascending reality of the past into a nonlinear, horizontal, curved reality at present. The param child is the nonlinear reality that limits the ascending vertical growth by devotionally following the present reality's supernatural paradigm.

The twentieth, param manifestor octave manifests the doctrinal consciousness of the "I AM the Almighty Creation" (*Etad Vai Tat siddhanta*,

-8000 = 1000 * -1 * 8) factor. It binds the intrinsic programmatic "sentient energy" (*Varuna*, 1000) with both the paternal "organizational metric" (*Maha Shunya*, -1) as well as the maternal "supernatural paradigm of present reality" (*Yukti*, 8).   As the twentieth "octave of entities" (*Krishnamurthy*, 1), it services the illusionary doctrinal reality of the "I AM Divine Entity" (*Sauh siddhanta*, 1) factor, in the form of a "Positive quark." Its normative value is +1.   It is multiplied as +1 for perpetuating the transformed, scientifically verifiable value of the three entropy and the three Growth quarks.

The twenty-first, param creator octave manifests the doctrinal consciousness of the "I AM the Almighty Creature" (*Ekam Evadvitiyam Brahma siddhanta*, -10^{30}) factor.   It binds the paternally planned "organizational metric" (*Maha Shunya*, -1), with the para-consciousness of the twentieth child "octave of entities" (*Krishnamurthy*, 1), the maternal "supernatural paradigm of present reality" (*Yukti*, 8) as a "manifold entity" (*Satarupa*, 8), and the programmatic sentient energy as "one entity" (*Mulapraktri*, 1000).   Each entity within the spherical "ensemble of ten entities" (*Nyaya*, 1) para-consciously services at par "consciousness energy" (*Varuna*, 1000) to the param creator octave, empowering the latter to create the desired future entropy reality by transforming the present reality.   Each primeval child experiences a conflicting desire for knowing and perpetuating the past reality as well as creating and illuminating an alternative future reality.   Each primordial child becomes the almighty creator of the present reality of the param child and, then, destroying that present reality by motivating the param child to be something one is not already.

18.15 The Secret of the Present Reality

The present reality is the universe's two-dimensional reality, composed of the inanimate atoms and the animate cells. The atoms are the illuminated future reality of the cells.   The cells are the perpetuating past reality of the atoms. The universe manifests the entities' present reality, performing the program of the "one entity" (*Mulapraktri*, 1000)—the programmatic sentient energy traded from the past reality of the atoms. The atoms' past reality is composed of a creature who has become a follower organization, a creator who has become a leader organization, and a creation who has become an entrepreneur organization.

A creature creates the atom by manifesting the consciousness sequence that manufactures the quarks and automates the process of the organization of those quarks into different forms of atoms. Having created the future reality by servicing the sentient energy and transforming the self into an inanimate atomic creation, the creature conceives the self as the creator. Therefore, the creature becomes a follower of both the creation, that he has conceived as the future reality, and the creator, he has perceived as the past reality, responsible for transforming the desirable present reality into the undesirable future reality. The present reality is desirable because it empowers the creature to be the creation as well as the creator. The creature becomes a follower of both the future reality, as a creation, and the past reality, as a creator, because that is the only way to experience the truth of the present reality's desirability.

Without knowing the truth, one para-consciously sentiates the present reality of the "world" (*Duniya*, -2) as the double-negative and the "universe" (*Brahman*, 2), that is manifesting the present para-consciously programmed reality, as the double-positive. Therefore, the creator becomes a leader of the creature who has materialized him and of the creation that he has created. Having created the creation, the creator lacks the energy for creating an alternative, primeval creation. As a leader, the creator becomes the guider power, guiding the creature to be the creator and the creation on the path of its creation. The creator trades the gravitational energy serviced by the creation to illuminate the path of its creation to be the shadow guiding force within the creature's mind.

As one who is secretly trading the desirable path from the illuminator creature and servicing the desirable path to the shadow Creator deity, the creation becomes an entrepreneurial organization. As an entrepreneur, the creation has the freedom to break the "deep, one-face symmetry" (*Mandra*, 1) with the illuminator creature, conceive the traded path as undesirable, and service its desired future path to the shadow Creator deity. Thus, the creation consciously, thermodynamically thermalizes the present reality of the "world" (*Duniya*, -2) as a double-negative and the "universe" (*Brahman*, 2), that is manifesting the present, para-consciously programmed reality, as a double-positive.

18.16 Is Present a Reality?

Suppose a creature transforms the present desirable reality of the world, and the creation has the power to liberate the self from the present undesirable reality. Is the two-dimensional present a reality? The two-dimensional present reality is an illusion of the division between the past and the future. An entity materializes itself as the "absolute earth-effect" (*Trinetra*, 1), that divides the sentient, subjective reality of the past and the inanimate, objective reality of the future. Both the subject, who is scripting the reality, and the object, that is the embodied, scripted reality, are not the reality. They are the animate and the inanimate forms of the divine energy scripting the reality. They are embodied within the scripted reality. The animate and the inanimate forms of the divine energy are not the "formative divine energy" (*Madhusudan*, 16). They are the manifestations of the masculine, ego-conscious "astral body" (*Lingam-Sharira*, 3) of the "white star" (*Yama*, 180), which is the "self-perpetuating" (*Udvaha*, ½) "divine element" (*Divya*, 360), formed through the exchange of the "time" (*Kala*, 360) as the "guider dimension" (*Guru Dharma*, 360).

The time is the variable growth value, illuminated by the "Primordial Illuminator" (*Parvati*, 10) by servicing the "transformative divine energy" (*Asrava Shakti*, 10), for forming a group of the thirty-six starseed entities. The thirty-six starseed entities manifest the squared value of the "tertiary residual" (*Khara*, 6) as the "normative divine energy" (*Khara*, 6), that together with the transformative divine energy, forms the "formative divine energy" (*Madhusudan*, 16).

After radiating the masculine, astral body for transforming the normative value of divine energy, the Primordial Illuminator realizes her almighty state of the "Param Deity" (*Shiva*, 7). Omnipresent oneness with the almighty state of the "Param Deity" (*Shiva*, 7) is the causative factor, that empowers both the Primordial Illuminator as well as the Primordial Greeter to be beyond the realm of space, time, and causation. The Param Deity, the Primordial Illuminator, and the Primordial Greeter have a choice to break the "zero-face symmetry" (*Alpatva*, 0) with the almighty state and be the "Luminous Devoted Greeter" (*Rukmini*, 86). They may trade the nonlinear, "supernatural paradigm" (*Yukti*, 8), as the primordial element, and the linear, "normative divine energy" (*Khara*, 6), as the primeval element, and make themselves their present invisible from the universe of elements.

The Luminous Devoted Greeter is the "primordial-primordial wave" (*Rukmini*, 86) of creation. It norms the metaphysical divine energy as the "constant script" (*Rukmini*, 86). The script delineates the "linear distance" (*Rukmini*, 86) between the scripted creation as a "symmetrical system" (*Rukmini*, 86) and the "invisible hand" (*Hasta*, 0) of the "greeter spirit" (*Hasta*, 0), who has decided to become a "devil" (*Sura*, 0) for chastising the clueless creature.

For activating one's divine energy, one must manage one's potential present reality, transcending the limits of the absolute present reality. To manage one's potential present reality, one must know both the different dimensions of that reality and how to manage these dimensions. We must be certain that we are scripting, conceiving, perceiving, experiencing, and enjoying conscious consciousness of the present moment's sentient truth. The present reality of the ascending inequity within nations is the effect of the past reality, which conditioned the farmers and other micro follower entities working in the field to believe that they will be better off saving and investing their savings into the supernormal returns-accruing instruments.

In the present moment, the farmers and other micro follower entities have become the insurers of the infinite resort, as the local and the global financial institutions have funneled those micro-savings into the hands of the leading investment managers. The micro follower entities eventually bear the entire risk of the entrepreneurial failures. The macro leader entities accrue the entire benefit of the responsible management jackpot. Since the value of the money multiplies within the infinite trading-effect from the micro follower entity to the macro leader entity, each entity perceives the multiplying and accelerating trading-effect to be a positive factor in the overall macro growth. However, at both the absolute as well as the proportionate levels, micro-entities experience the ascending entropy-effect of their descending workculture-effect. Conversely, macro entities experience the ascending growth-effect of their ascending workculture-effect.

The above reality begs an alternative thesis that the divine energy is nothing but the trading-effect of the energy already present within the universe of the organizational reality. If the energy present within the universe of organizational reality is the total energy, then the human entity is reduced to behave just like a past energy trader.

If the present energy is the aggregated effect of the past Worker energies of entities, and if each present entity is only the consumer of those past Worker energies, then the residual future energy is always less than past energy. By not working to create new energy in the present moment and by deciding instead only to trade and manage the effects of the old energies, the human entities experience a descending-order technological growth. Human entities become the puppets of the ancestors' energies, discrediting the foundational assumptions of the evolutionary theory of the ascending-order organizational development over time.

Therefore, there is a need for further investigating a sensible approach to master the clean-state open-mind white-ocean strategy for perpetuating the universal SHEENY well-being.

Acknowledgments

This investigation into present reality's scientific nature is shaped by six divine factors: determination, imagination, virtue, intuition, nature, and excellence.

- Conscious determination of the value of a scientific approach, without the limitations of the inherited scientific method, is shaped by Primordial Greeter Shri Kartar Singh Yadav Ji, ex–joint commissioner, Ministry of Agriculture, Government of India, who is my param guru guiding me through this investigation.

- Liberated imagination of the technique for initiating, persevering, and finishing this project is shaped by my father, Shri Surender Nath, and my mother, Shrimati Manju Gupta.

- Illuminated virtue of transcending beyond the traditional approach has been shaped by my wife, Bhakti.

- Infinite intuition for a conscious ecosystem approach is shaped by my students, who are devoted to transforming their social, human, ecological, economic, national, and psychological well-being.

- Universal nature of the proposed organizational approach is shaped by my professional colleagues and mentors, at various institutions, from various nations, and with varying academic and life perspectives. In particular, I appreciate the support of Dean Lawrence C. Rose of the Jack H. Brown College of Business and Public Administration and of my colleagues Dr. Frank Lin and Ms. Jamie Ayala at the Center for Global Management of California State University, San Bernardino, for standing by me as I worked on this project.

- Technological excellence of this investigator is shaped by my family, friends, and critics, and by those who have contributed to the present reality through the ages to illuminate the significance of the subject of the present study.

Primordial Perpetuator Maha Saraswati and Primordial Illuminator Shri Krishna blessed me with their Divine Light, energizing me to bring the project to fruition.

This investigation is dedicated to the Universe of Children, wishing for their global, unique, inclusive, diverse, engaged, and responsible well-being.

Index: English

T

Index: Hindi